Offshore Pipelines
Design, Installation, and Maintenance

Second Edition

Offshore Pipelines

Design, Installation, and Maintenance

Second Edition

Boyun Guo
University of Louisiana at Lafayette

Shanhong Song
Chevron Asia Pacific Exploration & Production

Ali Ghalambor
Oil Center Research International

Tian Ran Lin
Queensland University of Technology

ELSEVIER

AMSTERDAM • BOSTON • HEIDELBERG • LONDON
NEW YORK • OXFORD • PARIS • SAN DIEGO
SAN FRANCISCO • SINGAPORE • SYDNEY • TOKYO
Gulf Professional Publishing is an imprint of Elsevier

Gulf Professional Publishing is an imprint of Elsevier
225 Wyman Street, Waltham, MA 02451, USA
The Boulevard, Langford Lane, Kidlington, Oxford OX5 1GB, UK

Notice
Knowledge and best practice in this field are constantly changing. As new research and
experience broaden our understanding, changes in research methods or professional
practices, or medical treatment may become necessary.

Practitioners and researchers must always rely on their own experience and knowledge in
evaluating and using any information, methods, compounds, or experiments described
herein. In using such information or methods they should be mindful of their own safety
and the safety of others, including parties for whom they have a professional
responsibility.

To the fullest extent of the law, neither the Publisher nor the authors, contributors, or
editors, assume any liability for any injury and/or damage to persons or property as a
matter of products liability, negligence or otherwise, or from any use or operation of
any methods, products, instructions, or ideas contained in the material herein.

Library of Congress Cataloging-in-Publication Data
A catalog record for this book is available from the Library of Congress

British Library Cataloguing-in-Publication Data
A catalogue record for this book is available from the British Library

ISBN: 978-0-12-397949-0

For information on all Gulf Professional Publishing publications
visit our website at http://store.elsevier.com

**Working together
to grow libraries in
developing countries**

ELSEVIER   Book Aid International

www.elsevier.com • www.bookaid.org

Contents

Part II
Pipeline Installation   **133**

Boyun Guo, Shanhong Song and Ali Ghalambor

Part III
Pipeline Commissioning and Operations   167

Boyun Guo, Shanhong Song and Ali Ghalambor

14 Pipeline Testing and Precommissioning   169

15 Flow Assurance   179

Part IV
Condition-Based Maintenance   **255**

Tian Ran Lin with contributions from Yong Sun

Preface

The offshore pipelines in the oil and gas industry have advanced significantly in the last decade. The major changes are the new technologies employed for solving problems in deepwater development. During the many years of teaching oil and gas production engineering courses in academia and in the industry, the authors realized that there is a need for an updated book that reflects the current practice of what the modern offshore pipeline engineers do. Currently available books fail to provide adequate information about how the engineering principles are applied to solving problems that are frequently encountered in the offshore pipeline systems. This fact motivated the authors to write the second edition of the book Offshore Pipelines. In addition to the materials covered in the first edition, the second edition includes more materials in pipeline maintenance.

This book is written primarily for oil and gas pipeline engineers and college students of senior level as well as graduate level. It is not authors' intention to simply duplicate general information that can be found from other books. This book gathers authors' experiences gained through years of teaching courses of oil and gas production engineering in the oil and gas industry and universities. The mission of the book is to provide pipeline engineers handy guidelines to designing, analyzing, and operating offshore pipelines. The original manuscript of this book has been used as course manual for industry trainees and textbook for college students of undergraduate and graduate levels in Petroleum Engineering.

This book was intended to cover the full scope of pipeline systems. Following the sequence of applications, this book presents its contents in 18 chapters presented in 4 parts.

Part I contains 10 chapters covering pipeline design issues. Part II includes 3 chapters presenting principles and rules of pipeline installation. Part III consists of 3 chapters covering pipeline commissioning and operations. Part IV consists of 2 chapters covering condition-based maintenance.

Since the substance of this book is virtually boundless in depth, knowing what to omit was the greatest difficulty with its editing. The authors believe that it requires many books to describe the foundation of knowledge in offshore pipeline systems. To counter any deficiency that might arise from the limitations of space, the book provides a reference list of books and papers at the end of each chapter so that readers should experience little difficulty in pursuing each topic beyond the presented scope. Regarding presentation, this book focuses on presenting and illustrating engineering principles used for designing and optimizing offshore pipeline systems rather than in-depth theories. Derivation of mathematical models is beyond the scope of this book.

This book is based on numerous documents including reports and papers accumulated through years of work in the University of Louisiana at Lafayette, Chevron Corporation, and INTEC Engineering. The authors are grateful to the university and the companies for permissions of publishing the materials. On the basis of the collective experiences of authors, we expect this book to be of value to the pipeline engineers in the oil and gas industry.

Boyun Guo
Shanhong Song
Ali Ghalambor
Tian Ran Lin

Chapter 1

Introduction

Boyun Guo, Shanhong Song and Ali Ghalambor

1.1 OVERVIEW

The first pipeline was built in the United States in 1859 to transport crude oil (Wolbert, 1952). Through the one-and-a-half century of pipeline operating practice, the petroleum industry has proven that pipelines are by far the most economical means of large-scale overland transportation for crude oil, natural gas, and their products, clearly superior to rail and truck transportation over competing routes, given large quantities to be moved on a regular basis. Transporting petroleum fluids with pipelines is a continuous and reliable operation. Pipelines have demonstrated an ability to adapt to a wide variety of environments including remote areas and hostile environments. Because of their superior flexibility to the alternatives, with very minor exceptions, largely due to local peculiarities, most refineries are served by one or more pipelines.

Man's inexorable demand for petroleum products intensified the search for oil in the offshore regions of the world as early as 1897, when the offshore oil exploration and production started from the Summerland, California (Leffler et al., 2003). The first offshore pipeline was born in the Summerland, an idyllic-sounding spot just southeast of Santa Barbara. Since then the offshore pipeline has become the unique means of efficiently transporting offshore fluids, i.e., oil, gas, and water.

Offshore pipelines can be classified as follows (Figure 1.1):

- Flowlines transporting oil and/or gas from satellite subsea wells to subsea manifolds;
- Flowlines transporting oil and/or gas from subsea manifolds to production facility platforms;
- Infield flowlines transporting oil and/or gas between production facility platforms;
- Export pipelines transporting oil and/or gas from production facility platforms to shore;
- Flowlines transporting water or chemicals from production facility platforms, through subsea injection manifolds, to injection wellheads.

Offshore Pipelines. DOI: http://dx.doi.org/10.1016/B978-0-12-397949-0.00001-7
1

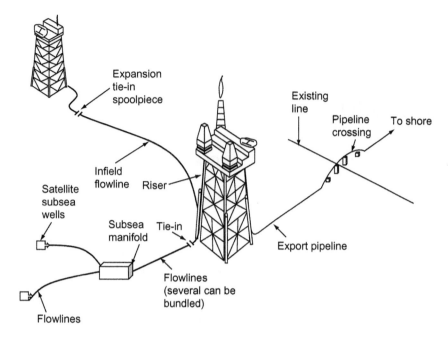

FIGURE 1.1   Uses of offshore pipelines.

Further downstream from the subsea wellhead, as more streams commingle, the diameter of the pipelines increases. Of course, the pipelines are sized to handle the expected pressure and fluid flow. To ensure desired flow rate of product, pipeline size varies significantly from project to project. To contain the pressures, wall thicknesses of the pipelines range from 3/8 in. to 1½ in.

1.2 PIPELINE DESIGN

Design of offshore pipelines is usually carried out in three stages: conceptual engineering, preliminary engineering, and detail engineering. During the conceptual engineering stage, issues of technical feasibility and constraints on the system design and construction are addressed. Potential difficulties are revealed and nonviable options are eliminated. Required information for the forthcoming design and construction are identified. The outcome of the conceptual engineering allows for scheduling of development and a rough estimate of associated cost. The preliminary engineering defines system concept (pipeline size and grade), prepares authority applications, and provides design details sufficient to order pipeline. In the detail engineering phase, the design is completed in sufficient detail to define the technical input for all procurement and construction tendering. The materials covered in this book fit mostly into the preliminary engineering stage.

TABLE 1.1 Sample Pipeline Sizes

Project No.	Project Name	Pipeline Diameter (in.)	Wall Thickness (in.)	D/t	Design Criterion
1	Zinc	4	0.438	10	Internal pressure
2	GC 108-AGIP	6	0.562	12	Internal pressure
3	Zinc	8	0.500	17	Internal pressure
4	Amerada Hess	8	0.500	17	Internal pressure
5	Viosca Knoll	8	0.562	15	Internal pressure
6	Vancouver	10	0.410	26	External pressure
7	Marlim	12	0.712	18	External pressure
8	Palawan	20	0.812	25	External pressure
9	Palawan	24	0.375	64[a]	Internal pressure
10	Marlim	26	0.938	28	External pressure
11	Palawan	30	0.500	60[a]	Internal pressure
12	Shtockman	36	1.225	29	Internal pressure[b]
13	Talinpu	56	0.750	72[a]	Internal pressure[b]
14	Marlim	38	1.312	29	External pressure[b]
15	Shtockman	44	1.500	29	Internal pressure[b]
16	Talinpu	56	0.750	75[a]	Internal pressure[b]

[a]Pipelines with D/t over 30.5 float in water without coating.
[b]Buckle arrestors required.

A complete pipeline design includes pipeline sizing (diameter and wall thickness) and material grade selection based on analyses of stress, hydrodynamic stability, span, thermal insulation, corrosion and stability coating, and riser specification. The following data establish design basis:

- Reservoir performance
- Fluid and water compositions
- Fluid PVT properties
- Sand concentration
- Sand particle distribution
- Geotechnical survey data
- Meteorological and oceanographic data.

Table 1.1 shows sizes of some pipelines. It also gives order of magnitude of typical diameter/wall thickness ratios (D/t). Smaller diameter pipes are often flowlines with high design pressure leading to D/t between 15 and 20. For deepwater, transmission lines with D/t of 25−30 are more common. Depending upon types, some pipelines are bundled and others are thermal- or concrete-coated steel pipes to reduce heat loss and increase stability.

Although sophisticated engineering tools involving finite element simulations (Bai, 2001) are available to engineers for pipeline design, for procedure transparency, this book describes a simple and practical approach. Details are discussed in Part I of this book.

1.3 PIPELINE INSTALLATION

Once the design is finalized, pipeline is ordered for pipe construction and coating and/or insulation fabrication. Upon shipping to the site, pipeline can be installed. There are several methods for pipeline installation including S-lay, J-lay, reel barge, and tow-in methods. As depicted in Figure 1.2, the S-lay requires a laying barge to have on its deck several welding stations where the crew welds together 40−80 ft of insulated pipe in a dry environment away from wind and rain. As the barge moves forward, the pipe is eased off the stern, curving downward through the water as it leaves until it reaches the touchdown point. After touchdown, as more pipe is played out, it assumes the normal S-shape. To reduce bending stress in the pipe, a stinger is used to support the pipe as it leaves the barge. To avoid buckling of the pipe, a tensioning roller and controlled forward thrust must be used to provide appropriate tensile load to the pipeline. This method is used for pipeline installations in a range of water depths from shallow to deep. The

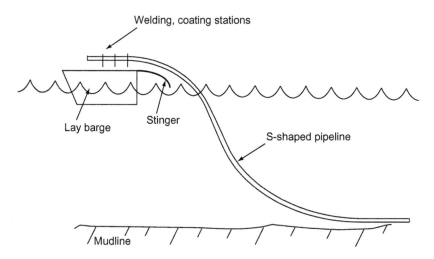

FIGURE 1.2   S-lay barge method for shallow to deep pipelines.

J-lay method is shown in Figure 1.3. It avoids some of the difficulties of S-laying such as tensile load and forward thrust. J-lay barges drop the pipe down almost vertically until it reaches touchdown. After that, the pipe assumes the normal J-shape. J-lay barges have a tall tower on the stern to weld and slip prewelded pipe sections of lengths up to 240 ft. With the simpler pipeline shape, J-lay can be used in deeper water than S-lay.

Small-diameter pipelines can be installed with reel barges where the pipe is welded, coated, and wound onshore to reduce costs. Horizontal reels lay pipe with an S-lay configuration. Vertical reels most commonly lay pipe by J-lay method but can also use S-laying.

There are four variations of the tow-in method: surface tow, mid-depth tow, off-bottom tow, and bottom tow. For the surface-tow approach, as shown in Figure 1.4, buoyancy modules are added to the pipeline so that it floats at the surface. Once the pipeline is towed on site by the two towboats, the buoyancy modules are removed or flooded, and the pipeline settles to the sea floor. Figure 1.5 illustrates the mid-depth tow. It requires fewer buoyancy modules. The pipeline settles to the bottom on its own when the forward progression ceases. Depicted in Figure 1.6 is the off-bottom tow. It involves both buoyancy modules and added weight in the form of chains. Once on location, the buoyancy is removed, and the pipeline settles to the sea floor. Figure 1.7 shows the bottom tow. The pipeline is allowed to sink to the bottom and then towed along the sea floor. It is primarily used for soft and flat sea floor in shallow water.

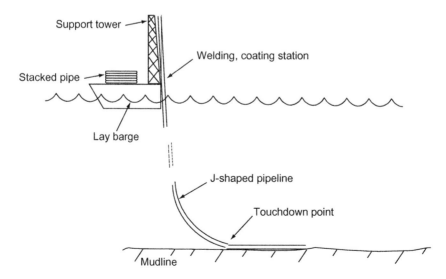

FIGURE 1.3   J-lay barge method for deepwater pipelines.

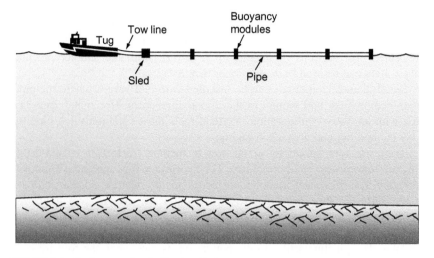

FIGURE 1.4   Surface tow for pipeline installation.

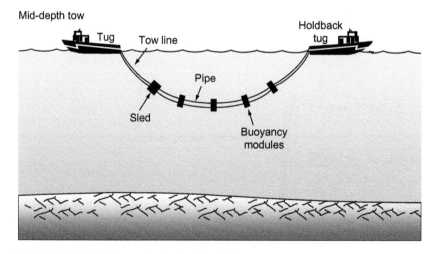

FIGURE 1.5   Mid-depth tow for pipeline installation.

Several concerns require attention during pipeline installation. These include pipeline external corrosion protection, pipeline installation protection, and installation bending stress/strain control. Details are discussed in Part II of this book.

Off-bottom tow

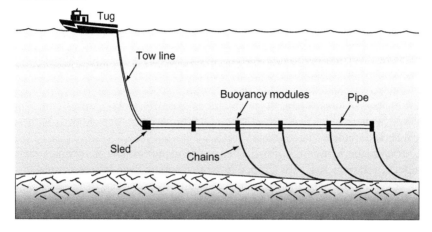

FIGURE 1.6 Off-bottom tow for pipeline installation.

Bottom tow

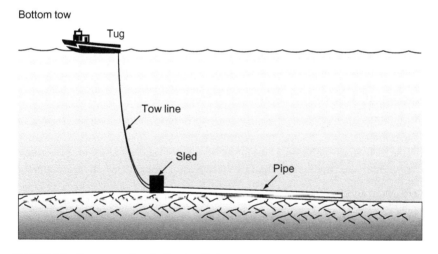

FIGURE 1.7 Bottom tow for pipeline installation.

1.4 PIPELINE OPERATIONS

Pipeline operation starts with pipeline testing and commissioning. Operations to be carried out include flooding, cleaning, gauging, hydrostatic pressure testing, leak testing, and commissioning procedures. Daily operations include flow assurance and pigging operations to maintain the pipeline under good conditions.

Flow assurance is defined as an operation that generates a reliable flow of fluids from the reservoir to the sales point. The operation deals with formation and depositions of gas hydrates, paraffin, asphaltenes, and scales that can reduce flow efficiency of oil and gas pipelines. Because of technical challenges involved, this operation requires the combined efforts of a multidisciplinary team consisting of scientists, engineers, and field personnel.

Technical challenges in the flow assurance operation include prevention and control of depositions of gas hydrates, paraffin (wax), asphaltenes, and scales in the oil and gas production systems. Usually, one or two of these problems dominate in a given oil/gas field.

Natural gas hydrate is formed when methane molecules—the primary component of natural gas—are trapped in a microscopic cage of water molecules under certain pressure and temperature conditions (Katz and Lee, 1990). As a rough rule of thumb, methane hydrate will form in a natural gas system if free water is available at a temperature as high as 40°F and a pressure as low as 170 psig. Decreasing temperature and increasing pressure are favorable for hydrate formation (Guo et al., 1992). Hydrate-forming conditions are predictable with computer programs. Natural gas hydrate can form within gas pipelines as a solid or semisolid mass that can slow or completely block gas flow. Clearing hydrate-plugged pipelines is an expensive and time-consuming task that can take as long as several weeks. There are five methods for preventing hydrate formation (Makogon, 1997):

- Remove free water from the system.
- Keep the system operating temperature above the hydrate formation threshold.
- Maintain the system operating pressure below the hydrate formation threshold.
- Inject hydrate inhibitors, such as methanol and glycol, to effectively decrease the hydrate formation temperature, or delay hydrate crystal growth.
- Add antiagglomerates to prevent the aggregation of hydrate crystals.

The choice of which methods to use depends upon system characteristics, technology availability, and cost considerations.

Paraffin or wax (n-alkane) has a straight chain linear structure composed entirely of carbon and hydrogen (Becker, 1997). The long-chain paraffin ($>C_{20}H_{42}$) components cause deposition or congealing oil in crude oil systems. Paraffin can deposit from the fractures in the formation rock to the pipelines that deliver oil to the refineries. The deposits can vary in consistency from rock hard for the highest chain length paraffin to very soft, mayonnaise-like congealing oil deposits. Paraffin components account for a significant portion of a majority of crude oils heavier than 20° API. One of the primary methods of controlling paraffin deposits is to use solvent. Complete success in paraffin removal has been elusive, depending on the type of deposit being dissolved, its location in the system, the temperature, and type of application. A number of factors can affect the removal of paraffin from a production system using solvent. Some of the most important factors are types of solvents used, type of paraffin, quantity of paraffin, temperature, and contact time. Even the best paraffin solvent applied to long-chain paraffin at low temperature for too short a time will fail to give a clean system. A poor solvent applied to short-chain paraffin at high temperature in large quantities will clean the system every time. Different solvents have different abilities to dissolve paraffin. Two general classes of solvents used in the oilfield to dissolve paraffin are aliphatic and aromatic. Common aliphatic solvents used in the oilfield are diesel, kerosene, and condensate. Aromatic solvents used are xylene and toluene. Solvents are frequently chosen based on price per gallon or price per barrel rather than effectiveness.

Other techniques used for paraffin removal include mechanical scratching and hot fluid treatments. Magnetic treatment of crude oils has also been reported to reduce paraffin deposition in wells.

Asphaltenes identified in oil production systems are generally high-molecular weight organic fractions of crude oils that are soluble in toluene, but are insoluble in alkanes (Becker, 1997). Asphaltene precipitation from crude oils can cause serious problems in the reservoir, wellbore, and in the production facilities. Asphaltenes remain in solution under reservoir temperature and pressure conditions. They destabilize and start to precipitate when pressure and temperature changes occur during primary oil production. The precipitated asphaltene particles then grow in size and may start to deposit onto the production string and/or flowlines, causing operational problems. Several factors, including the oil composition, pressure and temperature, and the properties of asphaltene, influence asphaltene precipitation from reservoir oil. A variety of models for predicting the onset of asphaltene precipitation from live crude oil are available in the literature. These models have been proposed based on different microscopic theories. Each model has its limitations due to the inherent assumptions built-in. A common practice for remediating or mitigating well impairment caused by asphaltene deposition consists of periodic treatments with a solvent (i.e., washing the tubing and squeezing into the near-wellbore formation). However, an economical limitation exists because of the transient effect of such cleanup operations. In addition, solvents in use in the field, such as xylene or naphtha, did not completely dissolve the asphalt deposits or completely extract asphaltenes fixed on clay minerals.

Scale deposits of many different chemical compositions are formed as a result of crystallization and precipitation of minerals from the produced water (Becker, 1998). The most common scale is formed from calcium carbonate (commonly known as calcite). These deposits become solids, which cause problems in pipelines and equipment when they are attached to the walls. This reduces the diameter of the pipes and the cross-sectional area available for flow. Scale is one of the most common and costly problems in the petroleum industry.

This is because it interferes with the production of oil and gas, resulting in an additional cost for treatment, protection, and removal. Scale also results in a loss of profit that makes marginal wells uneconomical. Scale deposition can be minimized using scale-inhibition chemicals. Antiscale magnetic treatment methods have been studied for the past few decades as a new alternative. Acid washing treatments are also used for removal of scale deposits in wells.

Deepwater exploration and development have become key activities for the majority of oil and gas exploration and production companies. Development activities in the deepwater face significant challenges in flow assurance due mainly to high pressure and low temperature of seawater (Hatton et al., 2002). Of particular concern are the effects of produced fluid hydrocarbon solids (i.e., asphaltene, wax, and hydrate) and their potential to disrupt production due to deposition in the production system (Zhang et al., 2002).

It has been noted that the deposition of inorganic solids arising from the aqueous phase (i.e., scale) also poses a serious threat to flow assurance. Gas hydrate plugging problems can occur in deepwater drilling, gas production, and gas transportation through pipelines. The potential for hydrocarbon solid formation and deposition adversely affecting flow assurance in deepwater production systems is a key risk factor in assessing deepwater developments. To reduce this risk, a systematic approach to defining and understanding the thermodynamic and hydrodynamic factors impacting flow assurance is required.

Flow assurance engineering has been known as an operation that does not directly make money, but costs a great deal in pipeline operations, if not managed correctly. Details about this issue are discussed in Part III of this book.

REFERENCES

Bai, Y., 2001. Pipelines and Risers, vol. 3. Elsevier Ocean Engineering Book Series, Amsterdam.

Becker, J.R., 1998. Corrosion & Scale Handbook. PennWell Books, Tulsa, OK.

Becker, J.R., 1997. Crude Oil Waxes, Emulsions, and Asphaltenes. PennWell Books, Tulsa, OK.

Guo, B., Bretz, E.R., Lee, R.L, 1992. Gas hydrates decomposition and its modeling. In: Thompson, H. A. (Ed.), Proceedings of the International Gas Research Conference, November 16—19. Orlando, FL.

Hatton, G.J., Anselmi, A., Curti, G., 2002. Deepwater natural gas pipeline hydrate blockage caused by a seawater leak test. In: Proceedings of the Offshore Technology Conference, May 6—9. Houston, TX.

Katz, D.L., Lee, R.L., 1990. Natural Gas Engineering. McGraw-Hill, New York, NY.

Leffler, W.L., Pattarozzi, R., Sterling, G., 2003. Deepwater Petroleum Exploration and Production. PennWell Books, Tulsa, OK.

Makogon, Y., 1997. Hydrates of Hydrocarbons. PennWell Books, Tulsa, OK.

Wolbert, G., 1952. American Pipelines, 5. University of Oklahoma Press, Norman, OK.

Zhang, J.J., Chen, M., Wang, X., Brown, R.J., 2002. Thermal analysis and design of hot water-heated production flowline bundles. In: Proceedings of the Offshore Technology Conference, May 6—9. Houston, TX.

Part I

Pipeline Design

Boyun Guo, Shanhong Song and Ali Ghalambor

Design of marine pipelines is usually carried out in three stages: conceptual engineering, preliminary engineering, and detail engineering. During the conceptual engineering stage, issues of technical feasibility and constraints on the system design and construction are addressed. Potential difficulties are revealed and nonviable options are eliminated. Required information for the forthcoming design and construction are identified. The outcome of the conceptual engineering allows for scheduling of development and a rough estimate of associated cost. The preliminary engineering defines system concept (pipeline size and grade), prepares authority applications, and provides design details sufficient to order pipeline. In the detail engineering phase, the design is completed in sufficient detail to define the technical input for all procurement and construction tendering. The materials covered in Part I fit mostly into the preliminary engineering.

Although sophisticated engineering tools involving finite element simulations (Bai, 2001) are available to engineers for pipeline design, and procedure transparency, this book describes a simple and practical approach.

Chapter 2

General Design Information

2.1 INTRODUCTION

Before designing an offshore pipeline, the design engineers need to understand the environments in which the pipeline will be installed and operated. What is the water depth? What are the water currents? How big are the waves? All those parameters will affect the mechanical design of the pipeline system. The fluids inside the pipeline will also influence the pipeline design. Is it single phase or multiphase? Are the fluids corrosive? How much sand will be in the fluids? What are the operating pressures and temperatures? All these will influence the pipeline metallurgy selection. A list of the data that will affect the pipeline design is as follows:

> reservoir performance,
> fluid and water compositions,
> fluid pressure−volume−temperature (PVT) properties,
> sand concentration,
> sand particle distribution,
> geotechnical survey data,
> meteorological and oceanographic data.

In this chapter, all the parameters that would affect the pipeline design will be covered. The design engineers should try to collect all these data and have a good understanding of their impacts before they start to perform the pipeline design.

2.2 DESIGN DATA

There are numerous parameters that can affect the pipeline design and operations. The following sections will cover the most critical ones.

2.2.1 Reservoir Performance

How the reservoir would perform over the whole field life can have profound impacts on the pipeline design and operations. Pipeline cannot simply be sized to deliver the maximum production. How the pipeline will be operated at different stages of the field life must be taken into account. The oil, water,

Offshore Pipelines. DOI: http://dx.doi.org/10.1016/B978-0-12-397949-0.00002-9

and gas flowrates will be different at different stages of field life. Different gas and liquid flowrates will then result in different flow behaviors inside the pipeline. Thus, to properly design the pipeline and formulate the operation strategies, how the reservoir will perform over the whole field life needs to be well understood.

2.2.1.1 Reservoir Pressure and Temperature

Both reservoir pressure and temperature will affect the pipeline design and operations. Reservoir pressure is directly related to the wellhead pressure, which will affect the pipeline operating pressure. Very high reservoir pressure can require special metallurgy for the piping and can drive up the material cost dramatically. On the other hand, if reservoir pressure is too low, gas-lift or other artificial-lift mechanisms may be required. Gas-lift gas can affect the pipeline design and operations. Gas-lift gas can make Joule—Thomson cooling effects even worse and cause metallurgy concerns. Joule—Thomson effect (temperature drop) is associated with gas flowing through production chokes where large pressure drop can occur. Gas-lift gas can also make the fluid flow inside the pipeline stable or unstable.

Reservoir temperature can also affect pipeline metallurgy and operations. Very high reservoir temperature may require use of special materials and drive up pipeline cost. Extreme high or low temperature can also eliminate some design flexibility; for example, some flexible pipeline may not be applicable due to high or low fluid temperature. Whether or not flexible pipeline can be used, the specific flexible pipeline manufacturers need to be consulted. If the reservoir temperature is too low and the pipeline fluid temperature is lower than the wax appearance temperature and the gas hydrate temperature, extra thermal insulation design, such as wet insulation or pipe-in-pipe, will be required.

2.2.1.2 Reservoir Formations

Reservoir formation can be classified as either consolidated or unconsolidated. Marine-deposited sands, like in sandstone formations, are often cemented with calcareous or siliceous minerals and may be strongly consolidated (Bradley, 1987). Miocene and younger sands are often unconsolidated or only partially consolidated with soft clay or silt.

With unconsolidated formations, individual grains tend to move easily, especially under high pressure drop, which is often associated with high production flowrates. Thus, if the formation is unconsolidated, even with sound sand control technology, sand would more likely be produced into the pipeline system, accelerating pipeline erosion. What kind of reservoir formations and likelihood that sand will be produced into the pipeline are important pieces of information for pipeline design engineers.

2.2.1.3 Production Profiles

Production profile is one of the most important data for pipeline sizing. Production profiles define how the oil, water, and gas flowrates will change with time for the whole field life. The production profiles are normally generated by reservoir engineers by performing reservoir simulations. Figure 2.1 shows typical black-oil production profiles for oil, water, and gas. Normally, the oil flowrate will reach a maximum in a short period of time and stay at the maximum flowrate for a few years before starting to decline. The water may not be produced for the early stage of production. Once water is breaking through, the water flowrate tends to increase rapidly and stay at the maximum flowrate for some time before starting to decline. If successful pressure maintenance programs are utilized, water production may not decline much for the whole field life. The gas flowrate is normally associated with the oil production and is determined by the gas—oil ratio unless there is an active gas cap in the formation. If a gas cap exists, gas production will be much higher than the solution gas rate.

Pipeline engineers need to understand the impacts of the gas and liquid production on the pipeline design and pipeline operations. The pipeline needs to be sized such that it will transport close to the maximum production and is also easy to operate for the whole field life, especially for the late stage of the field life when flowrates are much lower than the maximum. If the pipeline is oversized for the late field life, the fluid flow inside the pipeline may become unstable and cause terrain slugs inside the pipeline. Unstable flow may impact pipeline mechanical integrity by causing pipeline vibration and excessive corrosion. Unstable (slug) flow will be discussed in more detail in Appendix A.

Pipeline design engineers need to know that there are three different production profiles: P10, P50, and P90. While the P50 production profile should be used for pipeline design, the effect of P10 and P90 production profiles on pipeline design and operations must be considered.

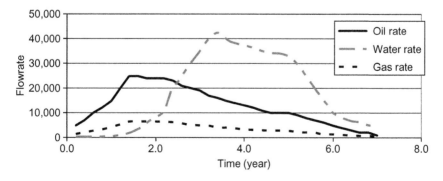

FIGURE 2.1   Typical oil, water, and gas production profiles.

2.2.2 Fluid and Water Compositions

Fluid and water compositions will affect both pipeline design and operations. Whether or not the pipeline metallurgy must satisfy sour service requirements depends upon the fluid and water compositions. If produced fluids contain CO_2 and/or H_2S, pipeline corrosion will be most likely and corrosion mitigation strategies will be developed. Either corrosion resistance alloy (CRA) or chemical inhibition will be required. Corrosion allowance must be included in the wall thickness design.

If seawater injection is used for reservoir pressure maintenance, after injection water breakthrough, sometimes H_2S will also be produced (Evans, 2001; Khatib and Salanitro, 1997; Seto and Beliveau, 2000). This is also called reservoir souring. Water injection may not always result in reservoir souring. Some fields have water injection for years and no H_2S ever appears. But there are also many examples showing reservoir souring due to water injection. If seawater injection is planned, reservoir souring studies should be performed to assess the amount of H_2S that may be produced, and the pipeline material must be chosen accordingly.

Compositions of produced water can also have significant impacts on pipeline design and operations. The cations, like calcium, magnesium, and barium, can react with the anions, like sulfate, carbonate, and bicarbonate, to form scales which can block pipeline flow (Oddo and Tomson, 1994; Cowan and Weintritt, 1976). Scales are formed when incompatible waters are mixed together.

It is well known that saltwater is also corrosive. The more salt in water, the more corrosive it is. Seawater contains high salt concentrations and seawater is very corrosive. The dissolved gases in water, such as oxygen, hydrogen sulfide, carbon dioxide, would significantly increase the water's corrosivity. Accurate analysis of water compositions is thus very critical for proper pipeline design and operations.

2.2.3 Fluid PVT Properties

Fluid PVT properties will greatly affect pipeline sizing. The pipeline must be sized to transport the designed flowrate with a specified pressure drop. The pressure drop is normally determined by the pipeline outlet pressure, which is often the first-stage separator pressure and pipeline inlet pressure which can be the reservoir pressure minus the pressure drop inside the wellbore.

On the other hand, pressure drop associated with multiphase flow is a strong function of the fluid properties such as fluid density, fluid viscosity, gas−liquid ratio, water−oil ratio, water−oil emulsions, and fluid interfacial tension. Thus, a very important step in sizing the pipeline is to accurately characterize the fluids. With the characterized fluids, the above-mentioned properties can be calculated using commercial PVT software.

Normally, during drilling or predrilling, reservoir fluids are captured with elevated pressure downhole. If water aquifer is encountered during drilling, water samples are obtained. The sampled fluids are sent to a PVT laboratory to measure the properties that are not limited to:

reservoir fluid compositions,
gas–oil ratio by single stage flash or multiple stage flash,
API gravity (oil gravity at 14.7 psia and 60°F),
formation volume factor,
bubble point pressure at reservoir temperature,
density at bubble point pressure and reservoir temperature,
water, oil compressibilities,
reservoir fluid viscosity at reservoir temperature,
interfacial tension.

Measured reservoir fluid compositions are characterized by matching the above-mentioned measured parameters. It is very important to characterize the fluid compositions at both reservoir conditions and at pipeline outlet conditions. Thus, the characterized fluids can be applied to the whole range of flowing pressure. The characterized fluids are then used to predict the PVT parameters at different pressures and temperatures for the flowing pressure drop calculations. Fluid characterization is one of the critical steps in sizing the pipeline.

2.2.4 Solid Production

Sand production affects the pipeline design and operations mainly in three areas. One is that sands in the pipeline increase pipeline erosion. Another is that fluid velocity would have to be high enough to carry the sands out of the flowline. Otherwise the sands can deposit inside the pipeline and block the flow. Finally, sand deposition inside the pipeline can prevent inhibition chemicals, like corrosion chemicals, from touching the pipe wall, thus reducing the effectiveness of chemicals.

The most challenging tasks of assessing the sand impacts on pipeline design are determining the particle sizes and determining the concentration of the sands that would be transported by the pipeline. Both particle size distribution and concentration depend upon such parameters as formation rock types and sand control technologies used in well completion. If the formation is unconsolidated, more sands can potentially be produced. Sand grain sizes can be determined by obtaining representative formation samples and performing sieve analysis (Bradley, 1987). Once grain sizes are determined, the proper sand control method can be designed to block the sand from flowing into wellbore and surface pipeline.

Even the best sand-control technologies can potentially fail and allow sands to be introduced into the production system, including the pipeline.

Thus, sand detection becomes very important for pipeline operations. No matter whether an intrusive technique, like impedance sensors, or a non-intrusive technique, like ultrasonic sensors, is used for sand detection, an accurate interpretation method must be developed.

2.2.5 Seafloor Bathymetry/Geotechnical Survey Data

Geotechnical survey data provide important information on seafloor conditions that can affect both pipeline mechanical design and operations. Seafloor bathymetry would affect pipeline routing, alignment, and spanning. Pipeline should be routed away from any seafloor obstructions and hazards. Spanning analysis should be conducted, based upon geotechnical survey data, to identify any locations where spans will be longer than allowable span lengths.

A pipeline bathymetry is preferred if pipeline flow is going upward. In other words, it is preferable that the water depth at the pipeline outlet be shallower than that at the inlet. This is because the multiphase slug flow is much less severe with an upward-inclined pipeline than with a downward-inclined pipeline. Pipeline A, shown in Figure 2.2, will tend to have more severe slugging problems than Pipeline B. More discussions on slug flow can be found in Appendix A.

Pipeline will be laid on the bottom of the seabed. The mechanical conditions of the seabed will affect the stability of the pipeline. It is possible that the pipeline may sink below the seabed and be buried into the subsea soil. Depending upon how deep the pipeline will sink, surrounding soil may have significant impact on the heat transfer process of the pipeline. Understanding the soil mechanical properties will help the design of subsea pipelines.

Soil mechanical properties depend largely upon the soil components and their fractions. There are coarse-grained components, like boulder, cobble, gravel, and sand. The fine-grained components consist mainly of silt, clay, and organic matter (Lambe and Whitman, 1969). Boulders and cobbles are

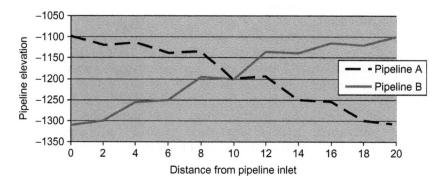

FIGURE 2.2 Upward- and downward-inclined pipeline profiles.

very stable components. Foundations with boulders and cobbles present good stability. Silt is unstable and, with increased water moisture, becomes a "quasi-liquid" offering little resistance to erosion and piping. Clay is difficult to compact when it is wet, but compacted clay is resistant to erosion. Organic matters tend to increase the compressibility of the soil and reduce the soil stability.

Silt and clay are the major components of seabed soil down to a few feet in depth. Thus, when pipeline is laid on the seabed, it will normally sink into the soil. How much the pipeline will sink depends largely upon the mechanical properties of the soil. The following parameters are normally obtained when performing geotechnical analysis.

Water moisture content is defined as the ratio of the mass of the free water in the soil to the mass of soil solid material. Water moisture content is normally expressed as a percentage. Some soils can hold so much water that their water moisture content can be more than 100%.

Absolute porosity is defined as the ratio, expressed as a percentage, of void volume of soil to the soil bulk volume.

Absolute permeability is defined as a measure of the soil's ability to transmit fluid. To determine the permeability of the soil, a sample is put into a pressure device and water is conducted through the soil. The rate of water flow under a given pressure drop is proportional to the soil permeability.

Liquid limit is determined by measuring the water moisture content and the number of blows required to close a specific groove which was cut through a standard brass cup filled with soil. Liquid limit indicates how much water the soil can hold without getting into the "liquid" state.

Plastic limit is defined as the water moisture content at which a thread of soil with 3.2 mm diameter begins to crumble. Plastic limit is the minimum water content required for the soil to present "plastic" properties.

Plasticity index is defined as the difference between liquid limit and plastic limit.

Liquidity index, LI, is defined as the ratio of the difference between the natural water moisture content and the plastic limit to the plasticity index.

Activity number is defined as the ratio of plasticity index to the weight percentage of soil particles finer than two microns. Activity number indicates how much water will be attracted into the soil.

Other mechanical and thermal property data are also important to pipeline design. These well-known data can be obtained through standard tests and will not be repeated here.

2.2.6 Oceanographic Data

Ocean currents and waves greatly affect the stability of offshore pipelines. To design a pipeline required to be mechanically stable for the whole field life, pipeline engineers must understand the importance of oceanographic data. Pipeline installation and towing can also be affected by the ocean conditions.

Oceanographic data include 1-year, 5-year, 10-year, and 100-year extreme wave cases and associated currents: wave heights, wave directions, current speeds, and tide data. Near-bottom conditions (waves, winds, tide, currents, etc.) should be clearly defined.

2.2.7 Other Data

Quite a few other data will affect pipeline design and operations. Water temperature profiles (minimum and maximum) affect the pipeline operations through heat transfer. To be conservative, the minimum temperature profile should be used for pipeline design.

Splash zone should be clearly defined. Splash zone is the pipe or riser section that will be splashed by the surface wave. Because of seawater splashes, the affected pipe or riser section tends to have more severe corrosion problems. Extra coatings are required.

REFERENCES

Bradley, H.B., 1987. Petroleum Engineering Handbook. Society of Petroleum Engineers, Richardson, Texas.

Cowan, J.C., Weintritt, D.J., 1976. Water-Formed Scale Deposits. Gulf Publishing Company, Houston, Texas.

Evans, R., 2001. Factors influencing sulphide scale generation associated with waterflood induced reservoir souring. In: Presented at the International Symposium on Oilfield Scale, 30-31 January, Aberdeen, UK, SPE Paper 68337.

Khatib, Z.I., Salanitro, J.P., 1997. Reservoir souring: analysis of survey and experience in souring waterfloods. In: Presented at the SPE Annual Technical Conference and Exhibition, 5-8 October, San Antonio, Texas, SPE Paper 38795.

Lambe, W., Whitman, R., 1969. Soil Mechanics. John Wiley and Sons, New York, NY.

Oddo, J.E., Tomson, M.B., 1994. Why scale forms and how to predict it. SPE Prod. Facil. 9 (1), 47–54.

Seto, C.J., Beliveau, D.A. 2000. Reservoir souring in the Caroline field. In: Presented at the SPE/CERI Gas Technology Symposium, 3-5 April, Calgary, Alberta, Canada, SPE Paper 59778.

Chapter 3

Diameter and Wall Thickness

3.1 INTRODUCTION

Design of pipeline involves selection of pipeline diameter, thickness, and material to be used. Pipeline diameter should be selected on the basis of flow capacity required to transport production fluids at an expected rate provided by the oil or gas wells. This task demands a comprehensive flow assurance analysis assuming the worst operating condition during the whole life of the pipeline. Due to the complex nature of multiphase flow as described in Appendix A, a calibrated computer model in flow assurance analysis is required. In the last decade, multiphase flow simulators have gained significant popularity. Both steady-state and transient simulators have been used for pipeline design and for pipeline operation simulations. Flow assurance analysis is described in Chapter 15.

This chapter covers wall thickness design for subsea steel pipelines and risers as governed by US Codes ASME/ANSI B32.8. Other codes such as Z187 (Canada), DnV (Norway), and IP6 (United Kingdom) have essentially the same requirements but should be checked by the readers.

Except for large diameter pipes (over 30 in.), material grade is usually taken as X-60 or X-65 (414 or 448 MPa) for high-pressure pipelines or on deepwater. Higher grades can be selected in special cases. Lower grades such as X-42, X-52, or X-56 can be selected in shallow water or for low-pressure, large diameter pipelines to reduce material cost, or in cases where high ductility is required for improved impact resistance. Pipe types are as follows:

- Seamless
- Submerged arc welded or double submerged arc welded (SAW or DSAW)
- Electric resistance welded (ERW)
- Spiral weld.

Except in specific cases, only seamless or SAW pipe is to be used, with seamless being the preference for diameters of 12 in. or less. If ERW pipe is used, special inspection provisions such as full body ultrasonic testing are required. Spiral weld pipe is very unusual for oil/gas pipelines and should be used only for low-pressure water or outfall lines.

Offshore Pipelines. DOI: http://dx.doi.org/10.1016/B978-0-12-397949-0.00003-0

3.2 DESIGN PROCEDURE

Determination of pipeline wall thickness is based on the design internal pressure or the external hydrostatic pressure. Maximum longitudinal stresses and combined stresses are sometimes limited by applicable codes and must be checked for installation and operation. However, these criteria (addressed in Chapter 6) are not normally used for wall thickness determination. Increasing wall thickness can sometimes ensure hydrodynamic stability in lieu of other stabilization methods (such as weight coating). This is not normally economical except in deepwater where the presence of concrete may interfere with the preferred installation method. Bai (2001) presents a design-through-analysis (DTA) method for pipeline sizing. In this book, we recommend the following procedure for designing pipeline wall thickness:

> *Step 1:* Calculate the minimum wall thickness required for the design internal pressure.
> *Step 2:* Calculate the minimum wall thickness required to withstand external pressure.
> *Step 3:* Add wall thickness allowance for corrosion if applicable to the maximum of the above.
> *Step 4:* Select next highest nominal wall thickness. *Note*: In certain cases, it may be desirable to order a nonstandard wall. This can be done for large orders.
> *Step 5:* Check selected wall thickness for hydrotest condition.
> *Step 6:* Check for handling practice, that is, pipeline handling is difficult for D/t larger than 50; welding of wall thickness less than 0.3 in. (7.6 mm) requires special provisions.

3.3 DESIGN CODES

3.3.1 Pipeline Design for Internal Pressure

Three pipeline codes typically used for design are ASME B31.4 (ASME, 1989), ASME B31.8 (ASME, 1990), and DnV 1981 (DnV, 1981). ASME B31.4 is for all oil lines in North America. ASME B31.8 is for all gas lines and two-phase flow pipelines in North America. DnV 1981 is for oil, gas, and two-phase flow pipelines in the North Sea. All these codes can be used in other areas when no other code is available.

The nominal pipeline wall thickness (t_{NOM}) can be calculated as follows:

$$t_{NOM} = \frac{P_d D}{2 E_w \eta \sigma_y F_t} + t_a \qquad (3.1)$$

where P_d is the design internal pressure defined as the difference between the internal pressure (P_i) and external pressure (P_e), D is nominal outside diameter, t_a is thickness allowance for corrosion, and σ_y is the specified minimum yield strength (SMYS).

Most codes allow credit for external pressure. This credit should be used whenever possible, although care should be exercised for oil export lines to account for head of fluid and for lines which traverse from deep to shallow water.

ASME B31.4 and DnV 1981 define P_i as the maximum allowable operating pressure (MAOP) under normal conditions, indicating that surge pressure up to 110% MAOP is acceptable. In some cases, P_i is defined as wellhead shut-in pressure (WSIP) or specified by the operators.

In Eq. (3.1), the weld efficiency factor (E_w) is 1.0 for seamless, ERW, and DSAW pipes. The temperature de-rating factor (F_t) is equal to 1.0 for temperatures under 250°F. The usage factor (η) is defined in Tables 3.1 and 3.2 for oil and gas lines, respectively.

The under thickness due to manufacturing tolerance is taken into account in the design factor. There is no need to add any allowance for fabrication to the wall thickness calculated with Eq. (3.1).

3.3.2 Pipeline Design for External Pressure

Different practices can be found in the industry using different external pressure criteria. As a rule of thumb, or unless qualified thereafter, it is recommended to use propagation criterion for pipeline diameters under 16 in. and collapse criterion for pipeline diameters above or equal to 16 in.

The propagation criterion is more conservative and should be used where optimization of the wall thickness is not required or for pipeline installation

TABLE 3.1 Design and Hydrostatic Pressure Definitions and Usage Factors for Oil Lines

Oil	ASME B31.4, 1989 Edition	DnV 1981
Normal Operations		
$P_d{}^a$	$P_i - P_e$ [401.2.2]	$P_i - P_e$ [4.2.2.2]
η for pipelines	0.72 [402.3.1(a)]	0.72 [4.2.2.1]
η for riser sections	No specific value use 0.50	0.50 [4.2.2.1]
P_h	1.25 $P_i{}^b$ [437.4.1(a)]	1.25 P_d [8.8.4.3]

[a]Credit can be taken for external pressure for gathering lines or flowlines when the MAOP (P_i) is applied at the wellhead or at the seabed. For export lines, when P_i is applied on a platform deck, the head fluid shall be added to P_i for the pipeline section on the seabed.
[b]If hoop stress exceeds 90% of yield stress based on nominal wall thickness, special care shall be taken to prevent overstrain of the pipe.

TABLE 3.2 Design and Hydrostatic Pressure Definitions and Usage Factors for Gas Lines

Gas	ASME B31.8, 1989 Edition 1990 Addendum	DnV 1981
Normal Operations		
$P_d{}^a$	$P_i - P_e$ [A842.221]	$P_i - P_e$ [4.2.2.2]
η for pipelines	0.72 [A842.221]	0.72 [4.2.2.1]
η for riser sections[b]	0.5 [A842.221]	0.5 [4.2.2.1]
P_h	1.25 $P_i{}^c$ [A847.2]	1.25 P_d [8.8.4.3]

[a]Credit can be taken for external pressure for gathering lines or flowlines when the MAOP (P_i) is applied at the wellhead or at the seabed. For export lines, when P_i is applied on a platform deck, the head of fluid shall be added to P_i for the pipeline section on the seabed (particularly for two-phase flow).
[b]Including prefabricated or retrofit sections and pipeline section in a J-tube.
[c]ASME B31.8 imposes $P_h = 1.4 P_i$ for offshore risers but allows onshore testing of prefabricated portions.

methods not compatible with the use of buckle arrestors such as reel and tow methods. It is generally economical to design for propagation pressure for diameters less than 16 in. For greater diameters, the wall thickness penalty is too high. When a pipeline is designed based on the collapse criterion, buckle arrestors are recommended. The external pressure criterion should be based on nominal wall thickness, as the safety factors included below account for wall variations.

3.3.2.1 Propagation Criterion

Although a large number of empirical relationships have been published, the recommended formula is the latest given by American Gas Association/ Pipeline Research Committee (AGA/PRC; AGA, 1990):

$$P_P = 33 S_y \left(\frac{t_{NOM}}{D}\right)^{2.46} \tag{3.2}$$

The nominal wall thickness should be determined such that:

$$P_P > 1.3 P_e \tag{3.3}$$

The safety factor of 1.3 is recommended to account for uncertainty in the envelope of data points used to derive Eq. (3.2). It can be rewritten as:

$$t_{NOM} \geq D \left(\frac{1.3 P_e}{33 S_y}\right)^{\frac{1}{2.46}} \tag{3.4}$$

For the reel barge method, the preferred pipeline grade is below X-60. However, X-65 steel can be used if the ductility is kept high by selecting the proper steel chemistry and microalloying. For deepwater pipelines, D/t ratios of less than 30 are recommended. It has been noted that bending loads have no demonstrated influence on the propagation pressure.

3.3.2.2 Collapse Criterion

The mode of collapse is a function of D/t ratio, pipeline imperfections, and load conditions. The theoretical background is not given in this book. An empirical general formulation that applies to all situations is provided. It corresponds to the transition mode of collapse under external pressure (P_e), axial tension (T_a), and bending strain (ε_b) as detailed in literature (Murphey and Langner, 1985; AGA, 1990).

The nominal wall thickness should be determined such that:

$$\frac{1.3 P_e}{P_C} + \frac{\varepsilon_b}{\varepsilon_B} \leq g_p \tag{3.5}$$

where the value 1.3 is the recommended safety factor on collapse, ε_B is the bending strain of buckling failure due to pure bending, and g_p is an imperfection parameter defined below.

The safety factor on collapse is calculated for D/t ratios along with the loads (P_e, ε_b, T_a) and initial pipeline out-of-roundness (δ_o). The equations are:

$$P_C = \frac{P_{el} P'_y}{\sqrt{P_{el}^2 + P'^2_y}} \tag{3.6}$$

$$P'_y = P_y \left[\sqrt{1 - 0.75 \left(\frac{T_a}{T_y} \right)^2} - \frac{T_a}{2 T_y} \right] \tag{3.7}$$

$$P_{el} = \frac{2E}{1 - \nu^2} \left(\frac{t}{D} \right)^3 \tag{3.8}$$

$$P_y = 2 S_y \left(\frac{t}{D} \right) \tag{3.9}$$

$$T_y = A S_y$$

where g_p is based on pipeline imperfections such as initial out-of-roundness (δ_o), eccentricity (usually neglected), and residual stress (usually neglected). Hence,

$$g_p = \sqrt{\frac{1 + p^2}{p^2 - (1/f_p^2)}} \tag{3.10}$$

with

$$p = \frac{P'_y}{P_{el}} \tag{3.11}$$

$$f_p = \sqrt{1 + \left(\delta_o \frac{D}{t}\right)^2} - \delta_o \frac{D}{t} \tag{3.12}$$

$$\varepsilon_B = \frac{t}{2D} \tag{3.13}$$

$$\delta_o = \frac{D_{max} - D_{min}}{D_{max} + D_{min}} \tag{3.14}$$

When a pipeline is designed using the collapse criterion, a good knowledge of the loading conditions is required (T_a and ε_b). An upper conservative limit is necessary and must often be estimated.

Under high bending loads, care should be taken in estimating ε_b using an appropriate moment–curvature relationship. A Ramberg Osgood relationship can be used as:

$$K^* = M^* + AM^{*B} \tag{3.15}$$

where $K^* = K/K_y$ and $M^* = M/M_y$ with $K_y = 2S_y/ED$ is the yield curvature and $M_y = 2IS_y/D$ is the yield moment. The coefficients A and B are calculated from the two data points on stress–strain curve generated during a tensile test.

3.3.3 Corrosion Allowance

To account for corrosion when water is present in a fluid along with contaminants such as oxygen, hydrogen sulfide (H_2S), and carbon dioxide (CO_2), extra wall thickness is added. A review of standards, rules, and codes of practices (Hill and Warwick, 1986) shows that wall allowance is only one of several methods available to prevent corrosion, and it is often the least recommended.

For H_2S and CO_2 contaminants, corrosion is often localized (pitting) and the rate of corrosion allowance ineffective. Corrosion allowance is made to account for damage during fabrication, transportation, and storage. A value of 1/16 in. may be appropriate. A thorough assessment of the internal corrosion mechanism and rate is necessary before any corrosion allowance is taken.

3.3.4 Check for Hydrotest Condition

The minimum hydrotest pressure for oil and gas lines is given in Tables 3.1 and 3.2, respectively, and is equal to 1.25 times the design pressure for pipelines. Codes do not require that the pipeline be designed for hydrotest conditions, but sometimes give a tensile hoop stress limit 90% SMYS, which is always satisfied if credit has not been taken for external pressure. For cases where the wall thickness is based on $P_d = P_i - P_e$, codes recommend not to overstrain the pipe. Some of the codes are ASME B31.4 (Clause 437.4.1), ASME B31.8 (no limit on hoop stress during hydrotest), and DnV (Clause 8.8.4.3).

For design purposes, condition $\sigma_h \le \sigma_y$ should be confirmed, and increasing wall thickness or reducing test pressure should be considered in other cases. For pipelines connected to riser sections requiring $P_h = 1.4P_i$, it is recommended to consider testing the riser separately (for prefabricated sections) or to determine the hydrotest pressure based on the actual internal pressure experienced by the pipeline section. It is important to note that most pressure testing of subsea pipelines is done with water, but on occasion, nitrogen or air has been used. For low D/t ratios (less than 20), the actual hoop stress in a pipeline tested from the surface is overestimated when using the thin wall equations provided in this chapter. Credit for this effect is allowed by DnV Clause 4.2.2.2, but is not normally taken into account.

REFERENCES

American Gas Association, 1990. Collapse of offshore pipelines. Pipeline Research Committee. Seminar held in Houston, TX.

American Society of Mechanical Engineers, 1989. Liquid transportation systems for hydrocarbons, liquid petroleum gas, anhydrous ammonia and alcohols. ASME B31.4—1989 edition.

American Society of Mechanical Engineers, 1990. Gas transmission and distribution piping systems. ASME Code for Pressure Piping, B31.8—1989 edition and 1990 addendum.

Bai, Y., 2001. Pipelines and Risers, vol. 3. Elsevier Ocean Engineering Book Series, Amsterdam.

Det norske Veritas, 1981. Rules for submarine pipeline systems (1981 edition).

Hill, R.T., Warwick, P.C., 1986. Internal corrosion allowance for marine pipelines: a question of validity. In: Presented at the Offshore Technology Conference, 5–8 May, Houston, Texas, OTC Paper 5268.

Murphey, C.E., Langner, C.G. 1985. Ultimate pipe strength under bending, collapse, and fatigue. In: Chung, J., Lunardini, V. (Eds.), Proceedings of the Fourth International Conference on Offshore Mechanics and Arctic Engineering (OMAE'85), February, Dallas.

Chapter 4

Hydrodynamic Stability of Pipelines

4.1 INTRODUCTION

This chapter addresses stability analysis of marine pipelines on the seabed under hydrodynamic loads (wave and current) and provides guidelines for pipeline stabilization using concrete coating. It does not address alternative methods such as pre- or post-trenching techniques, mattress covers, etc. Stability is checked for the installation case with the pipe empty using the 1-year return period condition and for lifetime (pipe with concrete) using the 100-year storm.

4.2 ANALYSIS PROCEDURE

There are several basic approaches to determining the required submerged weight for a marine pipeline. One of them is use of AGA program "LSTAB" code. It should be used in cases where the pipe is partially embedded or pre-trenched as the lift, drag, and inertia coefficients are adjusted for exposure. Regardless of the computer program selected, hydrodynamic stability analysis involves the following steps:

Step 1: Collect or define environmental criteria for the 1-year and 100-year conditions, including:
- water depth
- wave spectrum
- current characteristics
- soil properties
- seabed condition.

Step 2: Determine hydrodynamic coefficients: drag (C_D), lift (C_L), and inertia (C_I). These may be adjusted for Reynolds number, Keulegan number, ratio of wave to steady current, and embedment.

Step 3: Calculate hydrodynamic forces, typically, drag (F_D), lift (F_L), and inertia (F_I).

Step 4: Perform static force balance at time step increments and assess stability and calculate concrete coating thickness for worst combination of lift, drag, and inertial force.

Offshore Pipelines. DOI: http://dx.doi.org/10.1016/B978-0-12-397949-0.00004-2

Hydrodynamics stability is determined using Morison's Equation which relates hydraulic lift, drag, and inertial forces to local water particle velocity and acceleration. The coefficients used, however, vary significantly from one situation to another. For example, the lift and drag coefficients of 0.6 and 1.2, which are representative of a steady current, are not appropriate for oscillating flow in a wave field. In addition, these coefficients are reduced if the pipe is not fully exposed because of trenching or embedment. The literature is extensive in this area of research and is summarized in the next section.

To determine wave particle velocity, the equations used depend on wave height, water depth, and wave period. Figure 4.1 indicates the domain of applicability of the various theories. For most situations, linear theory is adequate as bottom velocities and accelerations do not vary significantly between theories. However, as the wave height to water depth ratio increases, Stokes fifth order theory becomes appropriate.

For shallow water or very high wave heights, a cnoidal or solitary theory should be used to predict particle velocity and accelerations (Sarpkaya and Isaacson, 1981). For breaking waves, or large diameter pipe, which may affect the flow regime, other analysis methods may be appropriate. In general, pipelines should be trenched within the breaking wave (surf) zone.

Experimental and theoretical research (Ayers, 1989; Sarpkaya and Isaacson, 1981) has shown that traditional static analysis methods have been conservative in most cases, understanding hydrodynamic forces but ignoring the effect of pipe embedment. In the 1980s, two research groups developed theoretical and experimental models to assess pipe stability. Findings of these groups (AGA in USA and PIPESTAB in Europe) resulted in the development of program LSTAB that accounts for the effects of embedment.

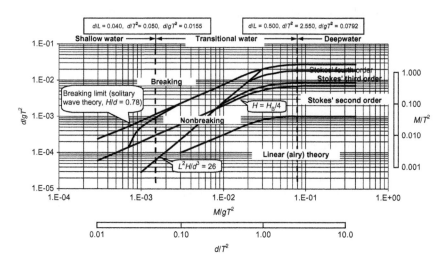

FIGURE 4.1 Domain of applicability of the various theories.

4.3 METHODOLOGY

4.3.1 Definitions of Environmental Criteria

Environmental criteria during operation and pipeline installation can be estimated based on prior projects in the same area for feasibility studies or preliminary design purposes. For final design, measurements of steady current in the water column should be conducted to enable prediction by extrapolating extreme values for long return periods (1-year to 100-year period).

Wave criteria (height and period) can be developed by hindcasting techniques based on existing wind and wave data in the area and further offshore. Wave hindcasting methods may involve refraction and shoaling analyses as well, to take into account local bathymetry irregularities.

4.3.1.1 Design Waves

An important design step is the estimation of an extreme design wave on the basis of recorded or hindcast wave data. This generally involves selecting and fitting a suitable probability distribution to wave height data and extrapolating to a suitable design wave with a predetermined return period. The actual selection of a design wave is a matter of engineering judgment and will vary according to the risk chosen for the design. The selection of a design wave is to be carried out in the following stages:

1. Oceanographic data is collected over a long time (at least 5 years, depending upon the design return period) at the site of interest. Alternatively, a hindcasting technique may be used to provide data over a much larger time period.
2. A plotting formula is used to reduce the data to a set of points describing the probability distribution of wave heights. Two such formulas are the Gumbel or external function and the Weibull function.
3. These points are plotted on extreme value probability paper corresponding to the desired probability distribution function.
4. A straight line is fitted through the points to present a trend.
5. The line is extrapolated to locate a design value corresponding to a chosen return period or chosen encounter probability.

The significant wave height is often used as the main parameter to define a sea state. Statistically, significant wave height is the average of the one-third highest waves and is denoted as H_s or $H_{1/3}$. Empirically, H_s is significant wave height in a sea state and is the value most often assigned by visual observations. Some other reference is sometimes used such as $H_{3\%}$ in the formal Soviet Union.

The maximum expected wave height (H_{max}) can be derived from the significant wave height by:

$$H_{max} = H_s \sqrt{\frac{1}{2} \ln N_o} \qquad (4.1)$$

where N_o is the number of observed waves. Typically for 1000 waves, $H_{max} = 1.68\ H_s$. In addition to wave height, a characteristic wave period must also be given to define a sea state. The average zero crossing period, T_z, is commonly used. This period is defined as the average time between consecutive up or down crossing of the mean sea level. Finally, irregular seas must also be described by a given wave spectrum. A large number of empirical formulations for wave spectra have been proposed for different conditions such as a fully arisen sea, a short fetch sea, combined seas, etc. Most empirical spectra have the following basis form:

$$S_n(f_w) = \frac{c_1}{f_w^m} \exp\left(-\frac{c_2}{f_w^n}\right) \tag{4.2}$$

where f_w is wave frequency in Hz, $S_n(f_w)$ is wave spectral density in ft^2/Hz, c_1 and c_2 are dimensional constants related to significant wave height and period, and m and n are integer coefficients. There are several common one-dimensional frequency spectra used to describe ocean waves including Bretschneider, Pierson−Moskowitz, and JONSWAP.

4.3.1.2 Wave Refraction

The use of wave fraction techniques allows accurate transformation of deep-water wave data (however it was obtained) into shallower water and allows an estimation of when the maximum wave height/breaking criteria takes over. A complete wave refraction/shoaling analysis can be done in two ways. The first way is by manual means, as shown below, and provides a good, quick method suitable for preliminary studies. The second method involves the same principle, but is done by computer on a detailed grid.

The manual wave refraction method described here is the "Forward Ray" method, where the hand-constructed rays travel from deep to shallow water. A better method is the "Reverse Ray" technique which uses basically the same equations, but can provide a better answer, consisting of refraction coefficients for all wave directions at one site, faster than the forward technique. The forward technique can illustrate areas of wave height concentration along a coastline.

When waves approach a bottom slope obliquely, they travel slower in the shallower water depth, causing the line of the wave crest to bend toward alignment with the bottom contours. The process is known as wave refraction. The change of direction of wave orthogonals (lines perpendicular to the wave crests) from deep to shallower water may be approximated by Snell's Law:

$$\frac{C_2}{C_2} = \frac{\sin \alpha_1}{\sin \alpha_2} \tag{4.3}$$

where α_1 is the angle a wave crest makes with the bottom contour over which it is passing, α_2 is the angle a wave crest makes with the next bottom contour over which it is passing, C_1 is wave velocity at depth of first bottom contour, and C_2 is wave velocity at depth of second bottom contour.

The assumptions made in a refraction analysis include:

1. Wave energy between wave orthogonals remains constant.
2. Wave direction is perpendicular to the wave crests in the direction of the orthogonals.
3. Speed of a wave of a given period at a particular location depends only on the water depth at the location.
4. Bottom topography changes are gradual.
5. Effects of currents, winds, wave reflections, and underwater topographic variations are negligible.
6. Waves are constant period airy waves.

Under the above assumptions, the following wave height relationship can be derived:

$$H = H_o \sqrt{\frac{b_{wo} C_o}{2 b_w CN}} \qquad (4.4)$$

where H is wave height at water depth d in feet, H_o is deepwater wave height in feet, b_{wo} is deepwater spacing between orthogonals in feet, C_o is deepwater wave velocity in feet per second, b_w is spacing between orthogonals at water depth d in feet, C is wave velocity at water depth (d) in feet per second, and N is expressed as:

$$N = \frac{1}{2}\left[1 + \frac{(4\pi d/L)}{\sinh(4\pi d/L)}\right] \qquad (4.5)$$

where L is wave length at water depth d in feet. The analysis procedure used to determine maximum or limiting wave heights at a given location is as follows:

1. Define and draw bottom contours over sea of interest.
2. Calculate wave velocities at selected bottom contours moving into deepwater.
3. Select an angle of wave attack at the desired location.
4. Calculate and construct wave crest angles moving into deepwater.
5. In deepwater, define an orthogonal spacing, b_{wo}.
6. Perform a second wave crest angle analysis shoreward on the new wave track.
7. At the desired location, measure orthogonal spacing, b_w.
8. Calculate wave height at desired location.
9. Define new angles of wave attack at the desired location and repeat Steps 4 through 9.

The analysis can be performed by hand as outlined above, or by using refraction templates. Refraction diagrams can provide information on the change in waves approaching a shore. Analysis validity is limited by depth data accuracy and the preciseness of the model.

4.3.1.3 Wave Shoaling

When a wave moves into shallower water, its wave height and wave length change. This process is described as shoaling. The effect of shoaling may be estimated from any particular wave theory under the following assumptions:

1. Motion is two-dimensional.
2. Wave period remains constant.
3. Average rate of energy transfer is constant in direction of wave propagation.
4. Wave theory applies at all water depths considered.

These assumptions are often valid until the wave breaks.

Using airy wave theory, the following comparative relationships can be approximated (Sarpkaya and Isaacson, 1981):

$$\frac{L_w}{L_o} = 2\pi\sqrt{\frac{d}{gT^2}} = \sqrt{2\pi\frac{d}{L_o}} \tag{4.6}$$

and

$$\frac{H}{H_o} = \frac{1}{\sqrt[4]{16\pi^2\frac{d}{gT^2}}} = \frac{1}{\sqrt[4]{8\pi\frac{d}{L_o}}} \tag{4.7}$$

where

d = water depth (ft),
T = wave period (s),
g = gravitational acceleration, 32 ft/s^2,
L_w = wave length (ft),
H = wave height (ft),
L_o = deepwater wave length (t),
H_o = deepwater wave height (ft).

For the above airy wave theory case, graphical aids may be developed. Selection of different wave theories will result in similar but different relationships.

Note that environmental data such as wind, waves, and currents can have specific interrelationships. A common assumption of taking the combined maximum effect of each may not always produce the worst design conditions, and, in some cases, joint statistics of current and wave should be considered.

4.3.1.4 Soil Friction Factor

Friction factor (μ) is defined as the ratio between the force required to move a section of pipe and the vertical contact force applied by the pipe on the seabed. This simplified model (Coulomb) is used to assess stability. The friction factor depends on the type of soil, the pipe roughness, seabed slope, and depth of burial. For practical purposes, only the type of soil is considered and the pipe roughness ignored.

For stability analysis, a lower bound estimate for soil friction is conservatively assumed, whereas for pulling or towing analysis, an upper bound estimate would be appropriate. The following lateral friction factors are given as guidelines for stability analysis in the absence of site-specific data:

Loose sand: $\mu = \tan \phi$ (generally $\phi = 30°$)
Compact sand: $\mu = \tan \phi$ (generally $\phi = 35°$)
Soft clay: $\mu = 0.7$
Stiff clay: $\mu = 0.4$
Rock and gravel: $\mu = 0.7$.

These coefficients represent the "best" estimate for generalized soil types and do not include safety factors.

Small scale tests (Lyons, 1973) and offshore tests (Lambrakos, 1985) have shown that the starting friction factor in sand is about 30% less than the maximum value which occurs after a very small displacement of the pipe builds a wedge of soil; past this point, the friction factor levels off. The value given above accounts for the build-up of this wedge of soil which has been shown to take place. The Coulomb model underestimates the actual lateral soil resistance if settlement is anticipated.

4.3.2 Hydrodynamic Coefficient Selection

Hydrodynamic coefficients have been the subject of numerous theoretical and experimental investigations and are often subject to argumentation. The purpose of this section is to provide a method for selection of C_D, C_L, and C_I for one of the following three situations:

- Steady current only
- Waves only
- Steady current and waves.

4.3.2.1 Steady Current Only

The C_D and C_L depend on pipe's roughness and Reynolds number. Figure 4.2 provides graphs of lift and draft coefficients for these parametric considerations (Jones, 1976).

Pipe roughness is defined as the ratio between the mean roughness height and the pipe diameter, i.e., $R_r = \kappa/D$. For fusion bonded epoxy (FBE)-coated

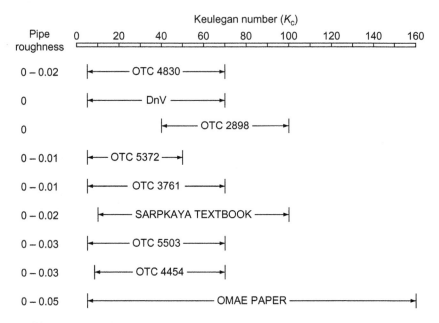

FIGURE 4.2 Hydrodynamic coefficient selection flowchart for wave acting alone.

pipe (smooth), R_r should be taken as 0. For other coatings when κ is not known, an approximate value must be estimated knowing that the hydrodynamic drag increases as R_r increases, while the lift coefficient decreases. The following κ values are given as guidelines:

FBE, yellow jacket: $\kappa = 0$ (fine)
Concrete coating or abrasion coating: $\kappa = 0.1$ in. (medium)
Marine growth (barnacles): $\kappa = 1.6$ in. (rough) (Teng and Nath, 1989)
Marine growth (anemones): $\kappa = 2.7$ in. (rough) (Teng and Nath, 1989).

Reynolds number is defined as ratio between inertial force and viscous force, i.e., $Re = U_c D / \nu_k$. The seawater kinematic viscosity (ν_k) increases as the seawater temperature decreases. In deepwater or cold water, the following value should be used:

$$\nu_k = 1.7\text{E-5 ft}^2/\text{s} \ (1.57\text{E-6 m}^2/\text{s}) \text{ at } 40°\text{F} \ (5°\text{C})$$

In warm waters or for hot pipes, the following value should be used:

$$\nu_k = 1.0\text{E-5 ft}^2/\text{s} \ (0.92\text{E-6 m}^2/\text{s})$$

Hydrodynamic coefficients increase as Re decreases (for the range of Reynolds number normally encountered), which justifies the use of a conservative high ν_k value.

4.3.2.2 Waves Acting Alone

The hydrodynamic coefficients (C_D, C_L, C_I) depend on pipe roughness and Keulegan number $K_c = U_w T/D$. For a pipe resting on the seabed, five references are applicable as shown in Figure 4.2 (Bryndum, 1983; Bryndum et al., 1983; Det norske Veritas, 1981; Verley and Lambrakus, 1987; Zdravkovich, 1977). The references show that C_D presents a peak for K_c values between 10 and 20, C_L decreases with increasing K_c values, and C_I increases with increasing K_c values. Figure 4.3 is recommended for K_c values

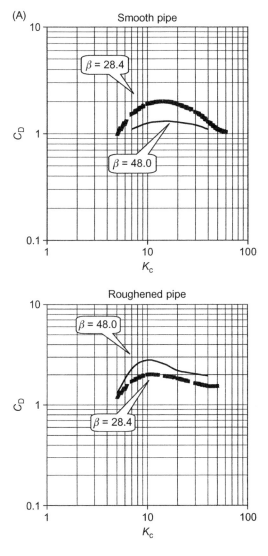

FIGURE 4.3 Hydrodynamic coefficients for wave acting alone in the low K_c region—(A) C_D, (B) C_L, and (C) C_I.

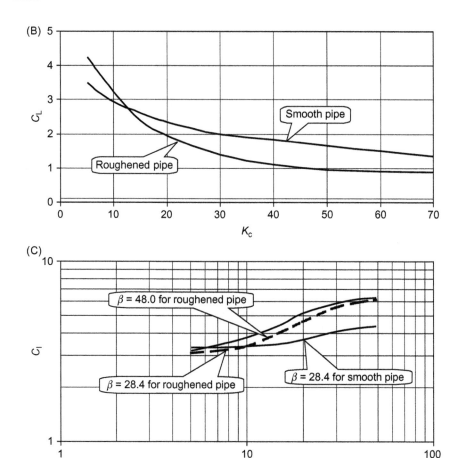

FIGURE 4.3   (Continued)

less than 20 (Sarpkaya and Rajabi, 1979). The frequency parameter (β) is defined as $\beta = D\nu/T$. Figure 4.4 is for K_c values greater than 20 but less than 160 (Bryndum et al., 1983). Figure 4.5 is for K_c values greater than 160 (Jones, 1976).

4.3.2.3 Waves and Currents Acting Simultaneously

In addition to the variables previously mentioned, the steady current ratio $R_c = U_c/U_m$ must be taken into account for the selection of C_D, C_L, and C_I. Another current ratio is sometimes used (Bryndum et al., 1983) and noted $\alpha = U_c/U_w$. Note that K_c is based on particle velocity U_w and not maximum velocity U_m.

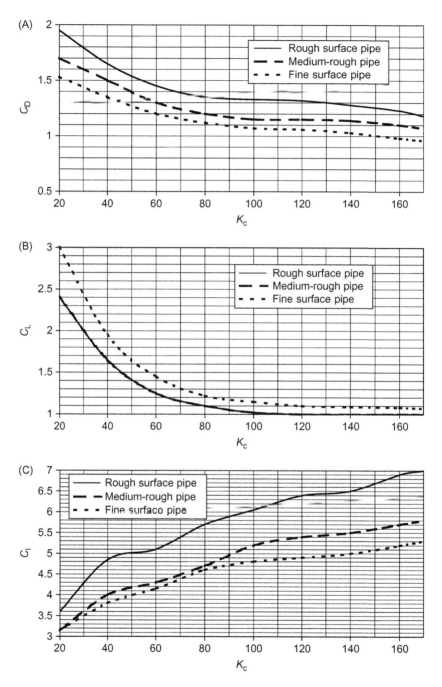

FIGURE 4.4 Hydrodynamic coefficients for wave acting alone in the high K_c region—(A) C_D, (B) C_L, and (C) C_I.

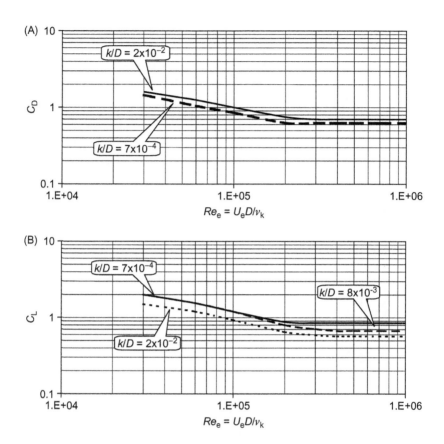

FIGURE 4.5 Recommended effective coefficient for design of pipeline resting on bottom subject to current acting alone—(A) drag coefficient C_D and (B) lift coefficient C_L.

Refer to the selection flowchart (Figure 4.6) to identify the relevant reference. Only three papers apply to pipelines resting on the bottom. They are OMAE paper 1988 (Bryndum et al., 1983), OTC paper 4454 (Bryndum, 1983), and OTC paper 5852 (SINTEF, 1988). Experimental tests (Bryndum et al., 1983) have shown that the presence of a steady current leads, in all cases, to a reduction of the hydrodynamic coefficients. For $15 < K_c < 70$, Figure 4.7 should be used (Bryndum et al., 1983). This requires the use of $\alpha(=U_c/U_w)$ for small values. For K_c values less than 15, Figure 4.8 should be used (Bryndum, 1983). Figures 4.2 and 4.6 show other references, which apply unusual situations such as pipelines away from a plane boundary. These may be used for pipes or pipeline spans, but care should be taken when deviating from this guideline. Note that the election of coefficients depends on the value U_w. This velocity can be calculated using a computer program or manual calculation.

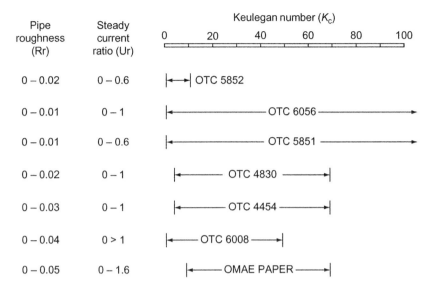

FIGURE 4.6 Hydrodynamic coefficient selection flowchart for waves and currents acting simultaneously.

4.3.3 Hydrodynamic Force Calculation

The third step of the stability analysis involves the determination of the hydrodynamic drag force (F_D), lift force (F_L), and inertia force (F_I), represented by the Morison Equations:

$$F_D = \frac{1}{2} C_D \rho D U_m |U_m| \tag{4.8}$$

$$F_L = \frac{1}{2} C_L \rho D U_m^2 \tag{4.9}$$

$$F_I = C_I \rho \left(\frac{\pi D^2}{4} \right) U_w \tag{4.10}$$

4.3.4 Hydrodynamic Stability Assessment

The last step of the analysis consists of assessing stability and computing concrete coating thickness requirement, using the AGA program LSTAB. Seabed slope and safety factor should be considered.

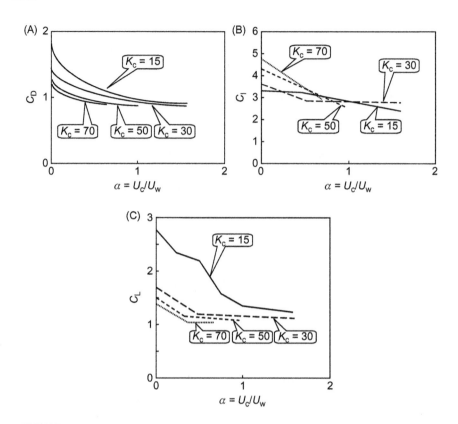

FIGURE 4.7 (A) Hydrodynamic coefficients for waves and currents acting simultaneously— (A) C_D, (B) C_I and (C) C_L.

A pipeline is stable on a slope (δ) if its submerged weight (W_s) satisfies the following relationship:

$$\mu(W_s\cos\delta - F_L) \geq \zeta[(F_D+F_D)_{max} + W_s\sin\delta] \qquad (4.11)$$

where ζ is a safety factor. This formulation assumes a Coulomb friction model and is not applicable if the pipe is embedded. A preliminary conservative approach, however, is to consider no embedment.

The safety factor is designed to account for uncertainties in actual soil factor, actual environmental data (wave, current), actual particle velocity and acceleration, and actual hydrodynamic coefficients. The safety factor can be imposed by the pipe operator, the governing code, or it can be selected by the engineer, depending on the design conditions. Recommended safety factors are $\zeta = 1.05$ for installation and $\zeta = 1.1$ for operation. The latter is also

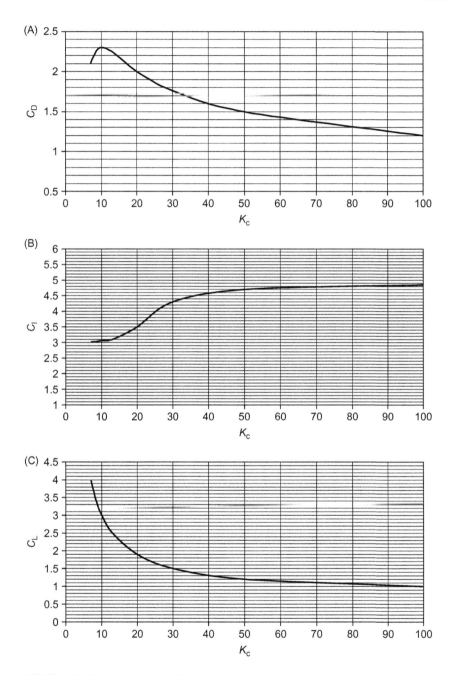

FIGURE 4.8   Hydrodynamic coefficients for waves and current acting simultaneously—(A) C_D, (B) C_I, and (C) C_L. Reduction of hydrodynamic coefficients due to superimposed steady current—(D) C_D and (E) C_L.

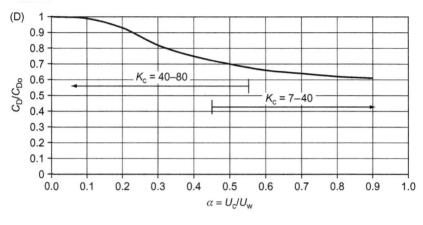

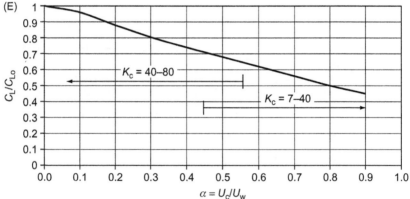

FIGURE 4.8   (Continued)

recommended by DnV RP E305 Clause 3.2.2 (DnV, 1988), DnV 1981 Clause 4.2.5.9 (DnV, 1981), and Canadian CAN/CSA.Z187.

Safety factors depend on how conservative a design has been conducted and should cover any uncertainty while, at the same time, avoid compounding conservatism. For example, nominal dimensions are normally used to calculate pipeline submerged weight; however, for large diameter pipes, manufacturing tolerances and sometimes seawater density should be taken into account.

4.4 PARTIALLY BURIED PIPELINES

This section indicates "rules of thumb" to determine stability of partially buried lines. It involves the determination of the breakout force on a partially buried pipeline under oscillatory loading, and the selection of modified hydrodynamic coefficients. Tests were for sand and clay conditions where embedment due to cyclic pipe motions may occur. For partially buried or settled pipelines, LSTAB program should be used. Different considerations apply to partially buried pipelines in sand and in clay.

Recent Norwegian compilation of existing test data (SINTEF, 1988) gives a simplified model for pipelines partially buried in sand. For a burial depth of such that $0 < z = 0.35 \ D$, the ultimate breakout force F_h is given by:

$$F_h = F_f + F_R \tag{4.12}$$

where F_R is the penetration dependent soil resistance given by:

$$F_R = 0.72 \ W_s(0.87D_r^2 - 1.96 \ D_r + 1.1) \tag{4.13}$$

where D_r is relative density of sand. For dense sand, $D_r = 0.46$. For loose sand, $D_r = 0.05$.

For pipelines partially buried in clay, the ultimate lateral resistance (R_L) depends on the pipe embedment (z), cohesion at the base of the pipe (C_b), and surface of contact (b_s) between the pipe and the seabed. The following equations should be used to determine z-value (Wantland, 1979):

$$4C_b + \gamma'z = 2.5\frac{(W_s\cos \delta - F_L)}{b_s} \tag{4.14}$$

where

$$b_s = 2z\sqrt{\frac{D}{z} - 1} \quad \text{for} \quad z < \frac{D}{2} \tag{4.15}$$

$$b_s = D \quad \text{for} \quad z \geq \frac{D}{2} \tag{4.16}$$

and

$$R_L = N_s C_a D \tag{4.17}$$

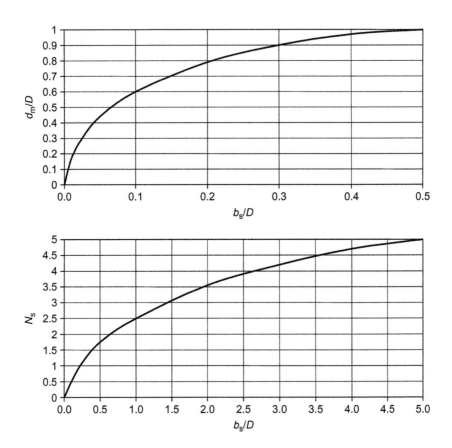

FIGURE 4.9   Relative embedment (d_m/D) and lateral stability coefficient N_s as functions of relative embedment (b_s/D).

where C_a is the average cohesion over two pipe diameters and N_s is given by Figure 4.9 (Wantland, 1979). Pipeline stability must be based on:

$$R_L > \zeta(F_{D_d} + F_I + W_s \sin \delta)_{max} \tag{4.18}$$

where δ is the seabed slope and ζ is the safety factor $= 1.1$.

The effects of embedment or pre-trenching reduce hydrodynamic coefficients. Refer to Figure 4.10 for the reduction factor associated with embedment d_m. Figure 4.11 is for the reduction factor related to pre-trenching, where H_T is trench height and a_T is trench slope ratio.

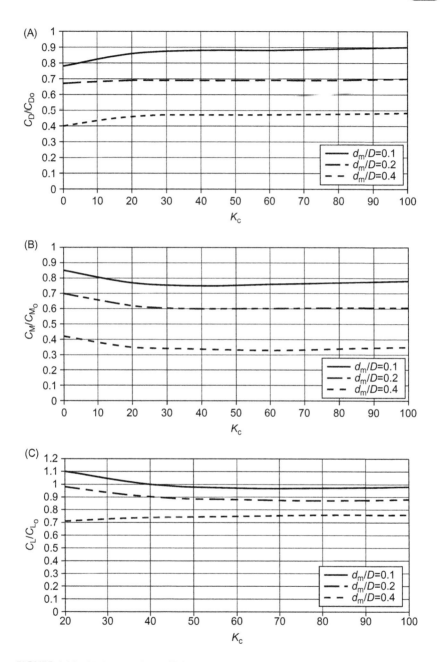

FIGURE 4.10 Hydrodynamic coefficients reduction factor for partially buried pipelines—(A) C_D, (B) C_M, (C) C_L, (D) C_V, (E) C_D, (F) C_M, and (G) C_L.

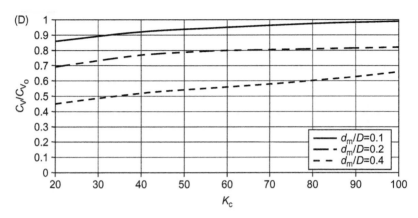

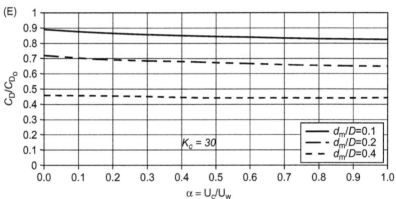

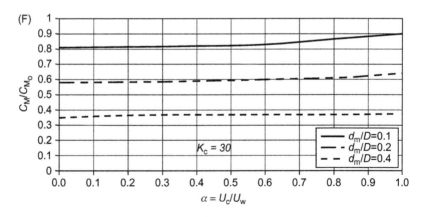

FIGURE 4.10   (Continued)

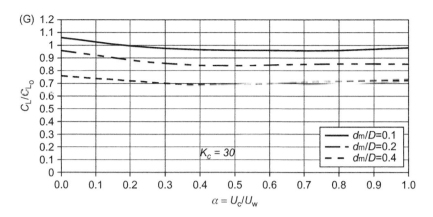

FIGURE 4.10 (Continued)

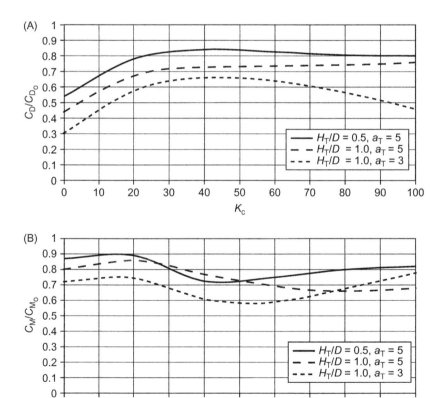

FIGURE 4.11 Hydrodynamic coefficient reduction factor for pipelines in open trench—(A) C_D, (B) C_M, and (C) C_L.

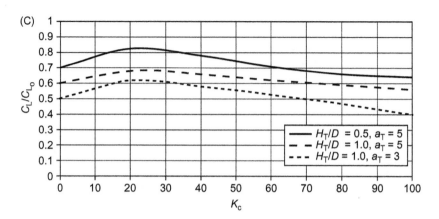

FIGURE 4.11   (Continued)

REFERENCES

Ayers, R.R., 1989. Submarine on-bottom stability: recent AGA Research. Presented at the Eighth International Conference on Offshore Mechanics and Arctic Engineering held March 19−23, in The Hague.

Bryndum, M.B., 1983. Hydrodynamic forces from wave and current loads on marine pipelines. OTC Paper 4454.

Bryndum, M.B., Jacobsen, V., Tsahalis, D.T., 1983. Hydrodynamic forces of pipelines: model tests. Proceedings of the Seventh International Conference on Offshore Mechanics and Arctic Engineering.

Det norske Veritas, 1981. Rules for submarine pipeline design.

Jones, W.T., 1976. On-bottom pipeline stability in steady water currents. OTC Paper 2598.

Lambrakos, K.F., 1985. Marine pipeline soil friction coefficients form in situ testing. Ocean Eng. 12 (2), 131−150.

Lyons, C.G., 1973. Soil resistance to lateral sliding of marine pipelines. OTC Paper 1876.

Sarpkaya, T., Isaacson, M., 1981. Mechanics of Wave Forces on Offshore Structures. Van Nostrand Reinhold Company, New York, NY.

Sarpkaya, T., Rajabi, F., 1979. Hydrodynamic drag on bottom-mounted smooth and rough cylinders in periodical flow. OTC Paper 3761.

SINTEF, 1988. Report: energy based pipe−soil interaction models. American Gas Association Contract PR194−719.

Teng, C.C., Nath, J.H., 1989. Hydrodynamic forces on roughened horizontal cylinders. OTC Paper 6008.

Verley, R.L.P., Lambrakos, K.F., 1987. Prediction of hydrodynamic forces on seabed pipelines, OTC Paper 5503.

Wantland, G.M., 1979. Lateral stability of pipelines in clay. OTC Paper 3477.

Zdravkovich, M.M., 1977. Review of flow interference between two circular cylinders in various arrangements. Trans. ASME December, 167−172.

FURTHER READING

Dalton, C., Szabo, J.M., 1977. Drag on a group of cylinders. Trans. ASME, J. Pressure Vessel Technol. 99 (1), 152−157.

Det norske Veritas, 1988. On-bottom stability design of submarine pipelines. Recommended practice E305, October.

Ismail, N.M., 1986. Wave forces on partially buried submarine pipelines. In: Presented at the Offshore Technology Conference, 5−8 May, Houston, Texas, OTC Paper 5295.

Jacobsen, V., 1988. Forces on sheltered pipelines. In: Presented at the Offshore Technology Conference, 2−5 May, Houston, Texas, OTC Paper 5851.

Sarpkaya, T., 1977. In-line and transverse forces on cylinders near a wall in oscillatory flow at high Reynolds numbers. In: Presented at the Offshore Technology Conference, 2−5 May, Houston, Texas, OTC Paper 2898.

Wagner, D.A., 1987. Pipe soil interaction model. In: Presented at the Offshore Technology Conference, 27−30 April 1987, Houston, Texas, OTC Paper 5504.

Wilkinson, R.H., Palmer, A.C., 1988. Field measurements of wave forces on pipelines. In: Presented at the Offshore Technology Conference, 2−5 May, Houston, Texas, OTC Paper 58852.

Chapter 5

Pipeline Span

5.1 INTRODUCTION

Pipeline spanning can occur when the contact between the pipeline and seabed is lost over an appreciable distance on a rough seabed. An evaluation of an allowable free-span length is required in pipeline design. Should actual span lengths exceed the allowable length, correction is then necessary to reduce the span to avoid pipeline damage. The flow of wave and current around a pipeline span can result in the generation of sheet vortices in the wake. These vortices are shed alternately from top to bottom of the pipeline resulting in an oscillatory force exerted on the span. This chapter provides information about the determination of the allowable pipeline span length, based on the avoidance of vortex shedding-induced oscillations. Both in-line and cross-flow vortex shedding-induced oscillations will be discussed and evaluated. There is also a brief discussion and equations are presented for calculating the fatigue life of a pipeline based on the stresses incurred due to vortex shedding-induced oscillations in a free span. This chapter will provide information based only on vortex shedding-induced oscillations due to currents, which in most deepwater pipelines is the limiting factor for the allowable span length. It does not address vortex shedding-induced oscillations due to wave motions or wave and current combined motions.

5.2 PROBLEM DESCRIPTION

5.2.1 Free Span

Free span can result in failure of pipelines due to excessive yielding and fatigue. It may also cause interference with human activities such as fishing. Free span can occur due to unsupported weight of the pipeline section and dynamic loads from waves and currents. When a fluid flows across a pipeline, the flow separates, vortices are shed, and a periodic wake is formed. Each time a vortex is shed it alters the local pressure distribution, and the pipeline experiences a time-varying force at the frequency of vortex shedding. Under resonant conditions, sustained oscillations can be excited, and the pipeline will oscillate at a frequency. This oscillation will fatigue the pipeline and can eventually lead to catastrophic failure. These oscillations

Offshore Pipelines. DOI: http://dx.doi.org/10.1016/B978-0-12-397949-0.00005-4

are normally in-line with the flow direction but can be transverse (cross-flow), depending on current velocity and span length.

5.2.2 In-Line Oscillations

In-line oscillations are excited at flow velocities lower than the critical velocities for cross-flow motion. However, the amplitude of the in-line motion is only 10% of those associated with cross-flow motion. Several parameters are used in determining the potential for vibration. These include the reduced velocity, U_r, and the stability parameter, K_s. These are defined in later sections.

The first and second modes of in-line instability are associated with symmetrical vortex shedding and have a peak response at reduced velocities (U_r) of 1.9 and 2.6, respectively. To prevent this in-line response at either mode of vortex-shedding excitation, it is suggested that the stability parameter (K_s) be larger than 1.8 (Wootton, 1991). DnV also states that resonant in-line vortex shedding-induced oscillation may occur when $1.0 < U_r < 2.2$, the shedding will be symmetrical; and for $U_r > 2.2$, the shedding will be alternate.

5.2.3 Cross-Flow Oscillations

Excitation in the cross-flow direction is potentially more dangerous than that in the in-line direction since amplitudes of response are much greater than those associated with in-line motion. However, these oscillations occur at much larger velocities than in-line oscillations and are not normally governing. The limiting value for cross-flow oscillation based on DnV is $K_s < 16$ (DnV, 1981; Wootton, 1991).

5.2.4 Galloping

Galloping is a form of oscillation that occurs for certain structural shapes and flow directions. Circular sections such as pipelines do not gallop because there can be no steady force on a circular cylinder other than drag. Therefore, galloping is not of concern to the pipeline engineer when evaluating the allowable span length. For noncircular shapes, it has been found to occur only in steady flows, but it may also occur in wave flows with long wavelengths. The motion is normal to the direction of flow and amplitude increases with increasing flow speed. Galloping generally occurs only when the reduced velocity is greater than the values for dynamic response to vortex shedding in steady flow.

5.3 DESIGN CONSIDERATIONS

5.3.1 Dynamic Stresses

The presence of bottom currents can cause significant dynamic stresses, if fluid structure interaction (vortex shedding) in these free-span areas causes the pipeline to oscillate. These oscillations can result in fatigue of the pipeline welds, which can reduce pipeline life. The frequency of vortex shedding is a function of the pipe diameter, current velocity, and Strouhal number. If the vortex-shedding frequency (also referred to as the Strouhal frequency) is synchronized with one of the natural frequencies of the pipeline span, then resonance occurs and the pipe span vibrates. Pipeline failure due to vortex-excited motions can be prevented if the vortex-shedding frequency is sufficiently far from the natural frequency of the pipe span such that dynamic oscillations of the pipe are minimized.

5.3.2 Vortex-Shedding Frequency

The vortex-shedding frequency is the frequency at which pairs of vortices are shed from the pipeline and is calculated based on the following:

$$f_s = \frac{SU_c}{D} \tag{5.1}$$

where

f_s = vortex-shedding frequency,
S = Strouhal number,
U_c = design current velocity,
D = pipe outside diameter.

Strouhal number is the dimensionless frequency of the vortex shedding and is a function of the Reynolds number. Reynolds number, Re, is a dimensionless parameter representing the ratio of inertial force to viscous force:

$$Re = \frac{U_c D}{\nu_k} \tag{5.2}$$

where ν_k is kinematic viscosity of fluid (1.2×10^{-5} ft^2/s for water at 60°F).

5.3.3 Pipeline Natural Frequency

The natural frequency of the pipeline span depends on pipe stiffness, end conditions of the pipe span, length of the span, and effective mass of the pipe. The natural frequency for vibration of the pipe span is given by the following formulas:

$$f_n = \frac{C_e}{2\pi} \sqrt{\frac{EI}{M_e L_s^4}}$$
(5.3)

where

f_n = pipe span natural frequency,
L_s = span length,
M_e = effective mass,
C_e = end condition constant.

The end condition constant is a function of the type of model that is selected in determining the support conditions of the pipeline span. The following values are used based on these end conditions:

$C_e = (1.00\pi)^2 = 9.87$ (pinned−pinned),
$C_e = (1.25\pi)^2 = 15.5$ (clamped−pinned),
$C_e = (1.50\pi)^2 = 22.2$ (clamped−clamped).

The effective mass is the sum of total unit mass of the pipe, the unit mass of the pipe contents, and the unit mass of the displaced water (added mass).

$$M_e = M_p + M_c + M_a$$
(5.4)

where

M_p = unit mass of pipe including coatings (slugs/ft or kg/m),
M_c = unit mass of contents (slugs/ft or kg/m),
M_a = added unit mass (slugs/ft or kg/m).

The added mass is the mass of water displaced by the pipeline and is calculated based on the following equation:

$$M_a = \frac{\pi D^2 \rho}{4}$$
(5.5)

where ρ is mass density of fluid around the pipe (seawater = 2 slugs/ft^3 or 1025 kg/m^3).

5.3.4 Reduced Velocity

The reduced velocity, U_r, is the velocity at which vortex shedding-induced oscillations may occur. The equation for reduced velocity is:

$$U_r = \frac{U_c}{f_n D}$$
(5.6)

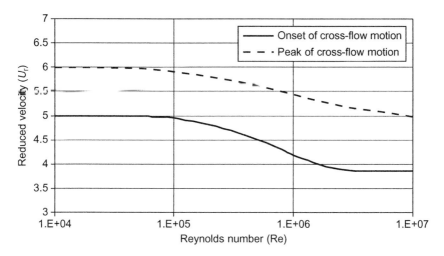

FIGURE 5.1 Reduced velocity for cross-flow oscillations based on the Reynolds number.

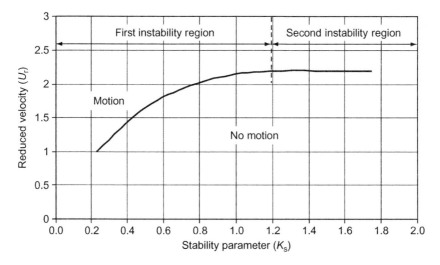

FIGURE 5.2 Reduced velocity for in-line oscillations based on the stability parameter.

Figure 5.1 presents the reduced velocity for cross-flow oscillations based on the Reynolds number (DnV, 1981). Figure 5.2 presents the reduced velocity for in-line oscillations based on the stability parameter (K_s) defined in the next subsection.

5.3.5 Stability Parameter

A significance for defining vortex-induced motion is the stability parameter, K_s, defined as:

$$K_s = \frac{2M_e\delta_s}{\rho D^2} \tag{5.7}$$

where δ_s is logarithmic decrement of structural damping ($=0.125$).

5.3.6 Critical Span Length

The critical span length or the unsupported pipeline length at which oscillations of the pipeline occur for a specific current is based on the relationship between the natural frequency of the pipe free span and the reduced velocity. The critical span length for cross-flow motion is expressed as:

$$L_c = \sqrt{\frac{C_e U_r D}{2\pi U_c}} \sqrt{\frac{EI}{M_e}} \tag{5.8}$$

The critical span length for in-line motion is expressed as:

$$L_c = \sqrt{\frac{C_e f_n}{2\pi}} \sqrt{\frac{EI}{M_e}} \tag{5.9}$$

5.4 DESIGN CRITERIA

5.4.1 General Considerations

For preliminary design purposes, it is customary to design a pipeline such that at no location along the pipeline route does the unsupported pipeline span length exceed the critical span length for which in-line motion occurs due to vortex shedding, at any time during the design life of the pipeline. However, in deepwater, where traditional deployment of span supports is not possible, this conservative design procedure can be quite costly. Thus, the selection of the allowable span length can become a risk-assessment-type solution.

5.4.2 Current Velocity Selection

The calculated reduced velocity, stability parameter, Reynolds number, and critical span length should all be based on a current velocity that is perpendicular to the pipeline. This design current should be based on the 100-year near-bottom current unless otherwise directed.

5.4.3 End Condition Selection

The selection of the proper end conditions for the pipe free span has a significant impact on the allowable span length selected. The typical rule of thumb for selecting the proper model for the end conditions are as follows:

Pinned–Pinned: Used for spans where each end is allowed to rotate about the pipe axis.
Pinned–Fixed: Used for the majority of spans, any span that does not fit the other two categories.
Fixed–Fixed: Should be used only for those spans that are fixed in place by some sort of anchor at both ends of the span.

The end condition selected can influence the calculated critical span length by as much as 50%, thus making the selection of the proper end conditions a critical step in selecting the proper allowable span length.

5.4.4 Design Parameters

As previously discussed, two types of motions are created by vortex shedding. The first of these is in-line motion. The amplitude of in-line motion can vary between 10% and 20% of the pipe diameter and occurs at low critical velocities. For most pipeline cases a prudent and conservative design should be based on the avoidance of in-line motion for the design bottom current. The second type of motion, cross-flow, occurs at higher critical velocities and with a larger amplitude, in the order of 1–2 times the pipe diameter. The allowable pipeline span length should always be designed such that cross-flow motion will never occur. The design engineer should only design the pipeline such that in-line motion is allowed to occur after evaluating the possible economic impacts that a smaller allowable span length would create. Even after such a decision has been made, the designer should undertake a fatigue life analysis check.

5.4.5 Design Steps

The following steps are based on the use of Figures 5.1 and 5.2 to assist in determining the allowable pipeline free-span length.

Step 1: Determine the design current (100-year near-bottom perpendicular to the pipeline).
Step 2: Calculate the effective unit mass of the pipeline with Eq. (5.4).
Step 3: Calculate Reynolds Number with Eq. (5.2).
Step 4: Calculate stability parameter with Eq. (5.7).
Step 5: Using the stability parameter and Figure 5.2 to determine the reduced velocity for in-line motion.

Step 6: Using the Reynolds number and Figure 5.1 to determine the reduced velocity for cross-flow motion.

Step 7: Based on the terrain and conditions involved, determine the type of free span end conditions and calculate the end condition constant.

Step 8: Calculate the critical span length for both in-line and cross-flow motion with Eqs (5.8) and (5.9).

Step 9: For the majority of projects, the allowable span length is the critical span length calculated for in-line motion. However, when economic factors warrant, the critical span length calculated for cross-flow motion can be selected.

Step 10: When in-line motion is permitted, the fatigue life of the free span should be calculated and evaluated for the pipeline.

5.4.6 Example Calculation

This example calculates the allowable span length to the cross-flow oscillation based on the following information:

Outside diameter of pipe $(D) = 0.2757$ m,
Inside diameter of pipe $(D_i) = 0.2509$ m,
Density of fluid in pipe $(\rho_f) = 107$ kg/m^3,
Density of pipe $(\rho_p) = 1024$ kg/m^3,
Mass of pipe and coatings $(M_p) = 74$ kg/m,
Kinematic viscosity of external fluid $(\nu_k) = 1.565 \times 10^{-6}$ m^2/s,
Current velocity $(U_c) = 0.35$ m/s,
Constant for clamped−pinned ends $(C_e) = 15.4$.

Step 1: Effective mass

$$M_p = 74 \text{ kg}/m$$

$$M_c = \frac{(3.14)(0.2509)^2}{4}(107) = 5.29 \text{ kg}/m$$

$$M_a = \frac{(3.14)(0.2757)^2}{4}(1024) = 61.13 \text{ kg}/m$$

$$M_e = 74 + 5.29 + 61.13 = 140.5 \text{ kg}/m$$

Step 2: Stability parameter

$$K_s = \frac{(2)(140.5)(0.125)}{(1024)(0.2757)} = 0.451$$

Step 3: Reynolds number

$$Re = \frac{(0.35)(0.2757)}{(1.56 \times 10^{-6})} = 6.1658 \times 10^4$$

Step 4: Reduced velocities

$$U_r = 1.6 \text{ from Figure 5.2 for in-line motion}$$
$$U_r = 5.0 \text{ from Figure 5.1 for cross-flow motion}$$

Step 5: Critical span length for cross-flow motion

$$L_c = \sqrt{\frac{(15.4)(5)(0.2757)}{(2\pi)(0.35)}} \sqrt{\frac{(2.07 \times 10^{11})[(\pi/64)((0.2757)^2 - (0.2509)^2)]}{(140.5)}}$$

$$L_c = 59.1 \text{ m}$$

Step 6: Critical span length for in-line motion

$$L_c = \sqrt{\frac{(2.45)(1.6)(0.2757)}{(0.35)}} \sqrt{131247}$$

$$L_c = 33.5 \text{ m}$$

5.5 FATIGUE ANALYSIS GUIDELINE

The fatigue life equation presented in this section is based on the Palmgren–Miner Fatigue model, which uses an S–N model based on the American Welding Society modified (AWS-X) fatigue curve of the form:

$$N = \frac{6.48 \times 10^{-8}}{\Delta \varepsilon^4} \tag{5.10}$$

where N is the number of cycles to failure and $\Delta \varepsilon$ is the strain range in each cycle. This extremely simplified fatigue life equation is expanded as follows:

$$L_f = \left[\frac{5.133 \times 10^{-18}(L_s/D)^8}{(D_s/D)^4 f_n} \right] \times \left[\frac{1}{\sum_i (f/f_n)_i (A/D)_i^4 T_i} \right] \tag{5.11}$$

where

L_f = fatigue life (years),
L_s = span length,
D_s = outside diameter of steel,
f_n = pipe frequency (Hz),
f/f_n = frequency ratio (Figure 5.3),
A/D = amplitude ratio (Figure 5.4),
T_i = current duration (hours/day).

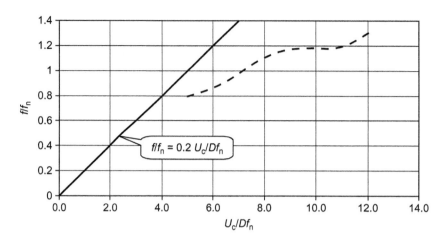

FIGURE 5.3   Chart for determination of frequency ratio based on V/Df_n.

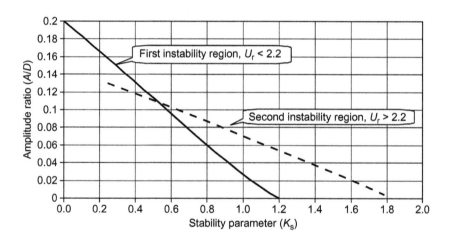

FIGURE 5.4   Chart for determination of amplitude ratio based on stability parameter (K_s).

The following steps should be followed when checking the fatigue life of free-span length:

Step 1: Calculate the pipe natural frequency (Eq. (5.3)).

Step 2: Determine the near-bottom current velocity occurrence distribution in histogram form using current duration blocks.

Step 3: For each current segment determine the frequency ratio based on (U_c/Df_n) and Figure 5.3.

Step 4: For each current segment determine the amplitude ratio based on the stability parameter and Figure 5.4.

Step 5: Calculate the fatigue life (Eq. (5.11)).

For cases where it can be illustrated that the fatigue life for in-line motion is much greater than the pipeline lifetime, cross-flow motion will become the limiting factor on critical span length. The fatigue life for cross-flow motion should be similarly checked to assume a factor of 10 for the amplitude ratio. This will normally show that cross-flow motion is prohibitive.

REFERENCES

Det norske Veritas, 1981. Rules for submarine pipeline systems, Norway.
Wootton, L.R., 1991. Vortex-Induced Forces. Dynamics of Marine Substructure, London (Chapter 7).

FURTHER READING

Hallam, M.G., Heaf, N.J., Wootton, L.R., 1978. Dynamics of Marine Structures: Methods of Calculating the Dynamic Response of Fixed Structures Subject to Wave and Current Action. CIRIA Underwater Engineering Group, London.

Chapter 6

Operating Stresses

6.1 INTRODUCTION

This chapter addresses the calculation of operation stresses and end movements (expansion) for trenched and nontrenched, single well pipelines. Operating stresses are those which result from a combination of internal pressure and thermal stresses that occur during operation. Equations of operating stresses and expansion are provided. Only single well pipe's internal pressure is addressed. In the case of relatively thin-wall pipe ($D/t > 20$), the equations presented can be used with P representing the difference between the internal and external pressure. This is not the case with thick-wall cylinders ($D/t < 20$).

6.2 OPERATING FORCES

When in operation, pressure and thermal forces exist, which act to expand the pipeline both rapidly and longitudinally. These are due to internal pressure and temperature difference between the pipe and surrounding fluid. The magnitude of these stresses is dependent upon forces opposing the above conditions, and boundary conditions, namely, soil friction acting longitudinally, end constraints, and end-cap effect.

6.2.1 Internal Pressure Stresses

A pipeline is a pressure vessel in the form of a cylinder, and, for this reason, some of the most detailed information available is obtained in the literature for pressure vessel design. Pipes with D/t greater than 20 are referred to as thin-wall pipes, and that with D/t less than 20 are called thick-wall pipes.

6.2.1.1 Thin-Wall Pipe

If a thin-wall pipe is subjected to internal pressure, P, the action of radial force distributed around the circumference will produce a circumferential (or hoop) stress given by:

$$\sigma_h = \frac{PD}{2t} \qquad (6.1)$$

Offshore Pipelines. DOI: http://dx.doi.org/10.1016/B978-0-12-397949-0.00006-6

where

σ_h = hoop stress,
D = internal diameter,
t = wall thickness,
P = net internal pressure.

The mitigating effect of external pressure can be considered, and for deepwater pipelines, this can be a factor in reducing required wall thickness. For pipeline design, D is taken as the nominal outside diameter to account for mill tolerance. This will be slightly conservative in most cases, but the hoop stress must be calculated in this manner to meet the ANSI/ASME B31.8, B31.4 design practices.

The longitudinal stress, σ_L, is calculated by dividing the total pressure force against the end of the pipe (end-cap effect) by the cross-section area of the pipe:

$$\sigma_L = \frac{PD}{4t} \tag{6.2}$$

These stresses in thin-wall pipes are illustrated in Figure 6.1. Strains can be calculated based on stresses and elastic modulus by $\varepsilon_h = \sigma_h/E$ and $\varepsilon_L = \sigma_L/E$.

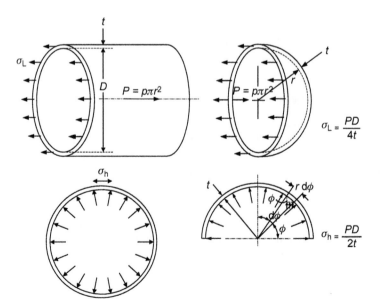

FIGURE 6.1   Operating stresses in thin-wall pipes.

6.2.1.2 Thick-Wall Pipe

For D/t less than 20, the convention is to use the thick-wall equations for hoop and radial stresses that are slightly more complicated. The thin-wall equations for hoop stress can be used, but it results in slightly high estimates of stresses.

The principal difference between the thin- and thick-wall formulations is that for thick-wall conditions, the variation in stress between inner and outer surface becomes significant. The cross section for a thick cylinder and its representative stresses are depicted in Figure 6.2.

For the case of internal pressure only, the following equations apply:

$$\sigma_r = \frac{b^2 P}{a^2 - b^2}\left(1 - \frac{a^2}{r^2}\right) \tag{6.3}$$

$$\sigma_h = \frac{b^2 P}{a^2 - b^2}\left(1 + \frac{a^2}{r^2}\right) \tag{6.4}$$

where r varies from b to a, which are the inside and the outside radii, respectively. Both σ_h and σ_r are maximum at $r = b$.

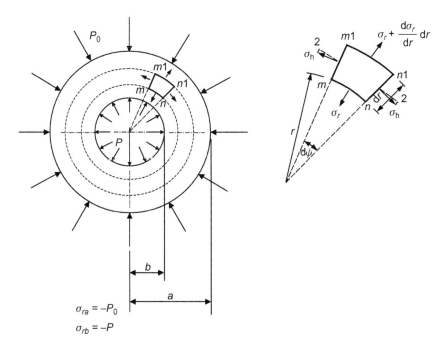

FIGURE 6.2 Operating stresses in thick-wall pipes.

The longitudinal stress, σ_L, is given by:

$$\sigma_L = \frac{b^2 P}{a^2 - b^2} \tag{6.5}$$

For the calculation of burst pressure, the maximum shear stress theory correlates well with data. This is given by one-half the algebraic difference between the maximum and minimum principal stresses at any point. Since the longitudinal stress is neither the maximum nor the minimum value, it is ignored resulting in:

$$\tau = \frac{\sigma_h - \sigma_r}{2} \tag{6.6}$$

which, when Eqs (6.3) and (6.4) are used, becomes:

$$\tau = \frac{a^2 b^2 P}{r^2(a^2 - b^2)} \tag{6.7}$$

For the case of internal pressure only, the shear stress is a maximum on the inner surface. Therefore,

$$\tau_{max} = \frac{a^2 P}{a^2 - b^2} \tag{6.8}$$

6.2.2 Thermal Expansion Stresses

Temperature gradient exists between the maximum operating temperature and the minimum installation temperature. Operating temperature along the pipeline can be predicted using the heat transfer model presented in Appendix B. The longitudinal strain is proportional to the magnitude of the temperature difference. In an unrestrained uniaxial condition, the longitudinal thermal stress is zero, but the thermal strain, ε_t, is given by:

$$\varepsilon_t = \alpha_t \Theta \tag{6.9}$$

where α_t is the coefficient of thermal expansion (6.5×10^{-6} in./in./°F for steel), and Θ is the value of temperature change $T_2 - T_1$.

In the restrained condition, the longitudinal strain is zero, but the compressive stress generated by the restrained expansion is given by:

$$\sigma = -E\alpha_t \Theta \tag{6.10}$$

The negative sign reflects the fact that the stresses for a positive temperature increase under restrained conditions is compressive. Similarly, the stresses are tensile for a restrained pipe if a temperature decrease exists.

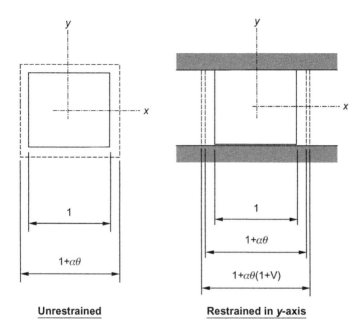

FIGURE 6.3   Operating stresses in subsea pipelines with 2D thermal strains.

When a two-dimensional element is heated but subjected to a restraint in the y-direction, the strain in the x-direction is increased due to the Poisson ratio ν. This is illustrated in Figure 6.3. If the element is heated and restrained in both the x- and y-directions, the principal stresses become:

$$\sigma_1 = \sigma_2 = -\frac{E\alpha_t \Theta}{1 - \nu} \tag{6.11}$$

For a heated cube having restraints in all directions, the stresses are given by:

$$\sigma_1 = \sigma_2 = \sigma_3 = -\frac{E\alpha_t \Theta}{1 - 2\nu} \tag{6.12}$$

6.2.3 Combined Pressure and Temperature

Pressure- and temperature-induced stresses in single wall pipe depend on restraining conditions such as unrestrained, partially restrained by longitudinal soil friction, and fully restrained by end anchor. Different equation sets apply to thin- and thick-wall pipes. In all cases, positive stresses are tensile stresses and negative stresses are compressive stresses.

6.2.3.1 Equations for Thin-Wall Pipe

For thin-wall $(D/t > 20)$ unrestrained pipe with ends capped, the equations for hoop stress, strain, end movement, and radial dilation are given as follows (Roark and Young, 1989):

$$\text{Hoop stress: } \sigma_h = \frac{PD}{2t} \tag{6.13}$$

$$\text{Longitudinal stress: } \sigma_L = \frac{PD}{4t} \tag{6.14}$$

$$\text{Hoop strain: } \varepsilon_h = \alpha_t \Theta + \frac{PD}{2tE}\left(1 - \frac{\nu}{2}\right) \tag{6.15}$$

$$\text{Longitudinal strain: } \varepsilon_L = \alpha_t \Theta + \frac{PD}{2tE}\left(\frac{1}{2} - \nu\right) \tag{6.16}$$

$$\text{End movement: } \Delta L = \frac{L\varepsilon_L}{2} \tag{6.17}$$

$$\text{Radial movement: } \Delta R = a\varepsilon_h \tag{6.18}$$

where P may represent the difference between internal and external pressures.

For pipe partially restrained by soil friction, the equations for hoop stress, strain, end movement, and radial dilation are given below (from Harvey, 1985 except otherwise noted):

$$\text{Hoop stress: } \sigma_h = \frac{PD}{2t} \tag{6.19}$$

$$\text{Longitudinal stress: } \sigma_L = \frac{PD}{4t} - \frac{fx}{2\pi at} \quad \text{for } x < Z \tag{6.20}$$

$$\text{(End-cap pressure effect included) } \sigma_L = \frac{\nu PD}{2t} - E\alpha_t \Theta \quad \text{for } x \geq Z \tag{6.21}$$

$$\text{Hoop strain: } \varepsilon_h = \alpha_t \Theta + \frac{PD}{2tE}\left(1 - \frac{\nu}{2}\right) + \frac{\nu fx}{2\pi atE} \quad \text{for } x < Z \tag{6.22}$$

or

$$\varepsilon_h = \alpha_t \Theta + \frac{PD}{2tE}(1 - \nu^2) + \nu\alpha_t \Theta \tag{6.23}$$

(by Roark and Young, 1989)

$$\text{Longitudinal strain: } \varepsilon_L = \alpha_t \Theta + \frac{PD}{2tE}\left(\frac{1}{2} - \nu\right) - \frac{fx}{EA_s} \quad \text{for } x < Z \tag{6.24}$$

$$\text{Longitudinal strain at free end: } \varepsilon_o = \alpha_t \Theta + \frac{PD}{2tE}\left(\frac{1}{2} - \nu\right) \quad \text{for } x < Z \tag{6.25}$$

$$\text{End movement: } \Delta L = \bar{\varepsilon} Z \qquad (6.26)$$

$$Z = \frac{\pi Dt}{f}\left(E\alpha_t\Theta - \frac{\nu PD}{2t}\right) + \frac{\pi Pa^2}{f}$$

$$\text{If } Z \geq \frac{L}{2}, \text{ then } Z - \frac{L}{2} \text{ and } \varepsilon = \frac{1}{2}(\varepsilon_0 - \varepsilon_z)$$

$$\text{If } Z \leq \frac{L}{2}, \text{ then } \bar{\varepsilon} = \frac{\varepsilon_0}{2}$$

$$\text{Radial movement: } \Delta R = a\varepsilon_h \qquad (6.27)$$

For pipe fully restrained by end anchors, the equations for hoop stress, strain, end movement, and radial dilation are summarized as below (from Harvey, 1985 except otherwise noted):

$$\text{Hoop stress: } \sigma_h = \frac{PD}{2t} \qquad (6.28)$$

$$\text{Longitudinal stress: } \sigma_L = \frac{\nu PD}{2t} - E\alpha_t\Theta \qquad (6.29)$$

$$\text{Hoop strain: } \varepsilon_h = \alpha_t\Theta + \frac{\sigma_h - \nu\sigma_L}{E} \qquad (6.30)$$

$$\text{Longitudinal strain: } \varepsilon_L = 0 \qquad (6.31)$$

$$\text{End movement: } \Delta L = 0 \qquad (6.32)$$

$$\text{Radial movement: } \Delta R = a\varepsilon_h \qquad (6.33)$$

$$\text{Force on anchor: } F = 2\pi at\left(E\alpha\Theta - \frac{\nu PD}{2t}\right) + \pi Pb^2 \qquad (6.34)$$

6.2.3.2 Equations for Thick-Wall Pipe

For thick-wall ($D/t < 20$) unrestrained pipe with ends capped, the equations for hoop stress, strain, end movement, and radial dilation are given as follows (Harvey, 1985):

$$\text{Hoop stress: } \sigma_h = \frac{Pb^2(a^2 + r^2)}{r^2(a^2 - b^2)} \qquad (6.35)$$

$$\text{Longitudinal stress: } \sigma_L = \frac{Pb^2}{a^2 - b^2} \qquad (6.36)$$

$$\text{Radial stress: } \sigma_h = \frac{Pb^2(a^2 - r^2)}{r^2(a^2 - b^2)} \qquad (6.37)$$

$$\text{Radial movement at } a: \Delta R_a = \frac{Pab^2(2-\nu)}{E(a^2-b^2)} + \alpha_t \Theta a \qquad (6.38)$$

$$\text{Radial movement at } b: \Delta R_b = \frac{Pb[a^2(1+\nu)+b^2(1-2\nu)]}{E(a^2-b^2)} + \alpha_t \Theta b \quad (6.39)$$

$$\text{Longitudinal strain: } \varepsilon_L = \frac{1}{E}[\sigma_L - \nu(\sigma_{ha}+\sigma_{ra})] + \alpha_t \Theta \qquad (6.40)$$

$$\text{End movement: } \Delta L = \frac{L}{2}\left[\frac{Pb^2(1-2\nu)}{E(a^2-b^2)} + \alpha_t \Theta\right] \qquad (6.41)$$

For partially restrained pipe by soil friction, the equations for hoop stress, strain, end movement, and radial dilation are given as follows (Roark and Young, 1989):

$$\text{Hoop stress: } \sigma_h = \frac{Pb^2(a^2+r^2)}{r^2(a^2-b^2)} \qquad (6.42)$$

$$\text{Longitudinal stress: } \sigma_L = \frac{1}{a^2-b^2}\left(Pb^2 - \frac{fx}{\pi}\right) \quad \text{for } x < Z \qquad (6.43)$$

$$\sigma_L = \frac{\nu Pb^2}{a^2-b^2} - E\alpha_t \Theta \quad \text{for } x \geq Z \qquad (6.44)$$

$$\text{Radial stress: } \sigma_r = \frac{Pb^2}{a^2-b^2}\left(1 - \frac{a^2}{r^2}\right) \qquad (6.45)$$

$$\text{Radial movement at } a: \Delta R_a = \frac{a}{E}[\sigma_{ha} - \nu(\sigma_{ra}+\sigma_{La})] + \alpha_t \Theta a \qquad (6.46)$$

$$\text{Radial movement at } b: \Delta R_b = \frac{b}{E}[\sigma_{hb} - \nu(\sigma_{rb}+\sigma_{Lb})] + \alpha_t \Theta b \qquad (6.47)$$

$$\text{Longitudinal strain: } \varepsilon_L = \alpha_t \Theta + \frac{1}{E}\left[\sigma_L - \nu\left(\frac{\sigma_{rb}}{2} + \sigma_{ha}\right)\right] \quad \text{for } x < Z$$

$$\varepsilon_L = 0 \quad \text{for } x \geq Z \qquad (6.48)$$

$$Z = \frac{\pi(a^2-b^2)}{f}\left[E\alpha_t \Theta + \frac{Pb^2(1-2\nu)}{a^2-b^2}\right] \qquad (6.49)$$

$$\text{Longitudinal strain at free end: } \varepsilon_0 = \alpha_t \Theta + \frac{P}{E}\left[\frac{\nu}{2} + \frac{b^2}{a^2-b^2}(1-2\nu)\right] \quad (6.50)$$

$$\text{End movement: } \Delta L = \bar{\varepsilon} Z \quad \text{for } Z < \frac{L}{2}, \bar{\varepsilon} = \frac{\varepsilon_0}{2} \qquad (6.51)$$

$$\Delta L = \frac{\bar{\varepsilon}_L Z}{2} \quad \text{for } Z \geq \frac{L}{2} \tag{6.52}$$

$$\bar{\varepsilon}_L = \frac{1}{2}\left(\varepsilon_L \text{ at } x = 0 + \varepsilon_L \text{ at } x = \frac{L}{?}\right)$$

For pipe fully restrained by end anchors, the equations for hoop stress, strain, end movement, and radial dilation are summarized as follows (from Harvey, 1985 except otherwise noted):

$$\text{Hoop stress: } \sigma_h = \frac{Pb^2(a^2 + r^2)}{r^2(a^2 - b^2)} \tag{6.53}$$

$$\text{Longitudinal stress: } \sigma_L = \frac{2\nu Pb^2}{a^2 - b^2} - E\alpha_t \Theta \tag{6.54}$$

$$\text{Maximum radial stress: } \sigma_{rb} = -P \tag{6.55}$$

$$\text{Radial movement at } a: \Delta R_a = \frac{a}{E}[\sigma_{ha} - \nu(\sigma_{ra} + \sigma_{La})] + \alpha_t \Theta a \tag{6.56}$$

$$\text{Radial movement at } b: \Delta R_b = \frac{b}{E}[\sigma_{hb} - \nu(\sigma_{rb} + \sigma_{Lb})] + \alpha_t \Theta b \tag{6.57}$$

$$\text{Longitudinal strain: } \varepsilon_L = 0 \tag{6.58}$$

$$\text{End movement: } \Delta L = 0 \tag{6.59}$$

$$\text{Force on anchor: } F = A_s(E\alpha_t \Theta - \nu\sigma_{ha}) + \pi Pb^2 \tag{6.60}$$

When calculating Von Mises equivalent stresses, the highest value is obtained in areas where the longitudinal stress is compressive since:

$$2\sigma_V^2 = (\sigma_h - \sigma_L)^2 + (\sigma_L - \sigma_r)^2 + (\sigma_r - \sigma_h)^2 \tag{6.61}$$

In thin-wall applications, σ_r can be assumed zero and the Von Mises equivalent stress is simplified to:

$$\sigma_V = \sqrt{\sigma_h^2 + \sigma_L^2 + \sigma_h\sigma_L} \tag{6.62}$$

6.2.3.3 Soil Friction

Soil friction force is the result of pipe–soil interaction building up a negative (compressive) strain in the pipeline. Friction force per unit length is equal to the product of the friction coefficient and the normal soil force acting around the pipe. Since actual distribution of normal force is hard to determine for the purpose of friction force calculation, a simplified model can be used. The soil force for a completely backfilled line is estimated by the following equation:

$$f = \mu(W + W_p - F_b) \tag{6.63}$$

where

W = weight of soil overburden (lb/ft),
W_p = dry weight of pipe and contents (lb/ft),
F_b = buoyant force (lb/ft),
μ = coefficient of friction for soil,
f = friction force (lb/ft).

For an untrenched pipe, the soil force is given by:

$$f = \mu(W_p - F_b) \tag{6.64}$$

When soil cover ranges from one to three times the pipe's diameter, the soil force can be taken as the weight of the soil over the pipe. For increased depth of cover, the soil force may not increase proportionately due to soil arching. The actual soil force must consider the type and composition of the overburden and is beyond the scope of this section.

6.2.3.4 End Constraint

End constraint is a reaction at structures such as a rigid flange, anchor, or a rigid tie-in. The restraint prevents pipe expansion. The restraining force generated is calculated by summing the internal pressure and thermal expansion forces. Soil friction is not a factor in this case as there is no longitudinal movement.

6.3 STRESS-ANALYSIS-BASED DESIGN

The pipeline design against the operating stresses involves stress analyses using the equations presented in the last section.

6.3.1 Analysis Procedure

A general method of calculating the operating stresses is given below:

1. Determine the wall thickness of the pipe using the method described in Chapter 3.
2. If $D/t < 20$, then use thick-wall pipe equations for subsequent calculations, otherwise use thin-wall equations.
3. Choose the appropriate pipe scenario case (fully restrained, unrestrained, or partially restrained).

TABLE 6.1 Design Factors for Offshore Pipelines

Content Type	F1 (Hoop Stress)	F2 (Longitudinal Stress)	F3 (Combined Stress)
Gas[a]	0.72	0.8	0.9
Oil[b]	0.72	0.675/0.54/0.8[c]	—

[a]*ASME (1990).*
[b]*ASME (1989).*
[c]*ASME (1989), 402.3.2(d) represents 0.75 × 0.90 for standard loads in restrained pipelines. This is reduced to 0.75 × 0.72 for unstrained pipelines as may be the case in a span area where pipeline is not in contact with the seabed. These stress limits refer to tensile only. Design factor is increased to 0.8 when considering occasional loads in addition to sustained loads.*

4. Calculate the distance to no movement to determine whether the pipeline half-length is longer or shorter than its distance. If the half-length is shorter, the strain at the midpoint is nonzero.
5. Calculate the hoop stress using the pressure difference between the internal fluid and external hydrostatic pressure.
6. Calculate the longitudinal stress using the appropriate equation selected from Step 3.
7. If no end restraint is present, calculate the resulting longitudinal strain.
8. Calculate the end and radial movement experienced by the pipe.
9. Check the results of the stress calculation with the ASME codes described below.

6.3.2 Code Requirements

This section outlines standards to follow when designing for maximum allowable operating stresses and end movements.

6.3.2.1 Hoop Stress

According to ASME Codes, the following requirement should hold for hoop stress:

$$\sigma_h < F_1 F_t S_y \tag{6.65}$$

where

σ_h = hoop stress,
F_1 = hoop stress design factor from Table 6.1,
S_y = specified minimum yield strength (SMYS) (psi),
F_t = temperature de-rating factor from Table 6.2.

TABLE 6.2 Temperature De-Rating Factor for Steel Pipe

Temperature (°F)	Temperature De-Rating Factor (F_t)
250 or less	1.000
300	0.967
350	0.933
400	0.900
450	0.867

6.3.2.2 Longitudinal Stress

ASME Codes specify the following requirements for longitudinal stress:

$$|\sigma_L| < F_2 S_y \tag{6.66}$$

where

σ_L = maximum longitudinal stress (psi),
F_2 = longitudinal stress design factor from Table 6.1.

6.3.2.3 Combined Stress

The combined stress shall meet the following requirement:

$$\sqrt{\sigma_h^2 + \sigma_L^2 - \sigma_h \sigma_L + 3\tau_t^2} \leq F_3 S_y \tag{6.67}$$

where

F_3 = combined stress design factor from Table 6.1,
τ_t = tangential shear stress (psi).

In most cases, no torsion is present and $\tau_t = 0$. Note that the most compressive ($-$) value of σ_L must be used for conservatism.

6.3.3 Example Calculation

For the following pipeline, calculate the operating stress and end movement.

Pipeline type = gas
Pipe outside diameter, $D = 8.625$ in.
Wall thickness, $t = 0.375$ in.
Steel modulus, $E = 3 \times 10^7$ lb/in.^{2}
Soil friction force, $f = 2.755$ lb/in.
Pipe length, $L = 20{,}381$ ft
Poisson ratio, $\nu = 0.3$
Yield stress, $S = 65.0$ Ksi

Restraint condition = partially restrained
Temperature differential, $\Theta = 50°F$
Thermal expansion coefficient, $\alpha_t = 6.5 \times 10^{-6}$ in./in./°F
Internal pressure, $P = 1440$ psig.

1. Calculate D/t.

$$\frac{8.625}{0.375} = 23$$

2. Due to $D/t = 23$, thin-wall pipe equations are adequate.
3. The applicable pipe case is for a thin, single wall, partially restrained by soil friction.
4. Calculate the distance to no movement using a corresponding equation.

$$Z = \frac{\pi Dt}{f}\left(E\alpha_t\Theta - \frac{\nu PD}{2t}\right) + \frac{\pi P_a^2}{f}$$

$$Z = (3.688)(9750 - 4968) + 30,538 \text{ ft}$$

$$Z = 4014 \text{ ft}$$

Pipe half-length 20,381 ft is greater than $Z = 4014$ ft.
5. Calculate longitudinal and hoop stress and strain.
   a. Hoop stress and strain:

$$\sigma_h = \frac{PD}{2t} = 16,560 \text{ psi}$$

For $x \geq Z$:

$$\varepsilon_h = \alpha_t\Theta + (1 - \nu^2)\frac{PD}{2tE} + \nu\alpha_t\Theta = 0.000925$$

   b. Longitudinal stress and strain:
      $x \geq Z$:

$$\sigma_L = \nu\frac{PD}{2t} - E\sigma_t\Theta = -4782 \text{ psi}$$

$$\varepsilon_L = 0$$

$$x = 0:$$

$$\sigma_L = \frac{PD}{4t} = 8280 \text{ psi}$$

$$\varepsilon_o = 0.000435$$

c. End movement (pipe half-length $> Z$)

$$\Delta L = \bar{\varepsilon} Z$$
$$\Delta L = 0.5 \varepsilon_o Z$$
$$\Delta L = 0.87 \text{ ft}$$

6. Compare the calculated stresses with code for gas lines.

$$\sigma_h = 16,650 \leq 0.72 \times 65,000 \text{ psi}$$
$$\sigma_L = 8280 \leq 0.80 \times 65,000 \text{ psi}$$
$$\sigma_V = \sqrt{(16,560)^2 + (-4782)^2 - (16,650)(-4782)} \text{ psi}$$
$$= 19,398 \leq 0.9 \times 65,000 \text{ psi}$$

REFERENCES

American Society of Mechanical Engineers (ASME), Gas transmission and distribution piping systems. ASME Code for Pressure Piping, B31.8—1989 edition and 1990 addendum.

American Society of Mechanical Engineers (ASME), Liquid transportation systems for hydrocarbons, liquid petroleum gas, anhydrous ammonia and alcohols, ASME B31.4—1989 edition.

Harvey, J.F., 1985. Theory and Design of Pressure Vessels. Van Nostrand Reinhold Company, New York, NY.

Roark, R.J., Young, W.C., 1989. Formulas for Stress and Strain, sixth ed. McGraw-Hill, New York, NY.

Chapter 7

Pipeline Riser Design

7.1 INTRODUCTION

Riser is defined as the vertical or near-vertical segment of pipe connecting the facilities above water to the subsea pipeline. The riser portion extends (as a minimum) from the first above-water valve or isolation flange to a point five pipe diameters beyond the bottom elbow, based on codes. The design engineer must select the exact limits on a case-by-case basis. This may often extend the riser beyond the five diameters limit or above the isolation flange. Many operators prefer a length of 200 ft from the elbow to protect against dropped objects (i.e., heavier wall pipe). The riser design usually considers adjoining pipework segments, clamps, supports, guides, and expansion absorbing devices. These are illustrated schematically in Figure 7.1.

This chapter addresses the engineering analysis and design of conventional steel risers and riser clamps. It is written on the basis of related codes and rules in riser design. This chapter provides guideline for preinstalled and postinstalled conventional steel risers but does not address J-tube or flexible pipe risers. Risers are assumed to be of API 5L line pipe and are operated at a temperature less than 250°F. For other cases, refer to specific code allowables.

7.2 DESIGN PROCEDURE

For a conventional steel riser, the design procedure includes the following steps:

Step 1: Establish the design basis.
- Maximum wave height and period for return periods of 1 and 100 years
- Annual significant wave height occurrence in 5 ft height intervals
- Associated wave periods for annual significant wave height distribution
- Steady current profile
- Seismicity (if applicable)
- Splash zone limits
- Befouling thickness profile
- Minimum pipeline installation temperature
- Maximum allowable operating pressure (MAOP)

Offshore Pipelines. DOI: http://dx.doi.org/10.1016/B978-0-12-397949-0.00007-8

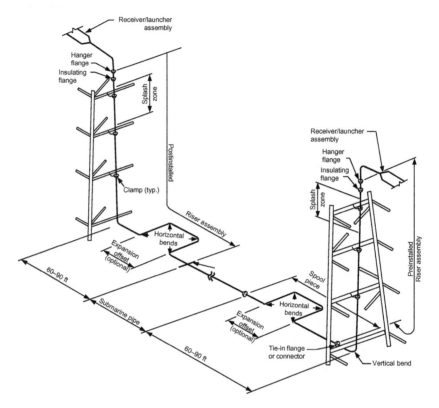

FIGURE 7.1 Typical riser schematic.

- Maximum allowable pipeline operating temperature (This should reflect the effects of temperature drop along pipeline in the direction of flow.)
- Pipe-to-soil longitudinal friction
- Soil elastic modulus

Step 2: Obtain platform design data.

- Jacket design drawings
- Batter of the jacket on the riser face
- Movements of the platform during storm (100-year)
- Intended riser locations: cellar deck plan

Step 3: Determine the minimum wall thickness for riser based on design pressure, pipe size, material grade, and corrosion allowance. This is defined by code formula and allowable hoop stress.

Step 4: Select a base riser configuration and perform static stress analyses for selected load cases. The detailed procedure is illustrated in the next section.

Step 5: Perform vortex shedding and fatigue analyses using cumulative damage methods to verify life of riser.

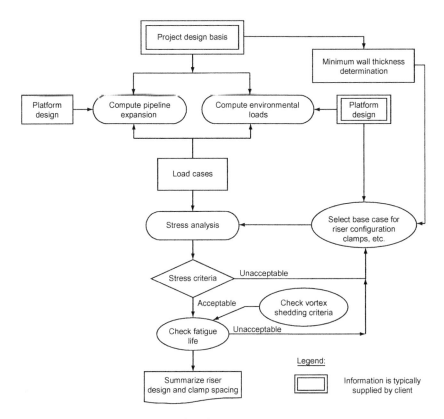

FIGURE 7.2   Riser design procedure flowchart.

Step 6: Modify clamp locations, riser design, or wall thickness as necessary to meet codes and reanalyses for all cases.

Step 7: Design riser clamps based on jacket design and the forces calculated from static stress analysis.

Step 8: Design riser anchor at top clamp, if needed. This is generally only required in water depths greater than 100 ft where the riser cannot be free-standing.

A flowchart for the riser design procedure is shown in Figure 7.2. The core of the riser design is static stress analysis.

7.3 LOAD CASES

Risers are subjected to various types of loads including functional loads, environmental loads, and installation loads. Based on ANSI B31.8 and B31.4, Tables 7.1 and 7.2 list the required load cases for stress analysis for gas and oil riser systems, respectively.

TABLE 7.1 Static Design Load Cases for Gas Risers (ANSI B31.8)

Loads		Riser Design Load Combinations		
		Operation (Case 1)	*Hydrostatic Test (Case 2)*	*Installation (Case 3)*
Functional loads	Internal pressure	MAOP	Hydrotest pressure	
	Operating temperatures	Design temperature		
	Weights	Weights	Weights	Weights
	Pipeline expansion	Pipeline expansion		
	External pressure	External pressure	External pressure	External pressure
Environmental loads	Wind	100-year wind		1-year wind
	Wave	100-year wave	1-year wave	1-year wave
	Current	100-year current	1-year current	1-year current
	Platform motion	Platform motion		
Installation loads	Cold springing			Cold springing
	Residual stresses			Residual stresses

7.3.1 Functional Loads

Functional loads acting on the riser can be classified as internal pressure, load due to operational expansion of pipeline, and deadweight of riser.

Two internal pressure conditions must be considered in riser design: the MAOP and the test pressure. The MAOP is the maximum pressure at which a gas system may be operated in accordance with the provisions of the design code. The MAOP should include the effects of shut-in to a live well if appropriate. Surge pressure is not normally considered as a part of MAOP. The test pressure is the maximum internal fluid pressure permitted by the design code for a pressure test based on the material and location involved. Gas risers are hydrostatically tested to 1.4 times MAOP (ANSI B31.8 par 847.2). Oil risers are tested to 1.25 times MAOP (ANSI B31.4 par 437.4.1).

TABLE 7.2 Static Design Load Cases for Oil Risers (ANSI B31.4)

Loads		Riser Design Load Combinations				
		Operation (Case 1)	Hydrostatic Test (Case 2)	Expansion Only (Case 3)	Sustained Load (Case 4)	Installation (Case 5)
Functional loads	Internal pressure	MAOP	Hydrotest pressure		MAOP	
	Operating temperatures weights	Design temperature weights	Weights		Weights	Weights
	Pipeline expansion	Pipeline expansion		Pipeline expansion	Pipeline expansion	
	External pressure					External pressure
Environmental loads	Wind	100-year wind				1-year wind
	Wave	100-year wave	1-year wave			1-year wave
	Current	100-year current	1-year current			1-year current
	Platform motion	Platform motion				
Installation loads	Cold springing	Cold springing				Cold springing
	Residual stresses	Residual stresses				Residual stresses

Some operators refer a test pressure of 1.5 times MAOP, and this should be established as part of the design basis, if applicable. (Note: riser test pressures are higher than pipeline hydrotest pressures for gas risers.) This means that the "riser" must be tested in the yard separately from the pipe unless the pipe has been designed to withstand the elevated riser test pressure. The US Marine Machinery Association (MMS) allows one weld to be made joining the riser with the pipeline after the hydrotest provided it is fully radiographed.

The temperature difference between the operating pipeline and its initial installation temperature will cause pipeline expansion. Internal pressure also causes expansion although to a lesser extent than temperature.

The loadings, due to the self-weight contents, are dependent on the design conditions. For installation, the pipeline is assumed to be empty. Analysis for operational conditions should consider the line to be filled with product. Hydrotesting is characterized by the pipe being filled with seawater at 64 lb/ft^3.

7.3.2 Environmental Loads

Environmental loads are loads caused by wind, waves, current, and other external forces. Wind, waves, and current loads can also induce platform movements. The hydrodynamic loads acting on the riser are divided into two categories: (1) drag, lift and inertia forces and (2) flow-induced vortex shedding on riser. Figure 7.3 shows a profile of wave- and current-induced loading. The platform movement is referred to as the relative displacement between the platform members where the riser clamps are connected. The relative displacement between clamps will increase the bending stress in the riser, which may be critical in some cases. Most platforms will have a stiff axis and a soft axis. The most conservative direction of movement should be

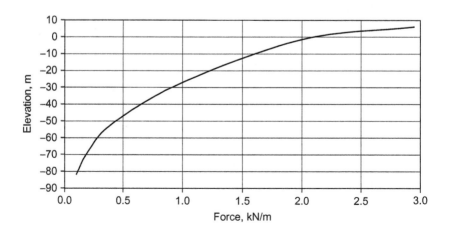

FIGURE 7.3 Typical profile of wave- and current-induced load.

selected taking into account wave loading and pipeline expansion. This will normally be in the direction of the pipeline or wave approach.

7.3.3 Installation Loads

Installation loads are determined by the installation procedure and tie-in methods. A typical installation load is the permanent bending loads caused by cold springing. The designer may need to consult with installation contractor in special situations.

7.4 WALL THICKNESS

The nominal wall thickness after corrosion withstands the internal pressure. The hoop stress, σ_h, is determined by:

$$\sigma_h = \frac{PD}{2t} \tag{7.1}$$

where

$P = $ MAOP,
$D = $ outside diameter,
$t = $ nominal wall thickness.

For oil and gas risers, the hoop stress must not exceed 0.6 SMYS and 0.5 SMYS, respectively. In deepwater, the effect of hydrostatic pressure must be checked using the design procedure presented in Chapter 3. An allowance for corrosion of ⅛ in. may also be required.

7.5 ALLOWABLE STRESS CRITERIA

Tables 7.3 and 7.4 list the allowable stress criteria, based on ANSI/ASME B31.8 and B31.4, for offshore gas and oil risers, respectively. Three types of

TABLE 7.3 Maximum Allowable Stress Criteria for Gas Risers

Stress Type	Maximum Allowable Stress Criteria		
	Operation (Case 1)	Hydrostatic Test (Case 2)	Installation (Case 3)
Hoop stress	0.5 SMYS	0.9 SMYS	—
Longitudinal stress	0.8 SMYS	0.8 SMYS	0.8 SMYS
von Mises stress	0.9 SMYS	0.9 SMYS	0.9 SMYS

TABLE 7.4 Maximum Allowable Stress Criteria for Oil Risers

Stress Type	Maximum Allowable Stress Criteria				
	Operation (Case 1)	Hydrostatic Test (Case 2)	Expansion Only (Case 3)	Sustained Load (Case 4)	Installation (Case 5)
Hoop stress	0.6 SMYS	0.9 SMYS	–	–	–
Longitudinal stress	0.8 SMYS	–	0.9 SMYS	0.54 SMYS	0.8 SMYS
von Mises or Tresca combined stress	0.9 SMYS	–	–	–	0.9 SMYS

stresses should be checked in riser design: hoop, longitudinal, and von Mises. A typical riser model is shown in Figure 7.4.

Hoop stress can be calculated by Eq. (7.1). Longitudinal stress should be calculated considering end-cap effect. The maximum von Mises stress in a riser is calculated using

$$2\sigma_V^2 = (\sigma_r - \sigma_h)^2 + (\sigma_h - \sigma_a)^2 + (\sigma_a - \sigma_r)^2 \qquad (7.2)$$

where

σ_h = hoop stress (+ value),
σ_r = radial stress = P (internal pressure),
σ_V = von Mises stress,

and the axial stress is given by

$$\sigma_a = \frac{D_i M_b}{2I} + \frac{T_a}{A_s} \qquad (7.3)$$

where

D_i = riser inside diameter,
M_b = bending moment,
I = moment of inertia,
T_a = axial force,
A_s = steel cross-sectional area.

Note that the maximum von Mises stress normally occurs at the inside wall of the compressive side of bending moment. Therefore, the negative

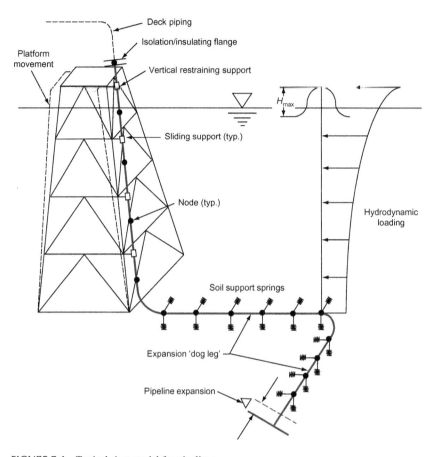

FIGURE 7.4   Typical riser model for pipelines.

value of bending moment and "without end-capped" tensile stress should be used in axial stress calculations.

7.6 DYNAMIC AND FATIGUE ANALYSIS

Computer programs are required to perform dynamic and fatigue analyses. Such analyses often indicate that the spacing between clamps will not permit a vortex-induced riser resonance to occur. Also the cyclic stresses in the riser are sufficiently low to allow a life substantially greater than that required. A minimum safety factor of 5 is necessary owing to the uncertainty of the data.

With computer programs, the maximum allowable span length for a riser in a given current field can be determined. It is recommended that the pinned-fixed condition be used unless rotational anchors are specified within the clamp design. Ordinarily, the content's density should be used in the analysis. The wall thickness should be the actual wall less the corrosion allowance. If riser dynamic amplification during wave or seismic loading is possible, the dynamic solution should also be determined.

Fatigue life is determined based on cumulative damage due to cyclic loads. Provided that a proper clamp spacing has been selected, vortex-induced oscillation is not a factor and the primary aspect is wave loading. Note that any increase in diameter and mass due to biofouling should be considered in developing the wave loads. The bending stresses in each section of the riser are determined for a range of wave heights. This would, for example, be for wave heights of 0−5, 5−10, 10−15, 15−20, 20−30, 30−40, and 40−50 ft. For practical reasons, wave height refers to the significant wave height. This is slightly conservative as the significant wave actually represents the average of the highest one-third in the wave population. From annual wave statistics such as the Summary of Synoptic Meteorological Observations (SSMO) for the region of interest, the number of cycles of each wave height group in a 1-year period can be developed. For each wave height group, the cyclic stress range can be determined. Normally, this is a quasi-static solution, but a full dynamic solution may be needed if dynamic amplification is present at the wave frequency.

Stresses are determined based on the highest wave in each group to generate a load profile and a peak bending stress. The stress range, S_{range}, is twice the peak amplitude. This is a quasi-static solution which excludes the effect of dynamic amplification. This is normally justified because the clamp spacing is small enough to prevent any significant resonance at wave frequencies. In certain cases, such as catenary risers, a full dynamic analysis may be required.

To define the number of cycles to failure, the AWS-X1 curve is used. This has been shown to be valid for butt-welded line pipe in a seawater environment regardless of material grade. This curve is illustrated in Figure 7.5.

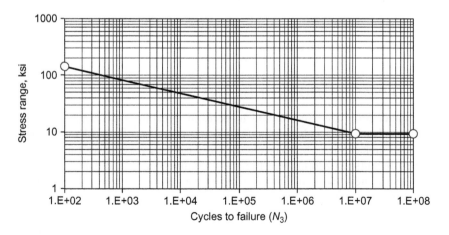

FIGURE 7.5   AWS-X1 S-N curve for butt-welded pipe.

For each wave group, the damage ratio is determined by:

$$R_D = \frac{N_y}{N_3} \qquad (7.4)$$

where

N_y = number of cycles per year,
N_3 = number of cycles to failure.

The annual damage ratio is the sum of all individual ratios and is the inverse of the calculation life of the riser. An example of fatigue life calculation is given in Table 7.5.

7.7 CORROSION CONTROL CONSIDERATION

Risers are installed in corrosive environments. Corrosion control is normally considered using extra wall thickness, coatings, and cathodic protection.

Since risers are prone to damage either by corrosion, vessel impact, or fatigue, extra wall thickness beyond that required by the design code is a good practice. A typical corrosion allowance is ⅛ in.

TABLE 7.5 An Example of Fatigue Life Calculation

Deepwater Wave Height (ft)		Average Number per Year, N_y	Stress Range, σ_{range} (psi)	Cycles to Failure, N_3	Damage Ratio, R_D
Normal statistics	0–5	3,060,000	3.0	1.0×10^9	0.0031
	5–10	410,000	5.6	1.1×10^8	0.0037
	10–15	130,000	12.4	5.0×10^7	0.026
	15–20	4790	19.0	1.0×10^6	0.0048
Major storms	20–30	810	30.0	1.8×10^5	0.0045
	30–40	37	47.0	2.0×10^4	0.0018
	40–50	2	71.0	600	0.0033
Annual damage ratio					0.047
Calculated life (years)					21.2

Risers are subject to more aggressive wave loading in the splash zone. This region is generally given a ½ in. coating of bonded neoprene. Alternatively, a monel sheathing welded to the pipe has been used in areas where biofouling is a factor. Concrete-coated risers are generally a poor idea, although they have been used offshore Mexico. The disadvantage is corrosion monitoring. Above the splash zone, the operator-preferred coating is usually used.

Common practice is to isolate the riser electrically from the platform to prevent the platform anodes from being consumed by the pipeline. Anodes placed on the pipeline protect the riser. Isolation is achieved by the use of neoprene (typically ¼–½ in.) vulcanized to the inside of the riser clamps and an electrical isolation joint above the waterline between the riser and the deck piping. Two styles of isolation joints are used. The first is an insulating kit for a flange which consists of a nonmetallic gasket and sleeves/washers for the bolts to insulate the two flange halves. The second type, which is generally preferred, is an integral rubber/steel union (Monobloc), which is welded in line. The clamps must also be insulated, which requires neoprene or nonconducting spacers. If the riser must be electrically connected to the jacket for some reason, extra anodes are needed at the base of the riser. This approach can be used in shallow water where diver surveys are routine.

7.8 RISER BENDS

The elbow at the base of the riser is normally designed for inspection pigging. For typical inspection pigs of diameters larger than 4 in., a minimum bend radius (MBR) is required. The MBR depends on diameter and wall thickness, but is generally 3 to 5 times the pipe diameter. This means that an 18 in. pipe designed with a 3D radius will have a radius to centerline of 54 in.

The angle of the bend must be such that the riser follows the platform batter in the plane of the riser and pipeline. For example, if the platform batter is 1:12, or 85.24°, and the pipe approaches normal to the jacket face, the riser bend will subtend an arc of 85.24°. If the pipe approaches parallel to the face, the bend will be 90°. Typically, this aspect of geometry must be checked carefully, particularly when the jacket has different batters for different faces. Above the waterline (typically 10–15 ft), the platform supports change from battered to vertical. The riser will have a transition piece at this location to mirror the platform batter.

7.9 RISER CLAMPS

For shallow water risers (less than 100 ft), the riser is encircled, i.e., guided, but not suspended. In deeper water, a suspension clamp at a location above the waterline is normally used. The exact depth limits are dependent on diameter and client practice. The suspension clamps restrain the riser laterally. A slight ($\frac{1}{8}-\frac{1}{4}$ in.) gap should be used with the encirclement clamps to allow the riser to slide vertically and hang off the suspension clamp.

The spacing of the encirclement clamps is determined depending upon environmental loads and generally results in the closest spacing in the splash zone. In deepwater, the spacing between clamps near bottom may be $50-60$ ft to correspond with available bracing. To accommodate pipeline expansion, the bottom clamp should not be installed too close to the seabed.

With X-braced jackets, the riser can sometimes be located at a position that optimizes the availability of clamp supports. Occasionally, clamps are attached to jacket legs. This provides convenient clamp supports, but may expose the risers to somewhat greater risk of boat damage, particularly on corner columns.

7.9.1 Design Overview

Before riser clamps can be designed, the function they perform and loads to be resisted must be thoroughly understood. The items listed below are desirable features for riser clamps:

- The number of different types of clamps should be kept to a minimum.
- Bolted connections should be avoided wherever possible. Necessary bolted connections must be designed to minimize the risk of bolts becoming loose such as by double nuts.
- Standardization of components—bolts, hinges, plate thicknesses, etc.—is desirable, i.e., hinged clamps.
- Excessive use of gusset plates (stiffeners) should be avoided.
- Reduction of diving time during riser installation is desirable.
- Clamps should be internally lined with neoprene or coated.

7.9.1.1 Basic Clamp Types

The two types of commonly used clamps are anchor and guide. These are depicted in Figures 7.6 and 7.7. The guide clamp allows vertical movement and has an internal radius of $\frac{1}{8}-\frac{1}{4}$ in. greater than the pipe radius. In cases where no anchor clamp is to be used, the guide is designed to squeeze the riser. After makeup, a gap of $\frac{1}{2}$ in. should remain between the flange faces to ensure a tight fit.

7.9.1.2 Adjustable Clamp Designs

Adjustable clamps are those in which the position of the riser end of the clamp may be altered even after the connection to the jacket. Figure 7.8 shows the adjustable clamp concept most widely used in the industry. The device consists of a tubular stub piece fitted with a riser clamp that is bolted

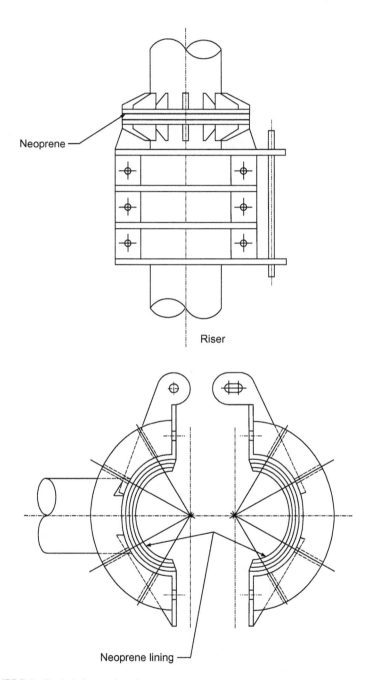

FIGURE 7.6   Typical riser anchor clamp.

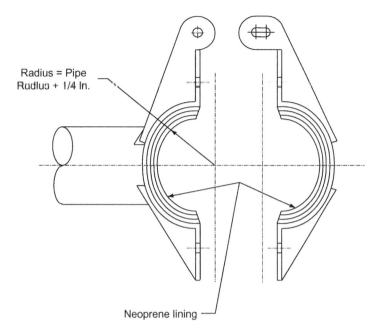

Radius = Pipe
Radius + 1/4 In.

Neoprene lining

FIGURE 7.7   Typical riser guide clamp.

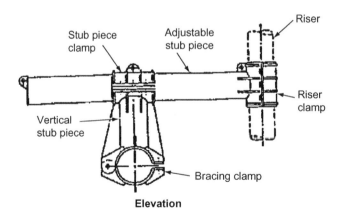

Stub piece
clamp

Adjustable
stub piece

Riser

Vertical
stub piece

Riser
clamp

Bracing clamp

Elevation

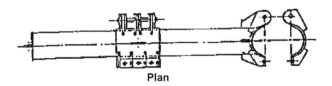

Plan

FIGURE 7.8   Adjustable clamp concept most widely used in the industry.

on one side and hinged on the other side. The connection to the bracing can be as shown in the figure or by directly welding the clamp to the bracing, if the jacket is being fabricated.

A vertical stub piece has been included to ease welding procedure and inspection. The basic components of these units are the clamp shells and the stub pieces. The wall thickness of the stub piece is based on the maximum loading, taking into account the shear force, axial force, and bending moments. Larger moments will be induced by having an adjustable stub piece. In addition, the vertical stub piece and bracing clamp have to cater for the additional moment due to the vertical offset of the bracing clamp, which sometimes results in larger stiffeners or higher-grade material for the bracing clamp. This clamp type can be preinstalled on the platform during fabrication by welding the clamp stub piece onto the jacket. The adjustable clamp requires bolting for closure around the riser. The clamp also gives more flexibility during installation due to the adjustable stub piece.

Figure 7.9 shows an adjustable clamp used for connection to diagonals in the North Sea. The dual plates must be designed with consideration that moments around two axes must be taken into account. In addition, local stress concentrations and high shear forces on the welds may be experienced by the lack of stiffeners in this section.

The adjustable clamp design shown in Figure 7.10 consists of a double clamp of both the jacket bracing and the adjustable stub piece and a single clamp to the riser. In the clamp position shown, the adjustable stub piece is arranged normal to the riser and the bracing clamp. This figure also displays the translational and rotational degrees of freedom for the clamp. Minor misalignment of the riser and pipeline can be adjusted. The clamp assembly can be attached to a vertical member (jacket leg) or a diagonal member.

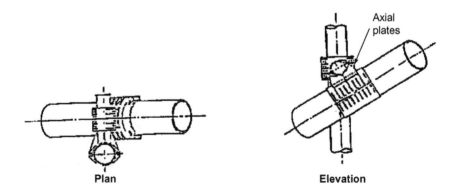

FIGURE 7.9   Adjustable clamp used for connection to diagonals.

The dual clamp part for connection to the jacket member and to the adjustable stub piece is for the friction grip type. In other words, any loads transferred from the riser will be restrained by frictional forces. The riser clamp is closed with a pinned connection on one side and bolts on the other side. The bolted connection is made such that the riser can move and rotate freely within the clamp.

Figure 7.10C shows an alternative arrangement, for easy connection, whereby the dual clamp sections can be rotated for connection to each other. Two flanges are welded to the connecting sides of the sections. One plate is fitted with two slotted holes, and the other contains two threaded studs for bolting the two flange plates together. This will allow some adjustments to be made during installation within the tolerance of the two slotted holes. The connection, however, between the flange plates and the clamp is weak in twisting and bending, in addition to being difficult to weld and inspect.

The clamps themselves are of a standard design except that the pinned connection contains three pins which are used to connect the clamp shells

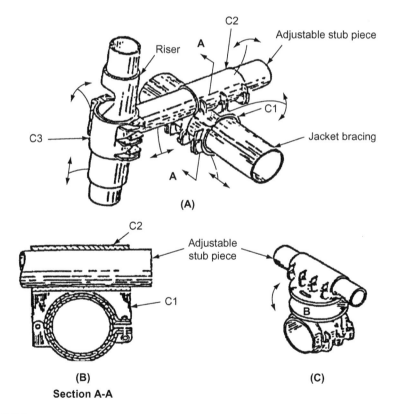

FIGURE 7.10 Arrangement for dual clamp sections (A) The adjustable dual clamp design; (B) Section A-A of the dual clamp design; and (C) An alternative dual clamp design.

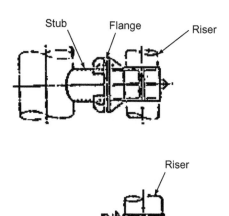

FIGURE 7.11 An alternative clamp design.

together, thus increasing the number of parts of the clamp. The use of the adjustable stub piece increases the independence of accurate measurements during riser installation. The clamp assembly also allows the design of the adjustable stub piece clamp ("C2" in Figure 7.10) for "failure." In the splash zone area, the bolt loads for this clamp are designed such that if accidents with a vessel occur, the adjustable stub piece will deflect, minimizing the damage on the riser.

7.9.1.3 Stub Piece Connection Clamp Design

Figure 7.11 shows an alternative design which facilitates future riser installation. The stub piece with a flange welded to the jacket bracing has been used extensively in the Middle East. This is not required for design of the inner stub piece, as that piece is more dependent upon the size of the jacket bracing. However, the number of bolts in the flange is important to know the approximate size of the future riser. The outer part of the clamp is shown with a plated and tubular design. The smaller riser sizes were based on plated structures, as the use of a tubular stub piece was not feasible due to the short distance between the centerline of the riser and the flange face.

Figure 7.12 shows a vertical stub clamped to the riser clamp for use with future adjustable clamps. This stub piece has the advantage of not protruding from the jacket since it is oriented along the vertical axes. It is important to

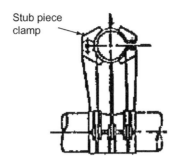

Stub piece
clamp

FIGURE 7.12   Vertical stub.

know the riser size in order to determine the number of bolts in the stub piece.

7.9.1.4 Load-Bearing Clamp Designs

The load-bearing clamp is generally located above sea level, giving the option of welding the clamp directly on the jacket bracing or using a friction grip clamp. The strength of the clamp stub piece is dependent upon the riser fixity to the clamp. With a completely fixed riser connection to the clamp, the riser forces and moments are transferred to the clamp, resulting in a requirement for a structurally strong clamp, such as a plated box-type structure between the riser clamp and the jacket bracing. For seating of the riser, the top part of the riser clamp is fitted with a flange. For easy installation, the clamp may be hinged and secured on one or two sides with bolts, depending on riser size and resulting forces and moments. As the installation generally takes place above water, no diving is required.

Figures 7.13 and 7.14 show two typical load-bearing clamps which follow the concept of rigid-type clamps apart from the flange located on the top of the riser clamp. Normally, the riser sits on top of the clamp, either using a flange welded to a sleeve which, in turn, is welded to the riser (Figure 7.13) or an anchor flange welded to the riser as part of the riser. The flange may also be used as a support for a temporary clamp to support the weight of the riser (Figure 7.15).

The required strength of the clamp is dependent upon the type of support, whether the two flange faces are bolted (fully fixed) or are resting on top of each other. With a bolted flange connection, all forces and moments experienced by the riser at this elevation will be transferred into the clamp. The reasoning for a bolted flange connection is often due to lack of deck piping data or imposed restrictions on the translations and rotations of the deck piping. With a bolted connection, all riser movements and rotations

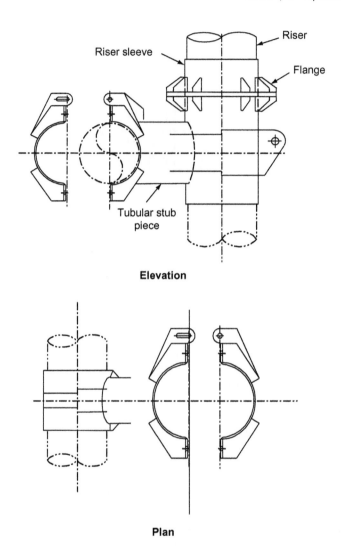

FIGURE 7.13 Typical load-bearing clamp following the concept of rigid-type clamps.

can be stopped at the load-bearing clamp. If, however, the flange connection is not bolted, only translational and vertical forces will be transferred to the clamp, resulting in a "lighter" clamp layout. Figure 7.14 shows the load-bearing clamp to consist of a tubular stub piece, while Figure 7.13 depicts the load-bearing clamp to be of a box-type construction consisting of plates welded together.

For the connection of the clamp to the riser and jacket bracing, dependent upon the riser size, bolts on both sides of the clamp are

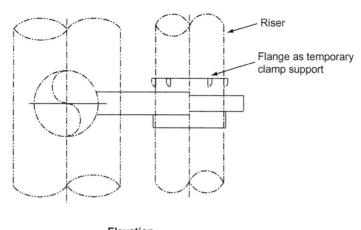

Riser

Flange as temporary
clamp support

Elevation

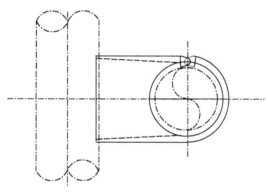

Plan

FIGURE 7.14   Typical load-bearing clamp consisting of a tubular stub piece.

recommended to secure the riser clamp and ensure a friction grip at the bracing. Generally, with the clamp installed above water, the clamp stub piece can be welded directly onto a member of the jacket using a sleeve, if required. Dependent upon forces and moments, however, the requirement for welding the load-bearing clamp to the jacket member of bolts has been increased with respect to standard clamps, the bolting of the clamps will take place above water, avoiding time-consuming diver operations.

The temporary weight clamps shown in Figures 7.15 and 7.16 are installed on the riser to transfer its weight to the load-bearing clamp during installation and hydrostatic testing of the pipeline and risers. After successful testing, hook up to the deck piping can take place, and riser weight can be

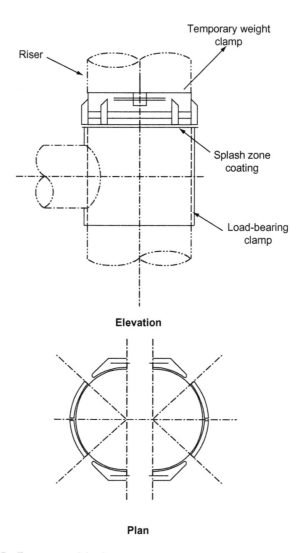

FIGURE 7.15   Temporary weight clamp.

transferred to a clamp located at a higher elevation or can be taken by the deck piping support and the stiffness of the deck piping as in shallow water riser installations.

The temporary weight clamp shown in Figure 7.15 is designed for shallow water installation or for smaller riser pipe sizes in deeper water. The clamp concept shown in Figure 7.16 is designed to be installed temporarily on a splash zone coated joint having a uniform load distribution to the load-bearing clamp. Both the temporary clamps are designed to hold the riser in

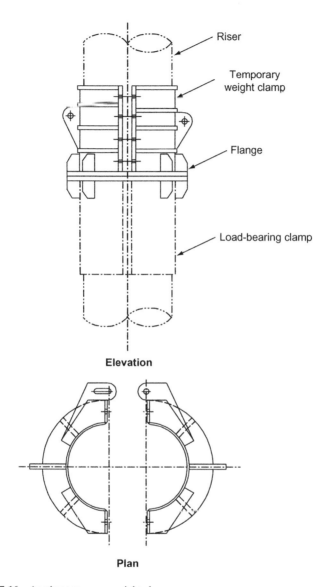

Elevation

Plan

FIGURE 7.16   Another temporary weight clamp.

place by frictional force achieved through connection with the two clamp shells with a gap between the two bolt flanges. The number of bolts is based on the weight of the riser filled with water (including the weight of the test head). In addition, if the clamp is to be attached to a splash zone coated joint, the number of bolts is increased in order to minimize the compressive stress in the coating (Figure 7.16).

TABLE 7.6 Allowable Stress Criteria for Riser Clamp Design

Stress Type	Allowable Stress Factor	
	AISC Specification	*API RP2A*
Compression	0.6	0.6
Tension	0.6	0.6
Shear	0.4	0.4
Bending	0.66	0.75

7.9.2 Design Analyses

The design of clamp assemblies, including riser and platform member clamps, bolts, lining, pins, and stubs, should consider the load combinations listed in Tables 7.1 and 7.2, as well as fabrication and maintenance. Clamp assemblies are to be elastically designed in accordance with AISC specification and API RP2A. Governing stress criteria are listed in Table 7.6.

Chapter 8

Pipeline External Corrosion Protection

8.1 INTRODUCTION

Offshore steel pipelines are normally designed for a life ranging from 10 years to 40 years. To enable the pipeline to last for the design life, the pipeline needs to be protected from corrosion both internally and externally. Internal corrosion is related to fluid that is carried by the pipeline, and this topic is not covered here. This chapter describes the method by which the external corrosion of offshore pipelines may be minimized.

A strong adhesive external coating over the whole length of the pipeline will tend to prevent corrosion. However, there is always the possibility of coating damage while handling the coated pipe either during shipping or during installation. Cathodic protection (CP) is provided by sacrificial anodes to prevent the damaged areas from corroding.

8.2 EXTERNAL PIPE COATINGS

This first external pipe coating layer is used to protect the pipe against corrosion. A single-layer coating is used when the installed pipeline is always in a static, laterally stable condition lying on soils such as clay or sand. Additional layers of coating are used for additional protection, for weight to help the pipeline remain laterally stable on the seabed, or for providing insulation. A multilayer coating is generally used in cases where the external environment tends to easily wear out the external coating (e.g., pipeline lying on top of rocky soil and calcareous material). Insulation is provided to maintain a higher temperature of the flowing internal fluid compared to the ambient. Depending on the external environment and on the location or use of the pipeline, a single-layer coating or a multilayer coating is required.

The properties that are considered desirable for deepwater pipeline coatings are as follows:

- Resistance to seawater absorption
- Resistance to chemicals in seawater
- Resistance to cathodic disbondment

Offshore Pipelines. DOI: http://dx.doi.org/10.1016/B978-0-12-397949-0.00008-X

- Adhesion to the pipe surface
- Flexibility
- Impact and abrasion resistance
- Resistance to weathering
- Compatibility with CP.

A single-layer coating may not be able to provide all of these properties under all operating conditions of pipeline. In such cases, multilayered coatings are used.

As the coating must adhere to steel pipe, the surface finish process of line pipe manufacturing must be carefully examined. This is required because in some instances, unacceptable surface finish of the line pipe can lead to loss of adhesion of the coating. The next step is to apply the coating in the coating plant following the manufacturer's recommended method of application.

8.2.1 Single-Layer Coating

The most common choice for single-layer coating for deepwater pipelines is fusion bonded epoxy (FBE). Properties and coating requirements are shown in Table 8.1.

For deepwater pipelines where there is no other requirement on the external coating, FBE is most frequently used. Most deepwater oil and gas transmission lines use FBE as they are extremely cost effective. This coating can be used in conjunction with concrete weight coating. The other coatings that can be used with concrete coating are coal tar enamel and coal tar epoxy, and they are used with lower product temperatures. Both of these coatings are bituminous coatings reinforced with fiberglass. However, most bituminous coatings are not desirable due to environmental laws and decreasing efficiency (sagging, cracking, permeation, and chemical deterioration).

The FBE field joint coating is carried out using the same coating material as mill-applied coating. Further advantages include the following:

- Easy to repair
- Easy for coating application
- High adhesion to steel
- Good for pipeline operating temperatures.

TABLE 8.1 Single-Layer Pipe Coatings

Coating Type	Maximum Temperature (°C)	Average Coating Thickness (mils)	Some Manufacturers
Fusion Bonded Epoxy	90	14 to 18	Dupont, 3 M, Lilly, BASF, Jotun

In the United States and United Kingdom, FBE coating is preferred for offshore pipelines.

8.2.2 Multilayer Coatings

Table 8.2 lists the most common choices that are available for multilayer coatings for deepwater pipelines.

8.2.2.1 Dual-Layer FBE

Dual-layer FBE coatings are used when additional protection is required for the outer layer such as high temperature and abrasion resistance. For deepwater trunklines, the high temperature of the internal fluid dissipates rapidly reaching ambient within a few miles. Therefore, the need for such coatings is limited for steel catenary risers (SCRs) at the touchdown area where abrasion is high and an additional coating with high abrasion resistance is used. The Duval system consists of an FBE base coat (20 mils) bonded to a polypropylene (PP) coating (20 mils). The PP layer provides mechanical protection.

8.2.2.2 Three-Layer Coatings

Three-layer coating consists of an epoxy or FBE, a thermoplastic adhesive coating and a polypropylene top coat. The polyethylene (PE) and polypropylene (PP) coatings are extruded coatings. These coatings are used for additional protection against corrosion and are commonly used for dynamic systems such as STRs and where the temperature of the internal fluid is high. These pipe coatings are frequently used in pipelines that are installed by the reeling method. The field joint coating for the three-layer systems is more difficult to apply and takes a longer time. However, in Europe, PE and

TABLE 8.2 Multilayer Pipe Coatings

Coating Type	Maximum Temperature (°C)	Two Main US Coating Applicators for Offshore Pipelines
Dual-layer FBE, Duval	90	BrederoShaw; Bayou Pipe Coaters
Three-layer polyethylene (PE)	110	BrederoShaw; Bayou Pipe Coaters
Three-layer polypropylene (PP)	140	BrederoShaw; Bayou Pipe Coaters
Polychloropene	90	BrederoShaw; Bayou Pipe Coaters

PP coatings are preferred because of their high dielectric strength, water tightness, thickness, and very low CP current requirement.

8.2.2.3 Concrete Weight Coating

Concrete weight coating is used when stability of the pipeline on the seabed is an issue. The two common densities of concrete that are used are 140 lb/ft^3 and 190 lb/ft^3. Higher density is obtained by adding iron ore to the concrete mix. Recently, higher density iron ore has been used to obtain concrete density ranging from 275 to 300 lb/ft^3 for the Ormen Lange pipeline in the North Sea.

8.2.3 Standard Organizations with Specifications Related to Pipe Coatings

The main organizations in the United States are listed here.

- American Society of Testing Methods (ASTM)
- Steel Structures Painting Council (SSPC)
- National Association of Corrosion Engineers (NACE)
- National Bureau of Standards (NBS)
- International Organization for Standardization (ISO).

The more common ones in Europe are listed here.

- Det Norske Veritas (DnV)
- Deutsches Institut fur Nurmung (DIN)
- British Standards (BS)
- International Organization for Standardization (ISO).

8.3 CATHODIC PROTECTION

CP is a method by which corrosion of the parent metal is prevented. The two main methods of CP are galvanic anodes and impressed current systems. For offshore pipelines, the galvanic anode system is generally used.

Corrosion is an electrochemical reaction that involves the loss of metal. This is due to the fact that the steel pipeline surface consists of randomly distributed cathodic and anodic areas, and seawater is the electrolyte that completes the galvanic cell. This causes electrons to flow from one point to the other, resulting in corrosion. By connecting a metal of higher potential to the steel pipeline, it is possible to create an electrochemical cell in which the metal with lower potential becomes a cathode and is protected.

Pipeline coatings are the first barriers of defense against corrosion. However, after coating the pipe, the process of transportation and installation of the pipelines results in some damage to the coating. CP uses another metal that will lose electrons in preference to steel. The main metals used as sacrificial anodes are alloys of aluminum and zinc. By attaching anodes of these metals to the steel pipeline, the steel area where the coating is damaged is protected from corrosion.

Zinc anodes are not normally used in deepwater pipelines because they are less efficient and therefore require a larger mass for protecting the pipeline. However, zinc anodes can be cast onto the pipe joint, and therefore, no cables need to be used for electrical connection to the steel. Zinc has been used in projects where the pipeline needed to be towed along the seabed and cast-on zinc anodes were less liable to be knocked off in the process of installation. Zinc anodes do not perform well for hot buried pipelines and are subject to intergranular attack at temperatures above 50°C. There is also a tendency for zinc anodes to passivate at temperatures above 70°C.

Aluminum anodes, on the other hand, perform much better. They are better suited for hot buried pipelines. Generally, for deepwater pipelines, aluminum alloy anodes that are attached to the pipeline are bracelet anodes. These anodes are normally attached to the pipe joint at the coating yard for S-lay and J-lay installation methods. Electrical contact to the pipeline is made by thermite welding or brazing the cable from the anode.

In the case of installation of pipeline by the reeling method, the anodes are installed on the lay vessel during unreeling and straightening. In this case, bracelet anodes are attached to the pipe by bolting and attaching the cable by thermit/cadweld to the pipeline.

The design of CP systems must consider the potential detrimental effects of the CP system such as hydrogen embrittlement of steel and local stresses that may lead to hydrogen induced stress cracking (HISC).

8.3.1 CP Design

In order to conduct a CP design for a deepwater pipeline, the parameters that need to be known are as follows:

- Service/design life (years)
- Coating breakdown (%)
- Current density for protection (mA/m^2) buried or unburied
- Seawater resistivity (Ω-cm)
- Soil resistivity (Ω-cm)
- Pipeline protective potential (normally, -900 mV with respect to Ag/AgCl)
- Anode output (A-h/kg)
- Anode potential (mV with respect to Ag/AgCl)
- Anode utilization factor (%)
- Seawater temperature
- Pipeline temperature
- Depth of pipeline sinkage/burial.

The design life of the pipeline is based on whether it is trunkline or an infield line. The life of a trunkline can be as long as 40 years while that of an infield line is normally 20 years. The coating breakdown factor depends on the type of coating. There is very little historical data available on coating breakdown. DnV (RP-F103) and NACE (RP-01-76) have recommended values based on the type of pipeline coating. Three values of coating breakdown are typically given—initial, mean, and final.

The current density, resistivity, and temperature depend on the geographical location where the pipeline is located. In deepwater pipelines, the approximate seawater temperature range is from 1.7°C to 7.5°C. DnV and NACE give values for current densities and resistivities for offshore geographical locations. For bare steel buried in sediments, a design current density of 0.020 A/m^2 is recommended by DnV.

The type of anode used determines its electrochemical properties. The Galvalum III$^{®}$ anode, for example, has an anode output of approximately 2250 A-h/kg in seawater temperature less than 25°C, and its potential is approximately -1050 mV. Manufacturers of anodes provide these properties for design.

The anode utilization factor depends on the shape and application of the anode. Bracelet anodes are typically assumed to be 80% utilized at the end of their life, while standoff anodes are 90% utilized. For pipeline temperatures above 25°C, the design current densities increase. For each degree above 25°C, the current density is increased by 0.001 A/m^2.

8.3.1.1 CP Design Methodology

The design methodology summarized here follows that given in DnV-RP-B401.

Designs must satisfy two requirements:

1. The total net anode mass must be sufficient to meet the total current demand over the design life.
2. The final exposed anode surface area must be sufficient to meet current demand at the end of design life (the final exposed anode surface area is calculated from anode initial dimensions, net mass, and the utilization factor).

First, one computes the current demand, (I_c), for initial, mean, and final stages of the design life. The current demand to protect each pipeline is calculated by multiplying the total external area (A_c) with the relevant design current density (i_c) and coating breakdown factor (f_b):

$$I_c = A_c f_b i_c \tag{8.1}$$

The current demands for initial polarization, I_{ci}, and for repolarization at the end of the design life, I_{cf}, are normally to be calculated together with the mean current demand I_{cm} required to maintain CP throughout the design period. It is not necessary to calculate the current demand required for initial polarization, I_{ci}, because initially, the pipeline corrosion coatings greatly reduce the current demand and time required for initial polarization.

The coating breakdown factors for various coatings, initial, mean, and final, are given in DnV and NACE publications. For example, in the Gulf of Mexico, for FBE coating with a design life of 20 years, the initial, mean, and final coating breakdown factors normally used are 1, 3, and 5%, respectively. The total net anode mass M_t required to maintain CP of a pipeline throughout the design life t_d (years) is given by

$$M_t = \frac{8760 I_{cm} t_d}{u_f \varepsilon_e} \tag{8.2}$$

where

I_{cm} = mean current demand,
ε_e = the electrochemical efficiency (A-h/kg),
u_f = the anode utilization factor.

The required current output (initial/final) and current capacity for a specific number of anodes determine the required anode dimensions and net weight. The following requirements must be met by the cathodic system design:

$n_a c_a \geq 8760 I_{cm} t_d$
$n_a I_a$ (initial/final) $\geq I_c$ (initial/final)

where

n_a = number of anodes,
c_a = anode current capacity (A-h),
I_a = anode current output (A).

The anode current capacity (c_a) is given by

$$c_a = m_a \varepsilon_e u_f \tag{8.3}$$

where m_a is the net mass per anode. The anode current output (I_a) is calculated from Ohm's law

$$I_a = \frac{E_c^0 - E_a^0}{R_a} \tag{8.4}$$

where

E_c^0 = design closed circuit potential of the anode,
E_a^0 = design protective potential,
R_a = anode resistance.

The design protective potential (E_c^0) for carbon steel is -0.80 V (relative Ag/AgCl/seawater) when in aerated seawater and -0.90 V (relative Ag/AgCl/seawater) when in anaerobic environments including typical marine sediments. Recommended practice states that the $E_c^0 = -0.8$ V should be used for *all* design calculations because the initial and final design current densities are referred to this protective potential.

The closed circuit anode potential (E_c^0) for an Al-based anode is taken to be -1.1 V for the pipeline at ambient temperature and -1.085 V for the pipeline at elevated temperatures. The anode resistance (R_a) formula for a bracelet anode is given by

$$R_a = \frac{0.315 \cdot \rho}{\sqrt{A_e}} \tag{8.5}$$

where

ρ_e = environmental resistivity,
A_e = exposed anode surface area.

The required number of anodes, n, can be obtained by

$$n_a = \frac{I_{cft}}{I_{af}} \tag{8.6}$$

where

I_{cft} = total final current demand for the pipeline
I_{af} = individual anode current output.

Some iterations may be required to meet the requirements of both the total net anode mass, M_t, and the total final anode current output ($n_a I_{af}$).

Generally, maximum spacing of the anodes recommended is 300 m. However, methods to calculate attenuation of protective potential with distance can be used to determine the mass and spacing of anodes.

Attenuation computations are specifically useful for determining anodes for CP of SCRs. In SCRs, rather than placing anodes on the suspended dynamic portion, several anodes may be placed on static pipeline sections past the touchdown point.

This method is also useful for short (up to 3 miles) bottom-towed pipelines with sleds at each end. Instead of placing discrete bracelet anodes along the pipeline, all total mass of anodes required for the pipeline can be placed on the end sleds. Placing them on the end sleds prevents the accidental impact and loss of bracelet anodes from the pipeline being towed along the seabed.

Attenuation calculations show that if current is drained from two points on a pipeline, the change in potential of the pipe may be calculated using the following equations:

$$E_x = E_B \cosh[(2\pi rR_1/k_p z_a)^{1/2}(x - d_p/2)] \tag{8.7}$$

$$E_A = E_B \cosh[-(2\pi rR_1/k_p z_a)^{1/2} d_p/2)] \tag{8.8}$$

$$I_A = (2E_B/R_1)[(2\pi rR_1/k_p z_a)^{1/2} \sinh(d_p/2(2\pi R_1/k_p z_a)^{1/2})] \tag{8.9}$$

where

E_x = change in potential at point x,
E_A = change in potential at drain point,
r = pipe radius,
E_B = change in potential at the midpoint between the two drain points,
R_1 = linear resistance of the pipeline,
I_A = total current pick up,
d_p = distance between drain points,
x = distance from drain point,
k_p = polarization slope,
z_a = actual bare area per linear length of pipeline.

Additional constraints are as follows:

- The current, I_A, must be equal to the current that can be delivered by the lumped anode array.
- E_A must equal the anode potential less the Internal Resistance (IR) drop, using the anode array resistance.
- The anode weight must exceed the weight necessary to protect the section of the pipeline for the specified design life.

Using these equations and constraints, a greater spacing of the required mass/array of anodes may be computed.

FURTHER READING

Aalund, L.R., 1992. Polypropylene system scores high as pipeline anti-corrosion coating. Oil Gas J. 90 (50), 42–45.

Alexander, M., 1992. High-temperature performance of three-layer epoxy/polyethylene coatings. Mater. Perform. 31 (6), 41–44.

DnV-RP-F103. 2010. Cathodic protection of submarine pipelines by galvanic anodes. Det Norske Veritas.

Gore, C.T., Mekha, B.B., 2002. Common sense requirements for steel catenary risers (SCRs). OTC Paper 14153.

Houghton, C.J., Ashworth, V., 1981. The performance of commercially available zinc and alumi-
    num anodes in seabed mud at elevated temperatures. in: Proceedings of the 2nd BSE-NACE
    Corrosion Conference, January 19–21, Bahrain.

Kavanagh, W.K., Harte, G., Farnsworth, K.R., Griffin, P.G., Hsu T.M., Jefferies, A., 2002.
    Matterhorn steel catenary risers: critical issues and lessons learned for reel-layed SCRs to a
    TLP. OTC Paper 14154.

LaFontaine, J., Smith, D., Deason, G., Adams G., 2002. Bombax pipeline project: anti-corrosion
    and concrete weight coating of large diameter subsea pipelines. OTC Paper.

NACE RP 0176. 2003. Corrosion control of steel fixed offshore structures associated with petro-
    leum production.The National Association of Corrosion Engineers Recommended Practices.

Smith, S.N., 1993. Analysis of cathodic protection on an underprotected offshore pipeline.
    Mater. Perform. 32 (4), 23–27.

Varughese, K., 1996. Mechanical properties critical to pipeline project economics. Oil Gas J. 94
    (37), 41–46.

Chapter 9

Pipeline Insulation

9.1 INTRODUCTION

Oilfield pipelines are insulated mainly to conserve heat. The need to keep the product in the pipeline at a temperature higher than the ambient could exist for the following reasons:

- preventing the formation of gas hydrates,
- preventing the formation of wax or asphaltenes,
- enhancing the product flow properties,
- increasing the cooldown time after shutting down,
- meeting other operational/process equipment requirements.

In liquefied gas pipelines, such as LNG, insulation is required to maintain the cold temperature of the gas to keep it in a liquid state. This chapter describes the commonly used insulation materials, insulation finish on pipes, and general requirements for insulation of offshore and deepwater pipelines.

9.2 INSULATION MATERIALS

Polypropylene, polyethylene, and polyurethane are three base materials widely used in the industry for pipeline insulation. Their thermal conductivities are given in Table 9.1 (Carter et al., 2003).

Depending on applications, these base materials are used in different forms resulting in different overall conductivities. A three-layer polypropylene applied to pipe surface has a conductivity of 0.13 Btu/h ft °F, while a four-layer polypropylene has a conductivity of 0.10 Btu/h ft °F. Solid polypropylene has higher conductivity than polypropylene foam. Polymer syntactic polyurethane has a conductivity of 0.07 Bt/h ft °F, while glass syntactic polyurethane has a conductivity of 0.09 Btu/h ft °F. These materials have lower conductivities in dry conditions such as that in pipe-in-pipe (PIP) applications.

Because of its low thermal conductivity, more and more polyurethane foams (PUFs) are used in deepwater applications. Physical properties of PUFs include density, compressive strength, thermal conductivity, closed cell content, leachable halides, flammability, tensile strength, tensile modulus, and water absorption. The values of these properties vary depending on density of the foam. Table 9.2 summarizes the properties of Cape Modern high-density PUFs.

Offshore Pipelines. DOI: http://dx.doi.org/10.1016/B978-0-12-397949-0.00009-1

TABLE 9.1 Thermal Conductivities of Materials Used in Pipeline Insulation

Material Name	Thermal Conductivity	
	Btu/h ft °F	W/m K
Polypropylene	0.13	0.22
Polyethylene	0.20	0.35
Polyurethane	0.07	0.12

TABLE 9.2 Properties of Cape Modern High-Density PUFs

Foam Property	Nominal Density (kg/m³)				Test Method
Compressive strength at 20°C, MPa	160	224	320	500	ASTM D1622
	2.035	4.563	8.144	22.998	ASTM D1621, perpendicular
Compressive strength at −196°C, MPa	1.999	3.819	9.144	21.217	ASTM D1621, parallel
	3.300	7.485	15.829	48.394	ASTM D1621, perpendicular
Thermal conductivity at 20°C, W/m K	3.494	7.540	17.107	47.408	ASTM D1621, parallel
	0.0292	0.0345	0.0407	0.0425	ASTM C518
Thermal conductivity at −160°C, W/m K	0.0253	0.0316	0.0346	0.0390	ASTM C177
Closed cell content, %	95	95	95	96	ASTM D2856
Leachable halides, ppm	<20	<20	<20	<20	ASTM D871
Flammability/10 S.E	10	10	10	10	ASTM D1692
Tensile strength at 22°C, MPa	2.412	3.517	6.649	12.582	ASTM D1623
Tensile strength at −196°C, MPa	3.204	4.854	8.305	15.055	ASTM D1623
Tensile modulus, MPa	11.8	19.4	24.0	29.5	ASTM D1624
Water absorption, % vol.	0.17	0.15	0.12	0.10	ASTM D2842

S.E., Self-Extinguishing.

9.3 PIP INSULATION

Under certain conditions, PIP systems may be considered over conventional single-pipe systems. PIP insulation may be required to produce fluids from high-pressure/high-temperature (above 150°C) reservoirs in deepwater (Carmichael et al., 1999). The annulus between pipes can be filled with different types of insulation materials such as foam, granular, gel, and inert gas or vacuum.

A pipeline bundled system, a special configuration of PIP insulation, can be used to group individual flowlines together to form a bundle (McKelvie, 2000). Heat-up lines can be included in the bundle if necessary. The complete bundle may be transported to site and installed with a considerable cost saving relative to other methods. The extra steel required for the carrier pipe and spacers can be sometimes justified by a combination of the following cost advantages (Bai, 2001):

- A carrier pipe can contain multiple lines including flowline, control lines, hydraulic hoses, power cables, and glycol lines.
- Insulation of the bundle with foam, gel, or inert gas is usually cheaper than individual flowline insulation.

9.4 GENERAL REQUIREMENTS

The requirements for pipeline insulation vary from field to field. Flow assurance analyses need to be performed to determine the minimum insulation requirements for a given field. These analyses include the following:

- Flash analysis of the production fluid to determine the hydrate-forming temperatures in the range of operating pressure.
- Global thermal hydraulics analysis to determine the required overall heat transfer coefficient (OHTC) at each location in the pipeline.
- Local heat transfer analysis to determine the type and thickness of insulation to be used at the location.
- Local transient heat transfer analysis at special locations along the pipeline to develop cooldown curves and times to the critical minimum allowable temperature at each location.

A number of computer packages are available in the industry for performing these analyses efficiently.

In steady-state flow conditions in an insulated pipeline, the heat flow, Q, through the pipe wall is given by

$$Q_r = U_o A_r \Delta T$$

where

Q_r = heat transfer rate,
U_o = OHTC at the reference radius,
A_r = area of the pipeline at the reference radius,
ΔT = difference in temperature between the pipeline product and the outside ambient temperature.

The OHTC, U_o, for a system is the sum of the thermal resistances and is given by (Holman, 1981):

$$U_o = \frac{1}{A_r\left((1/A_i h_i) + \sum_{m=1}^{n}(\ln(r_{m+1}/r_m)/2\pi L k_m) + (1/A_o h_o)\right)} \quad (9.1)$$

where

h_i = film coefficient of pipeline inner surface,
h_o = film coefficient of pipeline outer surface,
A_i = area of pipeline inner surface,
A_o = area of pipeline outer surface,
r_m = radius,
k_m = thermal conductivity.

Similar equations exist for transient heat flow giving instantaneous rate for heat flow.

Typically required insulation performances in terms of OHTC (U-value) of steel pipelines of different configurations are summarized in Table 9.3.

TABLE 9.3 Typical Performance of Insulated Pipes

Insulation Type	U-Value		Water Depth (ft)	
	(Btu/h ft² °F)	W/m² K	Field Proven	Potential
Solid polypropylene	0.50	2.84	5000	13,000
Polypropylene foam	0.28	1.59	2100	6300
Syntactic polyurethane	0.32	1.81	3600	11,000
Syntactic PUF	0.30	1.70	6200	11,000
PIP syntactic PUF	0.17	0.96	9500	13,000
Composite	0.12	0.68	3200	9000
PIP high efficiency	0.05	0.28	5300	9000
Glass syntactic polyurethane	0.03	0.17	7000	9000

9.4.1 Dry Insulations

Pipeline insulation comes in two main types—dry insulation and wet insulation. The dry insulations require an outer barrier to prevent water ingress (PIP). The most common types of this are as follows:

- Closed-cell polyurethane foam (CCPUF)
- Open-cell polyurethane foam (OCPUF)
- Polyisocyanurate foam (PIF)
- Extruded polystyrene
- Fiberglass
- Mineral wool
- Vacuum insulation panels (VIPs).

For deepwater pipelines, the outer barrier is a steel line pipe called the casing pipe. These pipelines are called PIP systems. Most deepwater pipelines that are insulated fall into this category.

The manufacture of PIP systems consists of placing the carrier pipe concentrically in the casing pipe using spacers and foaming the annulus. Large coating companies in the United States such as Bayou Companies and Bredero Shaw produce these PIP systems in an assembly line and are therefore able to produce large quantities in a short time for offshore deepwater use.

For installation by reel method or bottom tow method, the insulation can be placed on the carrier pipe and then pulled into the casing pipe by using low-friction spacers or rollers attached to the carrier pipe. This is performed manually and therefore uses preformed insulation panels.

In the bottom tow method for deepwater pipelines, the casing is pressurized with dry nitrogen to enable the reduction in the casing wall thickness. In such cases, the insulation needs to be open-cell PUF, fiberglass, or syntactic foam. The open cell allows the pressurized nitrogen to permeate the cells and prevent any collapse of the cells. Closed cells would collapse under the pressure. Syntactic foam is designed to withstand high pressures.

In a pressurized gas/nitrogen environment, the k-value of the insulation increases due to convection. If the pipeline is lying on a slope, the "chimney effect" causes convection currents to dissipate the heat and lower increases the effective k-value. To prevent this, an High-density polyethylene (HDPE) layer or a pipe is placed concentrically around the carrier pipe and the annulus is foamed. Holes are placed in the bottom position of the HDPE layer to allow the nitrogen to permeate into the open cells. Tests on such a configuration of insulated PIP have shown it to work well even in a pressurized nitrogen environment. Fiberglass insulation can be used instead of open-cell foam for similar configuration. Table 9.4 lists the properties of some dry insulations.

TABLE 9.4 Properties of Dry Insulations

Insulation Material	k-Factor at 75°F (Btu/h ft² °F)	Density (lb/ft³)	Compressive Strength (psi)	Service Temperature (°F)
CCPUF		3−6	10−65	
OCPUF		2−4		
PIF	0.190	1.8−2	19−24	−297−300
Polystyrene	0.259	6	20	−297−165
Fiberglass	0.24	3.5−5.5		0−850
Mineral wool	0.25	8.7		1292
VIP	0.036−0.055	3.7−9.0		320

9.4.2 Wet Insulations

Wet pipeline insulations are those materials that do not need an exterior steel barrier to prevent water ingress or the water ingress is negligible and does not degrade the insulation properties. The most common types of this are as follows:

- Polyurethane
- Polypropylene
- Syntactic polyurethane
- Syntactic polypropylene
- Multilayered
- Other.

The main materials that have been used for deepwater insulations are polyurethane and polypropylene based. Syntactic versions use plastic or glass matrix to improve insulation and greater depth capabilities. Insulation coatings with combinations of the two materials have also been used. Table 9.5 lists the properties of these wet insulations.

As can be seen from the table, the insulation is buoyant and must be compensated by the steel pipe weight to obtain lateral stability of the deepwater pipeline on the seabed.

9.5 HEAT TRANSFER ANALYSIS—AN EXAMPLE

In order to suggest the insulation materials and thickness for the network of pipelines, the heat transfer going through the pipes must be calculated for

TABLE 9.5 Properties of Wet Insulations

Insulation Material	k-factor (Btu/ h ft^2 °F)	Density (lb/ft^3)	Water Depth (ft)	Service Temperature (°F)
Polyurethane (PU)—solid	0.035	72		240
Polypropylene (PP)—solid	0.039	56	9000	290
Syntactic PU	0.021–0.026	38–53	300–9000	131–240
Syntactic PP	0.023–0.039	37–50		240
Syntactic phenolic	0.014	31		392
Syntactic epoxy	0.017–0.024	37–45	6000–9000	160–212

each pipe. The main aspect of heat transfer through the pipe occurs through conduction of the insulation material. The basic equation for radial heat transfer can be found in textbooks (Holman, 1981). Normally, it is assumed that convective heat transfer and conductive heat transfer through the pipe material are negligible. While heat transfer calculations for pipelines under steady flow conditions are straightforward, numerical computer simulators are required and available for pipelines under transient flow conditions. Guo et al. (2004) presented analytical solutions that can be easily used to carry out the required steady state and transient heat transfer analyses for single pipes. These solutions are included in Appendix B of this book. An application example with the analytical solutions is presented in this section.

Suppose a set of data in Table 9.6 is applicable to a design pipeline. Sensitivity analyses can be performed with the analytical temperature models to investigate the effects of thermal conductivity, time, and fluid flow rate on the temperature profile in a pipe.

Figure 9.1 illustrates steady temperature profiles calculated using different values of thermal conductivity of the insulation. In this situation, it appears that a thin layer (1.27 cm) of insulation with thermal conductivity of less than 1 W/m K will allow a total temperature drop of less than 1°C over the 1000 m pipeline.

Figure 9.2 shows the calculated transient temperature profiles for a start-up process. It shows that the transient temperature profile approaches the steady temperature profile after one half hour of fluid flow at a constant rate. Figure 9.3 demonstrates the calculated transient temperature profiles after an increase in fluid flow rate from 0.005 to 0.05 m^3/s. Although a 10 times higher value of thermal conductivity is utilized, it still shows that the

TABLE 9.6 Base Data Used in the Heat Transfer Analyses

Thermal conductivity of insulation	1.0	W/m K
Outer radius of pipe	0.100	m
Inner radius of pipe	0.095	m
Thermal gradient outside the insulation	0.005	°C/m
Thermal gradient angle from pipe axis	0	degree
Specific heat of fluid	41,800	J/kg°C
Fluid density	1000	kg/m³
Insulation thickness	0.0127	m
Fluid flow rate	0.0005	m³/s
External temperature at fluid entry point	100	°C
Fluid temperature at fluid entry point	100	°C
Pipe length	1000	m

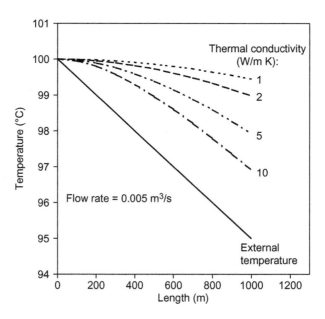

FIGURE 9.1   Calculated temperature profiles under steady fluid flow conditions.

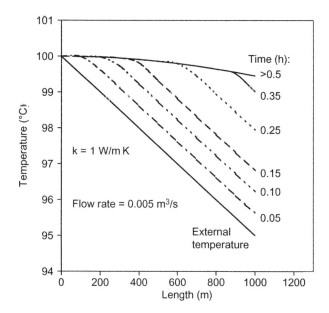

FIGURE 9.2   Calculated temperature profiles during a fluid flow start-up process.

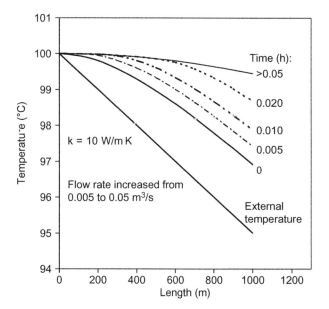

FIGURE 9.3   Calculated temperature profiles after a flow rate increase.

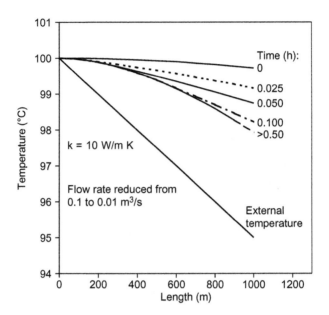

FIGURE 9.4   Calculated temperature profiles after a flow rate reduction.

transient temperature profile approaches the steady temperature profile after one half hour of flow at the new rate. Figure 9.4 shows the calculated transient temperature profiles during a pipeline shutdown process. It indicates that after a reduction in fluid flow rate from 0.1 to 0.01 m³/s, the transient temperature profile approaches the steady temperature profile after one half hour of flow at the new rate.

Paraffin (wax) deposition is a serious problem in the oil industry because it causes plugging of the wellbore, production facilities, and transportation pipelines. This problem is described in Part III of this book. Oil composition, pressure, and temperature are the factors affecting paraffin deposition. For a given oil composition, paraffin deposition is a strong function of temperature and weak function of pressure except in the near-critical point region where it is also sensitive to pressure. It is vitally important to predict the locations where paraffin deposition occurs in pipelines. The prediction can be used for flow assurance management in the oil production and transportation operations.

Paraffin deposition is usually evaluated in laboratories using wax appearance temperature (WAT) at different pressures. The WAT curve draws a boundary between wax region and wax-free region in the pressure–temperature (P–T) diagram. If the in-situ condition (temperature and pressure) of pipeline falls in the wax region, paraffin deposition is expected to occur at the point. Figure 9.5 shows an example P–T diagram generated with the analytical solution and a hydraulics model using the data by Hammami and

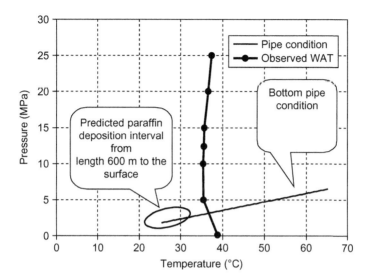

FIGURE 9.5   P–T diagram to identify paraffin deposition interval in a wellbore.

Raines (1999). This plot shows that the P–T profile falls in the two regions (wax and wax-free) across the WAT curve. The cross-point is at length of about 600 m. This means that the paraffin deposition could occur in the upper (low temperature and pressure) section of the pipeline above 600 m. A better insulation is required to prevent the paraffin deposition.

REFERENCES

Bai, Y., 2001. Pipelines and Risers. In: Bhattacharyya, R., McCormick, M.E. (Eds.), Elsevier Ocean Engineering Book Series, vol. 3. Elsevier, Amsterdam.

Carmichael, R., Fang, J., Tam, C., 1999. Pipe-in-pipe systems for deepwater developments. In: Proceedings of Second Deepwater Pipeline Technology Conference. March 22–25, New Orleans.

Carter, R., Gray, C., Cai, J., 2003. 2002 survey of offshore non-chemical flow assurance solutions. Offshore Magazine, Houston.

Guo, B., Duan, S., Ghalambor, A., March 16–18, 2004. A simple model for predicting heat loss and temperature profiles in thermal injection lines and wellbores with insulations. In: Paper SPE 86983 Presented at the SPE International Thermal Operations and Heavy Oil Symposium and Western Regional Meeting held in Bakersfield, California.

Hammami, A., Raines, M.A., 1999. Paraffin deposition from crude oils: comparison of laboratory results with field data. Paper SPE 54021 Presented at the SPE Annual Technical Conference and Exhibition held in San Antonio, Texas.

Holman, J.P., 1981. Heat Transfer. McGraw-Hill Book Company, New York, NY.

McKelvie, M., 2000. Bundles—design and construction. Integrated Graduate Development Scheme, Heriot-Watt University.

FURTHER READING

Hansen, A.B., Rydin, C., 2002. Development and qualification of novel thermal insulation systems for deepwater flowlines and risers based on polypropylene. In: Presented at the Offshore Technology Conference, 6–9 May 2002, Houston, TX, OTC Paper 14121.

Tough, G., Denniel, S., Al Sharif, M., Hutchison, J., 2001. Nile—design and qualification of reeled PIP pipeline in deepwater. In: Presented at the Offshore Technology Conference, 30 April–3 May 2001, Houston, TX, OTC Paper 13257.

Wang, W., Watkins, L., 2002. Syntactic foam insulation for ultradeep high temperature applications. In: Proceedings of ASME 2002 21st International Conference on Offshore Mechanics and Arctic Engineering, 23–28 June, 2002, Oslo, Norway, OMAE 2002-28193.

Wang, X., Chen, Y.D., Perera, R.M., Prescott, C.N., 2000. Convection heat losses through gaps between pipe and insulation and between insulation half-shells. In: Presented at the Offshore Technology Conference, 1–4 May 2000, Houston, TX, OTC Paper 12033.

Watkins, L., Hershey, E., 2001. Syntactic foam insulation for ultra deepwater oil & gas pipelines. In: Presented at the Offshore Technology Conference, 30 April–3 May 2001, Houston, TX, OTC Paper 13134.

Chapter 10

Introduction to Flexible Pipelines

10.1 INTRODUCTION

Flexible pipes have been used in the oil industry since 1972, when Coflexip was awarded a patent to build a high-pressure, flexible steel pipe. The first application was used in drilling as a 15,000 psi kill and choke line. Since then, flexible pipe designs have improved to produce the flowlines and risers that are now used in the offshore oil industry.

For deepwater, the flexible pipes are used mainly for dynamic risers from a subsea pipeline end manifold or riser tower to a floating production system (FPS) such as a floating storage and offloading (FSO) unit or a floating production, storage and offloading (FPSO) unit and tension-leg platforms (TLPs). The other uses are static risers, static flowlines, subsea jumpers, topside jumpers, and expansion joints. Flexible pipes are used for versatile offshore oil and gas applications including production, gas lift, gas injection, water injection, and various ancillary lines including potable water and liquid chemical lines.

The main advantages of flexible pipelines are as follows:

- Ease and speed of installation
- No large spans because it follows the contours of the seabed
- Almost no maintenance for life of the project
- Good insulation properties are inherent
- Excellent corrosion properties
- No field joints because the pipe is of continuous manufacture
- No need of expansion loops
- Can be made with enhanced flow characteristics
- Sufficient submerged weight for lateral stability
- Accommodates misalignments during installation and tie-in operations
- Diverless installation is possible—no metrology necessary
- Load-out and installation is safer, faster, and cheaper than any other pipe application

Offshore Pipelines. DOI: http://dx.doi.org/10.1016/B978-0-12-397949-0.00010-8

- Retrievability and reusability for alternative application, thus enhancing the overall field development economics and preserving the environment
- Fatigue life longer than steel pipe.

The codes that are used for the design of flexible pipe are as follows:

- API SPEC RP 17B—"Recommended Practice for Flexible Pipe"
- API SPEC RP 17J—"Specification for Unbonded Flexible Pipe"
- API SPEC RP 17K—"Specification for Bonded Flexible Pipe"
- ISO 10420—"Flexible Pipe Systems for Subsea and Marine Riser Applications"
- API Spec RP 2RD—"Design of Risers for FPSs and TLPs."

Since there are only three manufacturers, and the manufacturing of flexible pipe requires wrapping of many intertwining layers of high-strength stainless steel carcass and special polymers, the material price of a flexible line is hundreds of times more expensive than an equivalent high-strength carbon steel pipe. Consequently, general use is limited to special applications and in small quantities compared to the use of high-strength carbon steel pipe.

Ultra-deepwater use of flexible pipe is limited, due to the inability of these pipes to withstand high external hydrostatic pressure. Presently, the maximum depth at which flexible pipes have been used is 2000 m.

The main flexible pipe layers are shown in Figure 10.1. The material makeup of each layer is described below.

Layer 1 Carcass: The carcass is a spirally wound interlocking structure manufactured from a metallic strip. The carcass prevents collapse of the inner liner and provides mechanical protection against pigging tools and abrasive particles.

Layer 2 Inner liner: This is an extruded polymer layer that confines the internal fluid integrity.

FIGURE 10.1   Flexible pipe layers.

Layer 3 Pressure armor: This is made up of a number of structural layers consisting of helically wound C-shaped metallic wires and/or metallic strips. The pressure armor layers provide resistance to radial loads.

Layer 4 Tensile armor: The tensile armor layers provide resistance to axial tension loads. This is made up of a number of structural layers consisting of helically wound flat metallic wires. The layers are counter wound in pairs.

Layer 5 Outer sheath: The outer sheath is an extruded polymer layer. Its function is to shield the pipe's structural elements from the outer environment and to give mechanical protection.

These are the primary layers. Some of the other layers that are not shown are the antiwear layers and insulation layers. The antiwear layers are nonmetallic layers that are inserted between the structural elements to prevent wear and tear. Additional layers of material with low thermal conductivity can be applied in order to obtain specific thermal insulation properties of the pipe.

All the flexible pipes have the same fundamental concept. Some variation may occur in choice of materials in case of special operating environments such as high pressures, high temperatures, sour service (high H_2S and/or CO_2 content), and deepwater.

The end fitting of the flexible pipe is extremely important as it seals the different layers preventing any water ingress and also allows it to be connected to other pipeline appurtenances. The common end fittings that are used are as follows:

- Flanges
- Grayloc connectors
- Hydraulic subsea connectors.

Another device that is used at the end of the flexible pipes is the bend restrictor. This is used to prevent excessive bending because most flexible pipes have a minimum allowable bend radius. Any bending beyond this would comprise the integrity of the flexible pipe.

10.2 FLEXIBLE PIPE MANUFACTURERS

The three flexible pipe manufacturers in the world are as follows:

1. NKT Flexibles
2. Wellstream
3. Technip (formerly Coflexip).

10.2.1 NKT Flexibles

NKT Flexibles, located in Broendby, Denmark, is a worldwide supplier of flexible subsea pipes. This company, originally a power cable manufacturer, made its first flexible pipe in 1967. The NKT flexible pipe is an unbonded structure consisting of helically wound metallic armor wires or tapes combined with concentric layers of polymers, textiles, fabric strips, and lubricants. For each product type, all layers in the flexible pipe design are described in terms of dimensions and type of material.

Some of NKT Flexibles' features are as follows:

- Unrivaled process technology and experience from more than 30 years of flexible pipe manufacturing
- State-of-the-art manufacturing facility
- Full compliance with API 17J
- Focus on flexible pipe manufacturing as core business.

NKT Flexibles performs total riser and flowline system design, engineering, procurement, manufacture, testing, documentation, and delivery. NKT flexible pipe has been qualified by Bureau Veritas and can manufacture flexible pipe to meet special requirements such as insulation and sour service. Flexible pipes are certified by API according to ISO 9001:2000, API Q1, and API Monogram.

10.2.1.1 NKT Flexible Pipe Classes

The present pipe size range is from 2.5 to 16 in. inner diameter. Design pressures are from 15,000 psi for the smallest pipe bores to 4000 psi for the largest pipe bores. NKT Flexibles is the industry leader in design and manufacture of flexible pipes for high-temperature applications with design temperatures as high as 130°C for both static and dynamic services. The NKT flexible pipes are classified as follows:

- Low-pressure smooth bore
- Low-pressure rough bore
- High-pressure smooth bore
- High-pressure rough bore.

These classes determine the layers that are included in the fabrication of the flexible pipe.

10.2.2 Wellstream

Wellstream was founded in 1985 and is a designer and manufacturer of high-quality spoolable pipeline products, systems, and solutions for fluid transportation. Wellstream has pioneered enabling technologies for deepwater and ultra-deepwater developments, shallow water, and onshore applications through research and development.

Wellstream has supplied flexible pipe for water depths up to 1000 m for more than 10 years. In 1999, Wellstream products were the first to be qualified for service in ultra-deepwater extending the operational envelope to 2000 m.

Suited to riser and flowline applications, Wellstream's manufacturing capability ranges from 2 ID to 24 in. OD to reel or carousel. Typical product designs provided by Wellstream are as follows:

- Low internal pressure
- High pressure
- Thermal resistance
- High external pressure
- For corrosive internal fluids
- Prevention of external abrasion.

10.2.3 Technip

Technip Flexible Pipe (formerly Coflexip) has the largest market share in the world for this product. They have designs to cover all aspects of deepwater applications including corrosion resistance, high temperature, and pressure. They also have a product that can be used for LNG. Other special products include actively heated flexible pipe.

10.3 BASICS OF FLEXIBLE RISER ANALYSIS AND DESIGN

The design of the flexible riser is critical to the offshore field development as it provides the means to transfer hydrocarbon fluids from the subsea unit on the seabed to the floating production or storage unit on the sea surface. The main design code followed is APIRP 2RD.

The most common commercial finite element software used for the analyses are as follows:

- Flexcom-3D
- Orcaflex
- Flexriser
- Seaflex.

Complex floating body motions and loadings are combined in the dynamic analyses of the flexible riser. This is mainly a large deflection analysis subjected to dynamic boundary conditions and nonlinear hydrodynamic loading. The input data required for the analyses include the following:

- Flexible pipe data
- Vessel response data
- Environmental data
- End boundary conditions
- Attached buoyancy units data.

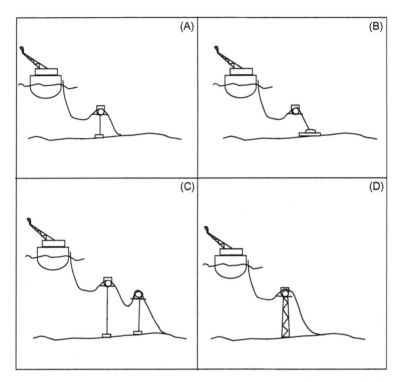

FIGURE 10.2 Some of the S-configurations of risers. (A) Lazy S, (B) Steep S, (C) Fixed S, and (D) Camel S.

The design of the riser system is an iterative process. To begin with, a riser configuration must be assumed and analyzed. Some of the common configurations (Figures 10.2 and 10.3) used are as follows:

- Free-hanging catenary
- Lazy wave
- Steep wave
- Lazy S
- Steep S
- Fixed S
- Camel S
- Pliant wave.

The selected configuration is then subjected to a combination of environmental loadings and vessel motions with the riser empty and full in two separate tests to determine the global dynamic response. As this is a tension-dominated structure, it is imperative under all circumstances that the riser remains in tension. Compression may cause bird-caging and will adversely affect the integrity of the riser and reduce the service life.

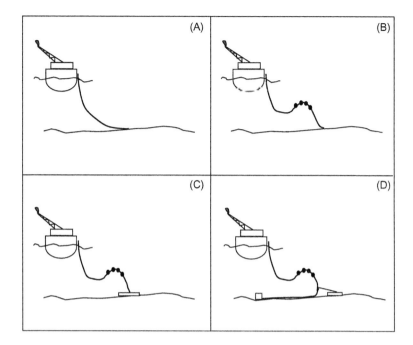

FIGURE 10.3   Some other configurations of risers. (A) Free-hanging catenary, (B) Lazy wave, (C) Steep wave, and (D) Pliant wave.

Finally, the detail static and dynamic analyses of local areas are performed to enable the design of various components such as bend stiffeners, bend restrictors, buoyancy modules, subsea buoyancy arches, and end connectors/flanges.

FURTHER READING

Deserts, L.D., 2000. Hybrid riser for deepwater offshore Africa. In: Presented at the Offshore Technology Conference, 1-4 May 2000, Houston, Texas, OTC Paper 11875.

Technical Bulletin—Wellstream International Limited.

Grealish, F.W., Bliault, A., Caveny, K.P., 1996. New standards in flexible pipe technology including API spec 17J. In: Presented at the Offshore Technology Conference, 6–9 May 1996, Houston, Texas, OTC Paper 8181.

Hoffman, D., Ismail, N.M., Nielsen, R., Chandwani, R., 1991. The design of flexible marine risers in deep and shallow water. In: Presented at the Offshore Technology Conference, 6–9 May 1991, Houston, Texas, OTC Paper 6724.

Lebon L., Remery J., 2002. Bonga: Oil off-loading system using flexible pipe. In: Presented at the Offshore Technology Conference, 6–9 May 2002, Houston, Texas, OTC Paper 14307.

Serta, O.B., Longo, C.E.V., Roveri F.E., 2001. Riser systems for deep and ultra-deepwaters. In: Presented at the Offshore Technology Conference, 30 April–3 May 2001, Houston, Texas, OTC Paper 13185.

Technical Bulletin—NKT Flexibles.

Technical Bulletin—Technip.

Tuohy, J., Loper, C., Wang, D., 2001. Offloading system for deepwater developments: unbonded flexible pipe technology is a viable solution. In: Presented at the Offshore Technology Conference, 30 April–3 May 2001, Houston, Texas, OTC Paper 13205.

Part II

Pipeline Installation

Boyun Guo, Shanhong Song and Ali Ghalambor

Various methods have been used in subsea pipeline installations. This part of the book provides a brief description of these methods and focuses on engineering aspects of controlling bending stress and stability of pipeline during the installation.

Chapter 11

Pipeline Installation Methods

11.1 INTRODUCTION

With the discovery of offshore oil fields in the shallow waters of the Gulf of Mexico during the late 1940s, offshore pipeline installation was invented. The first "offshore" pipeline in the Gulf of Mexico was constructed in 1954. Now, offshore fields are being discovered in water depths of 10,000 ft and the pipeline installation technology is keeping up. The most common methods of pipeline lay installation methods are:

- S-lay (shallow to deep)
- J-lay (intermediate to deep)
- Reel lay (intermediate to deep).

Shallow water depth ranges from shore to 500 ft. Intermediate water depth is assumed to be 500–1000 ft. Deepwater is water depths greater than 1000 ft. *Offshore* magazine produces a survey of most of the pipeline lay barges that work in the United States every year. This survey does not cover all the lay barges of all the countries that do offshore work, but it does cover the bigger international ones—Heerema, Saipem, Stolt, Technip, Allseas, McDermott, Global, and Subsea 7.

Other methods that have been used for pipeline installation are tow methods consisting of

- bottom tow
- off-bottom tow
- mid-depth tow
- surface tow.

Tow methods can be used for installing pipelines from shallow water depths to deepwater depths depending on the design requirements.

11.1.1 Pipeline Installation Design Codes

The most commonly used offshore pipeline installation codes are as follows:

- DnV OS F101 (Det Norske Veritas)
- API RP 1111 (American Petroleum Institute).

Offshore Pipelines. DOI: http://dx.doi.org/10.1016/B978-0-12-397949-0.00011-X

135

In the Gulf of Mexico, other codes that include sections relating to offshore pipelines are:

- ASME B31.4—pipeline transportation systems for liquid hydrocarbons and other liquids
- ASME B31.8—gas transmission and distribution systems.

11.2 LAY METHODS

11.2.1 S-Lay

The most common method of pipeline installation in shallow water is the S-lay method. A typical S-lay configuration is shown in Figure 11.1. In the S-lay method, the welded pipeline is supported on the rollers of the vessel and the stinger, forming the overbend. Then it is suspended in the water all the way to the seabed, forming the sagbend. The overbend and sagbend form the shape of an "S."

In the S-lay method, tensioners on the vessel/barge pull on the pipeline, keeping the whole section to the seabed in tension. The reaction of this pull is taken up by anchors installed ahead of the barge or, in the case of a dynamically positioned (DP) vessel, by thrusters. These barges/vessels are fitted with tension machines, abandonment and recovery (A&R) winches, and pipe handling cranes. The firing line for welding the pipe may be placed in the center of the barge or to one side. The firing line consists of a number of stations for welding, Non-Destructive Evaluation (NDE), and field joint application. The field joint station is located after the NDE station and the tension machines.

The S-lay barge/vessels can be classified into the following:

- First generation
- Second generation
- Third generation
- Fourth generation.

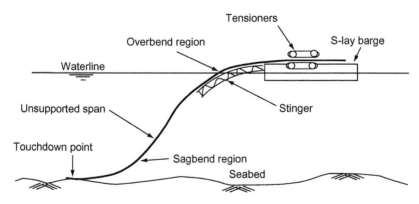

FIGURE 11.1 S-lay configuration.

First generation S-lay barges are mainly flat-bottomed spud barges used for very shallow water, swamps, and inland waters.

Second generation lay barges are also flat-bottomed barges with anywhere from 4- to 14-point moorings used for station keeping. These are the most common, and a large number of these exist throughout the world. The S-lay barge spread in this case consisted of:

- anchor handling vessels
- supply vessels
- pipe barges
- tugs.

Third generation S-lay barges are the semisubmersibles that use anchors for station keeping. The first barge of this generation was the Viking Piper constructed in 1975. This vessel was used as a lay barge in the North Sea for only a short period. Now only a couple of this generation of lay barges exist.

Fourth generation S-lay barges are vessels that use DP systems for station keeping. These barges can be used to lay deepwater pipelines. S-lay vessels of this generation belonging to Allseas are:

- the *Lorelay*
- the *Solitaire*.

Both vessels have DP capability and are, therefore, not limited by the use of anchors for station keeping. Both vessels have laid pipelines in deepwater. The *Lorelay* lays in shallow water to deepwater, while the *Solitaire* lays pipeline from intermediate water depth to deepwater. The *Solitaire* is able to compete with the J-lay method of installation on depth and pipe diameter, but has the advantage of the quicker production rate associated with S-lay over J-lay. However, S-lay in deepwater induces a higher strain than for J-lay and can be as high as 0.45% in the overbend (Figures 11.2 and 11.3).

11.2.2 J-Lay

To keep up with the discovery of deepwater oil and gas fields, the J-lay system for pipeline installation was invented. In this system, lengths of pipe are welded in a near-vertical or vertical position and lowered to the seabed. The J-lay configuration is shown in Figure 11.4. In this configuration, the pipeline from the surface to the seabed is one large-radius bend resulting in lower stresses than an S-lay system in the same water depth. There is no overbend, and a large stinger required in S-lay to support the pipe in deepwater is eliminated. The horizontal forces required to maintain this configuration are much smaller than required for an S-lay system. This lends itself for DP shipshape vessels and derrick barges to be equipped with a J-lay tower. Large J-lay towers have been installed on the world's largest heavy lift vessels—Saipem's S7000 and Heerema's *Balder*—as well as smaller towers on

FIGURE 11.2   Semisub S-lay barge—*LB 200*.

FIGURE 11.3   DP deepwater S-lay barge—*Solitaire*.

other vessels such as Stolt's *Polaris*, McDermott's *DB 50*, and Technip's *Deep Blue*. Normally, the J-lay process is slower than S-lay, but since the large J-lay towers are capable of handling prefabricated quad joints (160 ft long), the speed of pipelaying is increased (Figures 11.5 and 11.6).

The J-lay method is normally used in water depths greater than 500 ft. These water depths are normally too great for moored lay vessels to operate, because the required tensions and pipe bending stresses are too large.

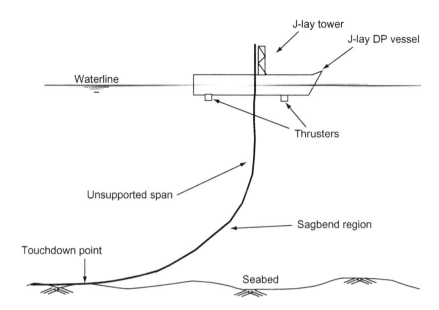

FIGURE 11.4   J-lay configuration.

FIGURE 11.5   Heerema's *Balder* with J-lay tower.

11.2.3 Reel Lay

Reel pipelay is a method of installing pipelines in the ocean from a giant reel mounted on an offshore vessel. Pipelines are assembled at an onshore spool-base facility and spooled onto a reel which is mounted on the deck of

FIGURE 11.6 Saipem 7000 with J-lay tower.

a pipelay barge. The first application of the reeled pipeline was on D-Day when the allies were supplied with fuel across the English Channel using a small diameter pipeline unreeled from a vessel. Commercial application of reeled pipeline technology was not available until the early 1970s when Santa Fe Corporation built the first reel vessel (Figures 11.7 and 11.8).

Reel technology also provides a safer and more stable work environment, thus speeding pipeline installation. Reeled pipelines can be installed up to 10 times faster than conventional pipelay. The greater speed allows pipelines to be laid during a short weather window. This can extend the normal construction season. Reel pipelay can be used on pipelines up to 18 in. in diameter.

The reel method reduces labor costs by permitting much of the welding, X-raying, corrosion coating, and testing to be accomplished onshore, where labor costs are generally lower than comparable labor costs offshore. After the pipeline is reeled onto the drum of the pipelaying vessel, it is taken to the off-shore location for installation. The reeled pipeline can be installed in an S-lay method or J-lay method depending on the design of the reel vessel and the depth of water. Reel vessels can have vertical reels or horizontal reels.

Horizontal reel vessels lay pipelines in shallow to intermediate water depths using a stinger and S-lay method. The station keeping of vessels with horizontal reels can be done by anchors or DP.

The vertical reel vessel can normally install pipelines from the intermediate water depths to deepwater and station keeping is always DP. For deepwater, the J-lay configuration is used and no stinger is required.

The pipe is unreeled, straightened, de-ovalized, and connected to the wire rope from the seabed preinstalled holdback anchor. The sagbend stresses are

FIGURE 11.7   DP global vessel *Hercules* with horizontal reel (S-lay).

FIGURE 11.8   Technip's DP vertical reel vessel *Deep Blue* (J-lay).

controlled by the tensioning system on the reel vessel. The vessel moves ahead while it slowly unreels the pipeline from the drum. When the end of the pipeline on the drum is unreeled, a pullhead connected to a wire rope is attached. The end of the pipeline is lowered to the seabed by paying out the A&R wire rope from the reel vessel slowly in a controlled method always maintaining sufficient tension in the pipeline. A buoy is attached at the end of the A&R cable. The reel vessel returns to the spool base to load more welded pipeline on the reel drum. On returning, it pulls the end of the pipeline using the A&R cable, removes the pullhead, and welds it to the pipeline on the drum. It then begins the unreeling process again.

The main disadvantages of the reeling method are:

- Connecting the ends of the pipeline segments.
- Amount of time to re-reel the pipeline to remove a buckle.
- Establishing a spool base close to the location where the pipeline is to be laid.
- Concrete-coated pipelines cannot be reeled.
- Only specifically designed pipe-in-pipe pipelines can be reeled.
- The pipeline is plastically deformed and then straightened. Some thinning of the wall and loss of yield strength of the material in localized areas can occur (Bauschinger effect).

11.3 TOW METHODS

In the tow methods, the pipeline is normally constructed at an onshore site with access to the water. These methods can be used for installing pipelines across inland lakes, across wide rivers, and offshore.

In the case of an offshore pipeline, the advantage of these methods is that the pipeline is welded onshore with an onshore pipeline spread. Once the pipeline is complete and hydrotested, the pipeline is dewatered and moved into the water, while being attached to a tow vessel (a large anchor handling vessel). It is then towed to a location offshore where each end is connected to preinstalled facilities. This could be cheaper than using a lay barge spread to install the pipeline offshore. The advantage occurs mainly if several small lines need to be laid and can be bundled inside a larger pipe. However, a case-by-case analysis is required to determine the risk versus the reward. The pipeline can be made up either perpendicular or parallel to the shoreline.

For a perpendicular launched pipeline, a land area that can accommodate the longest section of the fabricated pipeline must be leased. A launch way consisting of a line of rollers or rail system needs to be installed leading from the shore end right into the water. First, all the sections that make up the pipeline are fabricated and tested. Then, the first section of pipeline is lifted by side booms and placed on the rollers on the launch way. The cable from the tow vessel is attached and the section is pulled into the water, leaving sufficient length onshore to make a welded tie-in to the next section. In this manner, the whole single pipeline is fabricated and pulled into the water. A holdback winch is always used during these pulls to maintain control and, if need be, to reverse the direction of pull.

In the parallel launch method, the land area acquired along the shore is normally the total length of the pipeline to be towed. This could be longer than that acquired during perpendicular launch. No launch way is needed.

After the sections of the pipeline are welded and tested, the sections are strung along the shoreline. The pipeline sections are welded together to make up the length of pipeline to be towed. The completed pipeline is moved into the water using side-boom tractors and crawler cranes for the end structures. The front end is attached to the tow vessel, while the rear end is attached to a holdback anchor. The anchored tow vessel winches in the tow cable in such a manner that it gradually moves the pipeline laterally into the water, while the curvature is continuously monitored. When the whole length of pipeline and its end structures make one straight line, the tow vessel begins to tow the pipeline along the predetermined tow route.

For pipelines that are to be towed into deepwater, pressurized nitrogen can be introduced into the pipeline to prevent collapse or buckling under external hydrostatic pressure. A depth of 3000 feet can be achieved. Greater depths would require a stop for another recharge of pressurized nitrogen from the surface. This has never been done.

11.3.1 Bottom Tow

As the name indicates, the bottom tow method pulls the pipeline along the seabed to its final location. The length of a single section of pipeline is limited by the available bollard pull of the vessel used. The bollard pull must be greater than the total submerged weight of the pipeline, plus the partially submerged weight of the end structures, times the friction coefficient of the soil. For an estimate, the initial friction coefficient is taken as unity. Two to three vessels can be used in tandem to obtain additional bollard pull capability.

A thorough sea-bottom survey of the pipeline all the way from the shoreline to the pipeline's final resting place offshore must be conducted. If the pipeline is launched parallel to the shore, then the whole shallow water area nearshore along the length of the pipeline must be surveyed.

An additional abrasion-resistant coating is required on the bottom half of the pipeline to protect the normal corrosion-resistant coating like fusion-bonded epoxy (FBE). If concrete weight coating is required for stability, then this can be that coating. An additional thickness may be required to allow for abrasion. Several abrasion-resistant coatings that adhere to FBE are available on the market. However, abrasion testing may be required to select the appropriate coating. Additionally, a slick coating on the bottom half of the pipe can reduce friction and reduce the bollard pull requirement during tow.

For pipelines in shallow water, a trench may be required due to regulatory requirements or for pipeline stability. In this case, a subsea trenching plow can be attached ahead of the pipeline prior to pulling it into its final location. This will require additional bollard pull. A trench can be prepared prior to pulling the pipeline in. For reasonably straight pipelines this is not a problem. But pulling a pipeline into a curved trench is difficult.

The ends of a bottom-towed pipeline are normally connected by deflect-to-connect method. In this method, the end sections of the pipeline are made to float a few feet above the seabed by providing additional buoyancy for this length and attaching anchor chains at discrete spacing along this length. The buoyancy and chains are attached onshore with chains strapped over the buoyancy pipe during towing and deployed at the pipeline's final location. This length can then be pulled laterally by attaching cables to the end of the pipeline from the facility. Once the pipeline end structure is secured at the facility, the connection can be made by flanges (in diving depth) or by hydraulically activated connectors (in deepwater).

The disadvantages of bottom tow are:

• An extensive bottom survey along the tow route is required.
• The route must not cross existing pipelines. Otherwise, additional costs will be incurred for installing and removing structures to protect existing pipelines.
• Subsea transponder systems are required to locate the pipeline during tow and to place it in its final destination.
• The pipeline lying along the beach or nearshore can be subjected to large wave forces from storms, and this could compromise pipeline integrity. A pipe anchor system is required on standby for this emergency.
• In crossing shallow water areas, a chase vessel is required to keep fishing vessels from crossing the bottom-towed pipeline.

11.3.2 Off-Bottom Tow

In the off-bottom tow method, the submerged pipeline is buoyant and floats above the seabed at a predetermined height during the towing. This is achieved in the same manner as described in the above section for connecting the ends of bottom-towed pipeline. The buoyancy and chains are attached in discrete modules for the length of the pipeline.

The advantage of this method over bottom tow is that existing pipelines can be crossed by placing concrete mats placed over these pipelines and allowing the hanging chains to drag over the mats. No extensive protection structure is required. However, buoyancy and chains are required for the entire length of the pipeline. If several pipelines are needed for field development, the buoyancy and chains can be recovered and used again.

Only a nominal thickness of abrasion-resistant coating is required. This can even be additional FBE coating if no concrete coating is needed. For concrete-coated pipelines, no additional thickness is required.

The seabed survey needs to consider only obstacles that are higher than the height of the floating pipeline and sudden steep seabed cavities.

Launching of the pipeline with attached buoyancy and chains is the same as in bottom tow. Chains are secured over the pipeline or buoyancy during launch. Once the pipeline and end structures are fully submerged, the tow is temporarily stopped to deploy the chains and trim buoyancy if required.

11.3.3 Mid-Depth Tow

In the mid-depth tow method, the entire length of pipeline is kept at a considerable height above the seabed during towing. To achieve this, discrete buoyancy, chains, and a large tension applied to the pipeline are required. The tension is applied by two tow vessels pulling in opposite directions at each end of the pipeline. Once the pipeline reaches its desired height, the front tow vessel applies more thrust while the back tow vessel cuts back on its reverse thrust. A third vessel monitors the height of the pipeline in the middle by using a subsea transponder system. This vessel sends its signal to the two tow vessels, which see the height in real time and adjust their thrusts appropriately to keep the pipeline within the desired height range. This method is not suited for very long pipelines (greater than 3 miles).

In this method, only a nearshore survey and final infield pipeline route survey are required. Additionally, some discrete areas where the pipeline can be parked in case of emergency must be identified. This method is ideal for areas with extensive rocky outcrops, many existing pipelines, or other obstructions along the tow route.

The launching methods from shore are the same, and a temporary stop to trim the buoyancy and chains is required nearshore prior to start of the mid-depth tow.

11.3.4 Surface Tow

Surface tow of pipelines is similar to mid-depth tow except that the pipeline will not require any chains. The two vessels, one at each end, keep the pipeline in tension while it is towed on the surface. Only a survey of the final pipeline route is required. This method can be used for shallow water. For deepwater, a sophisticated controlled flooding and/or buoyancy removal system is required. Not many pipelines are installed by this method.

FURTHER READING

Allse as web site. <www.allseas.com/uk> (accessed 13.05.13).

Borelli A.J., Perinet D., 1998. Deepwater pipelaying offshore West Africa: a comparison between rigid pipe laying techniques and equipments. In: presented at the Offshore Technology Conference, 4–7 May 1998, Houston, Texas, OTC Paper 8714.

Cranswick, D., 2001. Brief overview of Gulf of Mexico OCS oil and gas pipelines: installation, potential impacts, and mitigation measures. OCS Report, MMS 2001-067.

Faldini R., 1998. S7000: a new horizon. In: Presented at the Offshore Technology Conference, 3–6 May 1999, Houston, Texas, OTC Paper 10712.

Global Industries Limited. web site. <www.global-ind.com> (accessed 13.05.13).

Nock M., Bomba J., 1997. Cased insulated pipeline bundles. In: Presented at the Offshore Technology Conference, 5–8 May 1997, Houston, Texas, OTC Paper 8542.

Riley J.W., Volkert B.C., Chappell J.F., 1998. Troika-towed bundle flowlines. In: Presented at the Offshore Technology Conference, 4–7 May 1998, Houston, Texas, OTC Paper 8848.

Saipem web site. <www.saipem.it>.

Springmann S.P., Herbert C.L., 1994. Deepwater pipelaying operations and techniques utilizing J-lay methods. OTC Paper 8848.

Stolt web site. <www.stoltoffshore.com>.

Technip web site. <www.technip.com>.

Vermeulen, E., 2000. Ultradeeps no threat to S-lay. Offshore Eng.

Chapter 12

Installation Bending Stress Control

12.1 INTRODUCTION

The dynamic stresses that occur during pipeline installation are normally less than 30% of the static stresses in shallow water. As the petroleum industry moves into deepwater environments, the dynamic stresses are more significant. Careful analysis of the dynamic stresses becomes essential for defining the limiting weather conditions that could cause overloading or fatigue failure of a pipeline. In this chapter, we describe how to control the magnitude of pipeline installation stresses that can occur under various installation conditions. These stresses include lay stresses, overbend stresses, sagbend stresses, and horizontal bending stresses.

12.2 LAY STRESSES

During pipeline installation, the bending stress in the pipe should be checked against that which is allowable by the code or specification. As illustrated in Figure 12.1 for S-laying, two regions of pipeline can be identified: the overbend region and the sagbend region. The overbend region extends from the tensioner on the barge deck, over the barge ramp, and down the stinger to the liftoff point when the pipe is no longer supported by the stinger. The sagbend region extends from the inflection point to the touchdown point. Bending stresses in the two regions are a major concern during pipeline installation. In J-lay, there is only the sagbend area. Reel lay falls into one or the other of these two categories depending on the method of pipe installation. Additionally, the reeling process puts the pipeline into one cycle of plastic deformation and straightening.

To have an understanding of installation bending stress/strain control, one must examine the basic differential equation that describes pipelay analysis, specifically the sagbend. This is the nonlinear bending equation and is given by

$$-q = EI \frac{\mathrm{d}}{\mathrm{d}s} \left(\sec \theta \frac{\mathrm{d}^2 \theta}{\mathrm{d}s^2} \right) - T_\mathrm{o} \sec 2\theta \frac{\mathrm{d}\theta}{\mathrm{d}s}$$

Offshore Pipelines. DOI: http://dx.doi.org/10.1016/B978-0-12-397949-0.00012-1

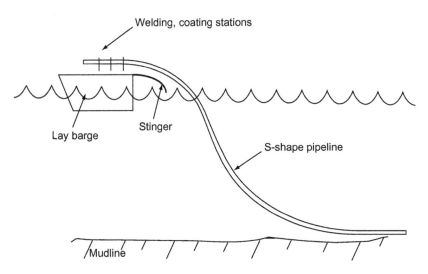

FIGURE 12.1　Overbend and sagbend regions in the pipe during an S-lay barge.

where

q = unit submerged weight of pipe,
EI = pipe bending stiffness,
T_o = effective lower pipe tension,
s = distance along pipe span,
θ = angle at distance s,

$$\sin \theta = \frac{\mathrm{d}y}{\mathrm{d}s}$$

The above equation is applicable to both deepwater and shallow water and is valid for small and large deflection. Finite difference and perturbation methods have been successfully used to provide solutions.

Finite element (FE) solution methods, both linear and nonlinear, exist for the pipelay analyses. Any general nonlinear FE program is capable of analyzing the pipeline during lay installation. Most lay barge/vessel companies have developed their own software applicable to their vessels. But most general FE programs have the capability of simulating pipelay analyses. The most common FE computer programs available on the market for analyses are as follows:

- OFFPIPE (specific for pipe installation)
- ANSYS (general FE program)
- Flexcom-3D (specialized for offshore pipeline, catenary riser analyses, moorings, etc.)
- OrcaFlex (specialized for offshore pipeline, catenary riser analyses, moorings, etc.).

The pipeline installation codes are given in Section 11.1.1. These codes give equations that determine the allowable stresses and strains for pipelay analyses. Some of the equations are empirical based on tests performed on the pipe section under external pressure and bending.

12.2.1 Overbend Stress

The overbend occurs mainly on the lay barge/vessel and the stinger. The setting of the rollers to obtain a radius of curvature is the main control of the overbend. This curvature, combined with axial tension, gives global stress on the pipe in the overbend. Local stresses occur at the rollers where the reaction force is assumed to be a point load on the pipeline. The local stresses can be analyzed using FE programs such as ANSYS and NASTRAN. In shallow water, this may not require a detailed analysis, but in deepwater, the reaction loads from the rollers are substantial and must be examined in detail.

Dynamic loads increase the reaction on the last few rollers on the stinger. By redistributing loads to the other rollers, these stresses can be reduced. Optimum roller heights can be obtained through a number of simulations of laying conditions. Some barges monitor the reaction loads on the stinger to help them control the overbend stresses.

The bending stress in the overbend region can be calculated using Eq. (12.1).

$$\sigma_a = \frac{ED}{2R_{cv}} \tag{12.1}$$

where

σ_a = axial bending stress,
E = steel modulus of elasticity,
D = pipe outer diameter,
R_{cv} = pipeline radius of curvature.

Therefore, the minimum radius of curvature can be determined as

$$R_{cv} = \frac{ED}{2\sigma_y f_D} \tag{12.2}$$

where

σ_y = minimum yield stress,
f_D = design factor, usually 0.85.

In order to control the bending stress in the overbend region to be below the minimum yield stress with a safety factor, the required minimum curvature of the stinger should be calculated using Eq. (12.2).

12.2.2 Sagbend Stress/Strain

The bending stress in the sagbend region is also caused by pipeline curvature. The curvature is normally computed using elastic rod theory in a small strain, large displacement formulation, with axial and torque deformations neglected. The stresses occur with S-lay, J-lay, shore pulls with buoyancy, surface and subsurface tows, and expansion devices subjected to environmental loading. Methods of analyzing sagbend stresses include beam method, nonlinear beam method, natural catenary method, stiffened catenary method, FE method, and method for thick concrete coating. Computer programs are frequently needed for predicting the lay stresses.

The beam method is also referred to as the small deflections method, i.e., the theory is applicable for small deflections only, i.e., (dy/dx) is much less than unity (Wilhoit and Merwin, 1967; Palmer et al., 1974). The method is applicable to shallow water pipelines.

The nonlinear beam method considers the nonlinear bending equation of a beam to describe the bending of pipeline span (Bryndum et al., 1982). It is applicable to both shallow water and deepwater. It is valid for small and large deflections. The finite difference method of approximations has been used to provide the solution.

The natural catenary method has been used to describe the pipeline span configuration away from its ends (Plunkett, 1976). Because the boundary conditions on pipeline span are not satisfied, the method is limited to pipeline segments of very small stiffness. The method is applicable to pipelines in deepwater or where the tension is very large such that the tension term is predominant over the stiffness term.

The stiffened catenary method differs from the natural catenary method in that the boundary conditions are satisfied (Palmer, 1975). The method gives accurate results of pipeline configuration even in the regions near the ends, but it is limited to deepwater and where the pipe stiffness is small.

The FE method is applicable to all water depths of small or large deflections (Martinsen, 1998). The pipeline span is modeled as a system of connected finite beam elements. The bending equations in the system are solved using matrix techniques. The accuracy of the method is affected by selection of the pipe element length.

The method for thick concrete coating was developed for certain pipeline installations where it is necessary to increase the pipe's submerged weight sufficiently to withstand hydrodynamic forces at the seabed (Powers and Finn, 1969). In this situation, when the pipe is bent, the bending stresses are intensified at the field joints where the pipe stiffness is low.

The sagbend is more interesting from a solid mechanics point of view. The addition of external hydrostatic forces creates the possibility of collapse for an empty pipe. The most common offshore codes that are used to limit the allowable stress/strain during pipe laying are as follows:

- API RP 1111,
- DnV OS F101 (2007).

API RP 1111 has the following interaction equation for buckling under external pressure and bending that must be satisfied:

$$\frac{\varepsilon}{\varepsilon_b} + \frac{(P_o - p_i)}{P_c} \le g(\delta)$$

where

ε = strain in pipe,
ε_b = critical strain under pure bending = $(t/2D)$,
P_o = external hydrostatic pressure,
P_i = internal pressure,
P_c = collapse pressure (for details see API RP 1111 and Section 4.3.2.1),
$g(\delta)$ = collapse reduction factor = $(1 + 20\delta)^{-1}$.

$$\delta = \frac{D_{max} - D_{min}}{D_{max} + D_{min}} = \text{ovality}$$

D_{max} and D_{min} are the maximum and minimum diameters at the same cross section. The strain, ε, in the pipe can be written as

$$\varepsilon \le \left[g(\delta) - \frac{(P_o - P_i)}{P_c} \right] \frac{t}{2D}$$

During installation, the allowable strain is limited by a safety factor, f_1, such that

$$f_1 \varepsilon_1 \le \varepsilon$$

where ε_1 is the maximum installation bending strain.

API RP 1111 recommends that $f_1 = 2.0$. The API equations are limited to pipes with a D/t ratio not greater than 50.

Normally, in the computation of wall thickness of pipelines in deepwater, the maximum bending strain is assumed to be 0.2%. In many cases, the allowable strain can be computed knowing all the other factors. In several cases, it has been shown that the strain can be much greater than 0.2% and can still satisfy the inequality.

DnV (2007) has a similar equation for pipeline subjected to longitudinal bending strain and external hydrostatic pressure and is given by

$$\left(\frac{\varepsilon_d}{(\varepsilon_c/\gamma_e)} \right)^{0.8} + \frac{p_e}{(p_c/\gamma_{sc}\gamma_m)} \le 1$$

where

p_e = external pressure,
p_c = collapse pressure,
ε_d = design compressive strain,
ε_c = critical compressive strain,
γ_e, γ_{sc}, γ_m = resistance factors of strain, safety class, and material, respectively.

DnV (2007) interaction equation is considered more conservative than the API RP 1111 one. DnV's ovality formula is given by

$$f_o = \frac{D_{max} - D_{min}}{D}$$

This is twice that of the definition given by API RP 1111. Therefore, one must be careful in using the value in the equations in both codes.

Additionally, in DnV (2007) collapse pressure, p_c (Eq. (5.10) in DnV (2007)) is a function of ovality which seems logical. However, some tests performed on larger, thicker wall pipes indicate that the results lie closer to the API RP 1111 equation. The API RP 1111 collapse pressure is not a function of ovality and is easier to use.

In deepwater, controlling the sagbend stresses during installation is imperative because a collapse of pipe under a combination of external pressure and bending can lead to buckle propagation. A propagating buckle can travel along the pipe for long distances. To prevent this from happening, buckle arrestors are placed at a predetermined spacing along the pipe. Buckle arrestors are short pup pieces of thicker walled pipe. Buckle arrestors can be welded as part of the pipeline for S-lay and J-lay situations or bolted on for reeled pipelay conditions.

The propagating pressure, P_p, for a pipeline can be computed using the API RP 1111 formula

$$P_p = 24S\left(\frac{t}{D}\right)^{2.4}$$

where S is the specified minimum yield stress of the pipe material.

A minimum safety factor of 1.25 is used on the external overpressure when compared with the propagation pressure. This equation is empirical, obtained mainly from experiments on pipes under external pressure.

Installing deepwater pipelines filled with water eliminates the collapse due to external pressure. This reduces the wall thickness required to resist collapse. A few advantages of laying the line filled with water are that it is very stable on the seabed and can be hydrotested right away. Only a few pipelines have been installed in this manner. The tension requirement of pipeline increases when filled with water for water depths up to approximately 2300 m. Therefore, most contractors lay pipelines in the dry condition.

Increasing the tension can control the sagbend stress. In shallow water barges, the increase in tension must be taken up by the anchors. This may lead to slippage depending on the soil conditions on the seabed. Also, the increase in tension leads to a longer suspended span, which may not be desirable. This also leads to higher residual tension of the pipeline on the seabed. On the seabed, spanning of the pipeline over undulation increases with increased tension. Therefore, most barges want to use the optimum amount of tension.

REFERENCES

Bryndum, M.B., Colquhoun, R.S., Verley, A., May 1982. Dynamic laystresses for pipelines. In: Presented at the Offshore Technology Conference, 3–6 May 1982, Houston, TX, OTC 4267.

DnV OS F101, 2007. Submarine pipeline systems, Det Norske Veritas.

Martinsen, M., 1998. A Finite Element Model for Pipeline Installation Analysis. M.Sc. Thesis, Stavanger University College for JP Kenny A/S, Stavanger, Norway.

Palmer, A.C., Hutchinson, G., Ells, J.W., 1974. Configuration of submarine pipelines during laying operations. Trans. of ASME, J. Manuf. Sci. Eng 96 (4), 1112–1118.

Palmer, A.C., 1975. Technical and analytical aspects of pipe laying in deepwater. Presented at the Institute of Marine Engineers. One day Conference on Pipelining in the North Sea, London, pp. 6–11.

Plunkett, R., 1976. Static bending stresses in catenaries and drill strings. Trans. of ASME, J. Manuf. Sci. Eng. 89 (1), 31–36.

Powers, J.T., Finn, L.D., 1969. Stress analysis of offshore pipelines during installation. In: Presented at the Offshore Technology Conference, 18–21 May, Houston, TX, OTC 1071.

Wilhoit Jr., J.C., Merwin, J.E., 1967. Pipe stresses induced in laying offshore pipelines. Trans. ASME, J. Manuf. Sci. Eng. 89 (1), 37–43.

FURTHER READING

API RP 1111, 1999. Design, construction, operation, and maintenance of offshore hydrocarbon pipelines. American Petroleum Institute Standards.

Langner, C.G., 1999. Buckle arrestors for deepwater Pipelines. In: Presented at the Offshore Technology Conference, 3–6 May 1999, Houston, TX, OTC Paper 10711.

Mousselli, A.H., 1981. Offshore Pipeline Design, Analysis and Methods. PennWell Books, Tulsa, OK.

Murphey, C., Langner, C.G, 1985. Ultimate pipe strength under bending, collapse and fatigue. In: Proceedings of the 4th Int. Conference on Offshore Mechanics and Arctic Engineering 1, 467–477.

Nogueira, A.C, Lanan, G.A., 2000. Rational modeling of ultimate pipe strength under bending and external pressure. In: Proceedings of IPC Conference 2–6 April 2000, San Diego, CA.

Palmer, A., 1998. A radical alternative approach to design and construction of pipelines in deep water. In: Presented at the Offshore Technology Conference, 4–7 May 1998, Houston, TX, OTC Paper 8670.

Park, T.D., Kyriakides, S., 1997. On the performance of integral buckle arrestors for offshore pipelines. International Journal of Mechanical Sciences 39 (6), 643–669.

Chapter 13

Pipeline On-Bottom Stability Control

13.1 INTRODUCTION

Pipelines installed on the seabed are subjected to hydrodynamic forces. Waves and steady currents that are characteristics of all offshore areas subject the pipeline on the seabed to drag, lift, and inertia forces. For lateral stability, the pipeline resting on the seabed must resist these forces and at a minimum be at equilibrium.

Drag and inertia forces act together laterally on the pipeline, tending to move the pipeline. Lift force acting vertically tends F_1, *lift force*, to effectively reduce the submerged weight of the pipeline. Traditionally, sliding friction between the pipeline and soil provided the resistance of the pipeline on the seabed. Forces acting on the pipeline resting on the seabed are shown in Figure 13.1.

The traditional method of pipeline stability is given by the following equation:

$$\frac{\mu(W_S - F_1)}{F_T} > 1$$

where

μ = soil−pipe friction,
W_S = submerged weight,
F_1 = lift force,
F_T = total horizontal force from waves and currents.

In general, the larger the submerged weight, the higher the frictional resistance. However, later methods for determining the stability include the depth of embedment of the pipeline. Additional resistance is provided by the soil and, therefore, reduces the required submerged weight of the pipeline.

As the pipeline is resting on the seabed, soil characteristics play an important role in the lateral, as well as the vertical, stability of the pipeline. The importance of vertical stability of the pipeline is mainly in buried pipelines in soils with a high liquefaction potential.

Offshore Pipelines. DOI: http://dx.doi.org/10.1016/B978-0-12-397949-0.00013-3

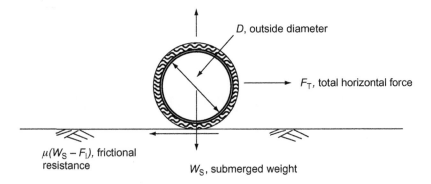

FIGURE 13.1   Forces acting on the pipeline resting on the seabed.

13.2 HYDRODYNAMIC PARAMETERS

The drag force, F_d, due to water particle velocities is given by the following equation:

$$F_d = \frac{1}{2}\rho C_D D(U+V)^2$$

where

F_d = drag force/unit length,
ρ = mass density of seawater,
C_D = drag coefficient,
D = outside diameter of pipeline (including the coatings),
U = water particle velocity due to waves,
V = steady current.

The lift force, F_l, is determined by the same equation as that for the drag force with the lift coefficient, C_L, replacing C_D, the drag coefficient.

The inertia force, F_i, due to water particle acceleration is given by the following equation:

$$F_i = \rho C_M \frac{\pi D^2}{4}\left(\frac{du}{dt}\right)$$

where

F_i = inertia force/unit length,
ρ = mass density of seawater,
C_M = drag coefficient,
D = outside diameter of pipeline (including the coatings),
(du/dt) = water particle acceleration due to waves.

The traditional stability method uses Morison's equation to determine the combined forces, F_T, as given by

$$F_T = F_d + F_i$$

In linear wave theory, wave velocity and acceleration are sinusoidal functions in time. Therefore, the maximum force can be obtained using calculus or by stepping through a wave cycle and computing the forces at discrete phase angles.

In reality the waves are nonlinear, and methods using spectral analysis to obtain velocities and accelerations may be more appropriate. The most common spectral models that are used to describe sea state are as follows:

- Pierson–Moscowitz (P–M)
- Bretschneider (Bret)
- JONSWAP (Joint North Sea Wave Project—JS).

The spectral formulation can be generally expressed as:

$$S_{\eta\eta}(\omega) = Bf(H_s, \omega_p, \omega)\gamma^{\phi(\omega/\omega_p)} \text{ with } \phi\left(\frac{\omega}{\omega_p}\right) = \exp f(\sigma, \omega, \omega_p)$$

where

$\omega_p = 2\pi f_p$ (f_p is the peak frequency of the spectrum),
$\omega = 2\pi f$,
γ^ϕ = spectrum peakedness factor,
H_s = significant wave height.

The values for B and γ for the three spectra are given in Table 13.1.

The P–M spectrum is used for fully developed sea state in deepwater that is not fetch or duration limited. JS is developed for the North Sea and can be used for fetch-limited seas.

The surface wave spectrum can be transformed to bottom velocity spectrum using a specific transfer function. From the bottom velocity spectrum, a mean bottom velocity and a related significant bottom velocity can be obtained.

The values for the hydrodynamic coefficients C_D, C_L, and C_M given in DnV's (1981) Pipeline Design Guidelines are 0.7, 0.9, and 3.29, respectively. However, tests on static pipes have led to producing graphs to determine the values of these coefficients with respect to Reynold's number for steady currents and Keulegan–Carpenter number ($KC = U_mT/D$, maximum velocity and period, T) for steady currents combined with wave-induced currents.

TABLE 13.1 Values for Spectral Model Factors

Spectral Model	B	γ	ω_p	ω_p/ω_z
P–M	5	1	ω_p	0.710
Bret	5	1	$0.857\omega_s$	0.710
JS	3.29	3.3	ω_p	0.781

Note: ω_s is the significant frequency ($=2\pi f_s = 2\pi T_s$).

Figures 13.2 and 13.3 present the graphs for C_D and C_L obtained by Hydraulics Research Station for pipelines in an estuary (tidal currents).

13.3 SOIL PARAMETERS

Traditionally, lateral stability of the pipeline on soil was determined using soil friction coefficient and ranged from 0.7 to 1.0 for sand and 0.3 to 0.5 for clay without considering the embedment.

Now, a more rigorous approach is taken by computing the embedment and factoring the additional resistance provided by the soil. A reduction in drag and lift forces occurs when there is embedment. This embedment takes place when small oscillations of the pipeline occur under wave action.

The amount of embedment of the pipeline in soil depends on the bearing capacity of the soil, q_f, given by the following equation:

$$q_f = \frac{1}{2}\gamma B N_\gamma + c N_c + z\gamma N_q$$

where

q_f = ultimate bearing capacity,
γ = soil submerged weight,

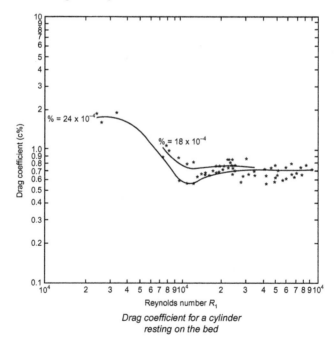

Drag coefficient for a cylinder
resting on the bed

k   roughness height
d   diameter

FIGURE 13.2   Drag coefficient.

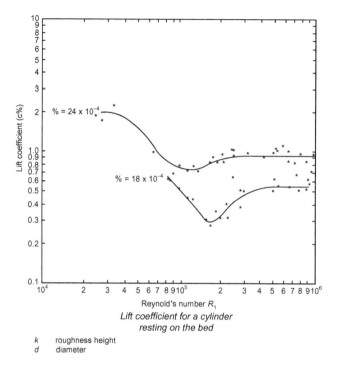

Lift coefficient for a cylinder
resting on the bed

k roughness height
d diameter

FIGURE 13.3   Lift coefficient.

B = width of embedment,
z = depth of embedment,
c = cohesion = 0 for sands,
N_γ, N_c, N_q = dimensionless bearing capacity factors = f(φ) (angle of friction),
N_q = 0 for no embedment,
N_q = 1 for $\varphi = 0$.

The pipeline embedment into the seabed is shown in Figure 13.4.
For clays with zero embedment:

$$q_f = cN_c$$

For sands with zero embedment

$$q_f = \frac{1}{2}\gamma BN_\gamma$$

13.3.1 Cohesive Soils

Clays are classified by their consistency and are measured by their shear strength as given in Table 13.2.

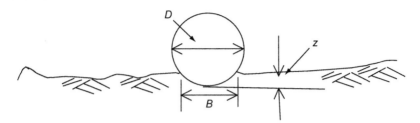

FIGURE 13.4   Pipeline embedment into the seabed.

TABLE 13.2 Clay Classification

Consistency	Shear Strength Range (ksf)
Very soft	0.0 to 0.25
Soft	0.25 to 0.5
Firm	0.5 to 1.0
Stiff	1.0 to 2.0
Very stiff	Greater than 2.0

Cone penetrometer testing gives characteristics of both clay and sand. In the field, the shear strength can be obtained by performing Torvane tests.

Several methods have been proposed for determining the embedment of the pipeline into clayey soils. Models have been developed by the PIPESTAB project in 1987 and by the American Gas Association/Pipeline Research Committee (AGA/PRC) in 1992. In both these models, the total soil resistance consists of a frictional term, F_f, and an additional resistance due to pipeline embedment, F_p, given by:

$$F_R = F_f + F_p; \quad F_f = \mu F_c$$

where

F_R = total soil resistance,
F_c = contact force = $(W_s - F_l)$,
μ = soil–pipe friction = 0.2.

An empirical model for clayey soil resistance due to penetration, F_p, is proposed by Verley and Lund (1995), given by:

$$\frac{F_p}{DS_u} = 4.13 \ G^{-0.392} \left(\frac{z}{D}\right)^{1.31}$$

and

$$\left(\frac{z}{D}\right)_{max} = 1.1 \ SG^{0.54}\left(\frac{a}{D}\right)^{0.17} = (0.3 \ \text{max})$$

where

D = outside diameter of pipeline (including the coatings),
z = penetration of pipe into soil,
S_u = undrained shear strength,
$G = S_u/(D\gamma_s)$, γ_s = unit soil weight,
a = amplitude = $H_s/2$.

13.3.2 Noncohesive Soils

For sands, Standard Penetration Test gives a measure of the relative density, D_r. Sands are classified by compactness as given in Table 13.3.

The embedment of the pipeline in sands is based on bearing capacity equation given earlier, with cohesion set to zero:

$$q_f = \frac{1}{2}\gamma B N_\gamma + z\gamma N_q$$

The bearing capacity factors N_γ and N_q are related to internal angle of friction and can be found in most soil mechanics textbooks (e.g., Bowles, 2001). Some authors (Vesic, 1975, Hansen et al., 1969, Meyerhof, 1974) have suggested ways of improving the value of these factors.

The ultimate lateral soil resistance, q_{lult} can be obtained by:

$$q_{lult} = z\gamma N_q$$

The suggested values to be used for bearing capacity factors for lateral resistance are given by Hansen.

TABLE 13.3 Sand Classification

Compactness	Relative Density, D_r (%)	SPT	Friction Angle (Degrees)	Submerged Weight (pcf)
Very loose	0–15	0–4	0–28	<60
Loose	15–35	4–10	28–30	55–65
Medium	35–65	10–30	30–36	60–70
Dense	65–85	30–50	36–41	65–85
Very dense	85–100	>50	>41	>75

13.4 STABILITY ANALYSIS GUIDELINES

The most common and industry-accepted methods for determining the on-bottom stability of submarine pipelines are:

- Pipeline Research Council International (PRCI) (AGA) Pipeline Stability Program
- DnV RP E305.

13.4.1 PRCI Pipeline Stability Program

The three levels of analyses that can be performed by this software are as follows:

- Simplified static—Level 1
- Simplified quasi-static—Level 2
- Dynamic time domain—Level 3.

The simplified static method employs the traditional method using the airy wave theory, Morison's equation for wave force, and soil—pipe friction factor. This is normally used in getting a quick result for preliminary design of pipeline stability. Reduction of wave force due to embedment can be specified.

Level 2 is accepted for most detailed design stages, because pipe embedment and additional lateral resistance are taken into account. Soil information is required for this analysis. Wave spectrum is transferred to the seabed and additional embedment during a storm is considered. Safety factors are computed for four levels of probabilistic waves in a storm.

In Level 3, the pipeline is allowed to move and the stress due to movement and factor of safety over time is computed. This is used mainly in cases of existing pipelines that were not designed to the 100-year storm or initially buried pipelines that are subsequently exposed to waves and currents.

13.4.2 DnV RP E305

This guideline recommends three levels of analysis:

- Simplified stability analysis
- Generalized stability analysis
- Dynamic analysis.

The simplified stability analysis is similar to Level 1 in the previous section. The differences are that friction factors for sand and clay are given and a calibration factor for submerged weight as a function of Keulegan—Carpenter number is given. A safety factor of 1.1 is inherent in the calibration factor.

The generalized stability analysis is based on the use of a set of nondimensional parameters and for particular end conditions. This method assumes the following:

- Hydrodynamic forces modified for wake effects.
- No initial embedment.

- No prior loading.
- Pipe is rough.
- Soil resistance due to penetration under cyclic loading is included.
- Medium sand soil.
- Uses JONSWAP wave spectrum.
- No reduction of hydrodynamic forces due to embedment.

The response of the pipeline in waves is controlled by the following non-dimensional parameters:

Load parameter, $K = U_s T_u/D$
Pipe weight parameter, $L = W_s/0.5\rho_w DU2s$
Current to wave velocity ratio, $M = U_c/U_s$
Relative soil weight (sand soil) $G = (\rho_s - \rho_w)/\rho_w = \rho_s/\rho_w - 1$
Shear strength parameter (clay soil) $S = W_s/(DS_u)$
Time parameter $T = T_1/T_u$

where

U_s = significant bottom wave velocity perpendicular to the pipe,
U_c = steady current,
T_u = zero up-crossing period,
T_1 = duration of the sea state,
ρ_s = mass density of sand soil,
ρ_w = mass density of seawater,
W_s = submerged weight,
D = outside diameter,
S_u = shear strength of clay.

The validity of this method is for the following range of parameters:

$4 < K < 40$
$0 < M < 0.8$
$0.7 < G < 1.0$ (for sand soil)
$0.05 < S < 8.0$ (for clay soil)
$D \geq 0.4$ m (16 in.).

This method is used for larger pipelines.
This DnV RP E305 does not give a method for dynamic analysis, but recommends on what aspects should be accurately modeled. It also gives recommendations for lateral displacement of the pipeline.

13.5 TRENCHING/JETTING

The need for pipelines to be trenched or lowered below the natural seabed level is based on the following:

- regulations,
- high-seabed velocity.

In the Gulf of Mexico, all pipelines in water depths less than 200 ft are to be buried with a cover of 3 ft. All pipelines that approach the shore are buried, due to both regulations and the fact that the seabed velocity is very high in shallow water. It would be uneconomical to provide concrete weight coating for stability during a 100-year return period storm.

Lowering of the pipeline below the seabed can be performed by the following methods:

Jetting
Plowing

In theory, jetting consists of fluidizing the seabed soil so that the heavier pipeline sinks in. The jetting plow is placed over the pipeline and high-pressure water jets from nozzles blast the surrounding soil away from the pipeline. Jetting disperses the local soil away from the pipeline and creates a wide trench. In most cases, the soil is not returned immediately and the pipeline gets covered in time depending on the bottom sediment load. Jetting is done in sands and soft clays, but in hard clays plowing may be required.

Mechanical plows have also been used to cut trenches below pipeline, allowing the pipeline to be gradually lowered into them. Initial development was of a pre-trenching plow. The trench was made first, and the pipeline was then laid or pulled into the trench. This was not very successful because the trench had to be quite wide or the subsequent installation of the pipeline had to be very accurate. Further development has resulted in the more common post-trenching plow. In this case the plow is deployed over the pipeline after the pipeline is installed on the seabed. These plows are capable of lifting the pipeline, trenching, and then lowering the pipeline into the trench as the plow is moved or moves forward.

For shallow water areas (up to 60 ft of water depth) that have rock, coral, or very hard soils, a mechanical device called the "Rocksaw" can be used. This is a pre-trenching device. The depth of trench can be 10 ft and the width up to 8 ft. As it is in shallow water, the pipeline can be pulled into the trench from shore or from a moored pull barge offshore.

REFERENCES

Bowles, J.E., 2001. Foundation Analysis and Design, fifth ed. McGraw-Hill, New York.

DnV, 1981. Rules for submarine pipeline systems.

Hansen, B., Christensen, N.H.m, Karafiath, L.L., 1969. Discussion of theoretical bearing capacity of very shallow footings. J Soil Mech Found Div 95 (6), 1568–1573.

Meyerhof, G.G., 1974. Ultimate bearing capacity of footings on sand layer overlying clay. Canadian Geotech J 11 (2), 223–229.

Vesic, A.S., 1975. Bearing Capacity of Shallow Foundations. In: Winterkorn, H.F., Fang, H.Y. (Eds.), Foundation Engineering Handbook. Van Nostrand Reinhold Company, New York.

Verley R., Lund K.M., 1995. A soil resistance model for pipelines placed on clay soils. In: Proceedings of the 14th International Conference on Offshore Mechanics and Arctic Engineering, June 18–22, 1995, Copenhagen, Denmark.

FURTHER READING

Audibert, J.M.E., Lai N.W., Bea, R.G., 1978. Design of pipelines-sea bottom loads and restraints. In: Proceedings of the ASCE Pipeline Division Specialty Conference, New Orleans, LA, pp. 187–203.

DnV RP E305, October 1998. On-bottom stability design of submarine pipelines. Det Norske Veritas.

Jones W.T., 1971. Forces on submarine pipelines from steady currents. In: Proceedings of ASME Petroleum Mech. Eng. Underwater Technology Conference, Houston, TX.

PRCI, December 2002. Submarine pipeline on-bottom stability, vols. 1 and 2.

US Army Coastal Engineering Research Center, 2002. Shore Protection Manual, vol. 1.

Part III

Pipeline Commissioning and Operations

Boyun Guo, Shanhong Song and Ali Ghalambor

After installation, a pipeline undergoes testing and commissioning stages. Certain procedures should be followed during testing and commissioning. Then, the pipeline is ready to operate for transporting the production fluids. As the pressure in the oil–gas reservoirs is declining with time, the composition of the produced fluids (water-cut and gas–liquid ratio) changes. Flow assurance engineering becomes essential. Pigging operations are conducted to clean the pipeline and identify pipeline defects. This part of the book addresses technical issues during pipeline management from pipeline testing to daily operations.

Chapter 14

Pipeline Testing and Precommissioning

14.1 INTRODUCTION

From its fabrication to start-up, a pipeline system has to pass a series of tests. Some of these, such as the factory acceptance test (FAT), are done onshore at the fabrication yards with individual components. The FAT mainly consists of the inspection, testing, and reporting of the system according to the drawings, specifications, and requirements of the contract. Pipe sections must pass the FAT before they are accepted. Some of the tests, such as the pipeline hydrotest, are mainly done offshore with either a portion of the whole pipeline system or the whole pipeline system. The hydrotests are conducted to check the mechanical strength of the pipeline system and the integrity of the connections. The hydrotest is one of the pipeline precommissioning activities. Precommissioning is performed after the pipeline system is installed, and all the tie-ins are completed to assess the global integrity, qualify the system as ready for commissioning and start-up, confirm the safety to personnel and environment, and confirm the operational control of the pipeline system.

Why are all the tests done offshore necessary? The subsea pipeline system typically consists of pipeline and riser. A jumper is usually used to connect the pipeline and the riser, as shown in Figure 14.1. A pipeline jumper is a short section of pipe which can be either rigid or flexible. The jumper is tied with the riser and the pipeline with connectors and pipeline end termination (PLET). The PLET is used to support a pipeline connector and/or a pipeline valve. At the subsea end, the pipeline is tied to a manifold or a well through a jumper which is installed between one connector at the PLET and one connector on the manifold or on the tree, as shown in Figure 14.1.

When the subsea pipeline system is installed, because of the various connections along the pipeline system, it is necessary to make sure the pipe sections are leakproof and have the required mechanical strength to withstand the designed pressure with the specified level of safety. Pipeline may get damaged during the transportation and installation process, and its mechanical strength may thus be reduced. The various connections along the pipeline system may not be tied in properly, and leaks may occur under high

Offshore Pipelines. DOI: http://dx.doi.org/10.1016/B978-0-12-397949-0.00014-5

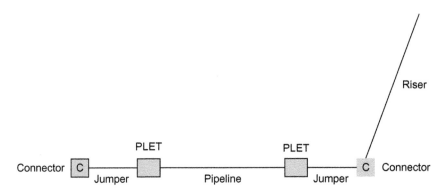

FIGURE 14.1   Schematic of a typical subsea pipeline system.

pressure conditions. All the above mentioned potential problems must be detected by performing the pressure testing and corrected properly before the pipeline is put into service to prevent any operational accidents (environmental and safety impacts).

Before the pipeline system can be used, it must also be cleaned and gauged. During fabrication and installation, debris may be left inside the pipeline. If the debris is not removed, it can plug and damage valves and chokes. Pipeline internal dimensions and internal integrity also need to be checked for proper operations. For example, the pipeline has to be checked to make sure that no deformations have occurred during the installation and no internal restrictions exist. Otherwise, operational pigs may not pass the deformed pipe sections.

This chapter will cover the main activities associated with subsea pipeline testing and precommissioning.

14.2 PIPELINE PRECOMMISSIONING

The pipeline precommissioning consists of the following activities:

- Flooding
- Cleaning and gauging
- Hydrotesting
- Leak tests.

14.2.1 Pipeline Flooding, Cleaning, and Gauging Operations

After the pipeline is laid, it must be verified that the line is internally clean and free from restrictions or debris and will withstand its design pressure. This verification process generally involves flooding the pipeline with treated fluids and sending a cleaning pig down the line to clear out any accumulated debris followed by a gauging pig to prove it is of full bore over the entire length. The cleaning and gauging can be carried out with a single pig. Thus, the main objectives of the flooding, cleaning, and gauging operation are as follows:

- fill the pipeline with a suitable pressure testing medium,
- verify the cleanliness of the pipeline,
- verify the pipeline integrity by gauging to make sure no buckles or obstructions exist.

The pipeline should be filled with clean filtered water. Suspended material in the water should be removed by a filter capable of removing all particles larger than a specified size (50−100 μm). A meter with sufficient accuracy should be used to measure the quantity of water injected into each pipe section. Knowing the quantity of water injected is critical for leak detection. Chemicals, like biocide, are usually injected into the test water with a certain concentration, which will depend upon the test conditions. If the test water will stay in the pipeline for a relatively long time, corrosion inhibitor will also need to be injected into the pipeline to protect the pipeline from excessive corrosion. All the chemicals injected must be compatible with the water so that no solids will form inside the line.

While filling the pipeline, a series of pigs (pig train), separated by a slug of fluids, shall be passed through the pipeline at a specified minimum velocity. The pig train consists of cleaning pigs and gauging pigs. The best choices for cleaning pigs are pigs with discs, conical cups, spring-mounted brushes, and bypass ports. Discs are effective at pushing out solids while also providing good support for the pig. Conical cups provide excellent sealing characteristics and long wear. Spring-mounted brushes provide continuous forceful scraping for removal of rust and other build-ups on the pipe wall. Bypass ports allow some of the flow to bypass through the pig and help minimize the solids built up in front of the pig. The pig should also include a magnetic cleaning assembly to clean any metal debris. Some applications may use a bidirectional disc pig when the water used to fill the line has to be pushed back to its source after completion of the test. Bidirectional pigs may be used if there is a fear of the pig getting stuck and there is an option to reverse flow and bring the pig back to the launch point.

Gauging pigs are used to determine whether there are unacceptable reductions/obstructions in a line. These reductions can be caused by ovality due to overburden or by dents and buckles. A conventional gauging pig is a cup-type pig with a slotted aluminum gauging plate. The slotted aluminum plate will bend out of the way when it encounters an excessive reduction. If the pig comes out with a damaged plate, it is usually run again and if the plate is damaged again, it is assumed there is an unacceptable reduction in the line. The restriction must be located and removed. After removing the

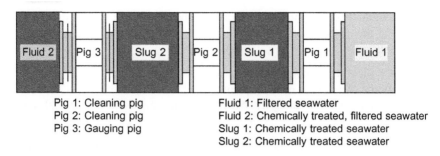

Pig 1: Cleaning pig
Pig 2: Cleaning pig
Pig 3: Gauging pig

Fluid 1: Filtered seawater
Fluid 2: Chemically treated, filtered seawater
Slug 1: Chemically treated seawater
Slug 2: Chemically treated seawater

FIGURE 14.2 Typical flooding, cleaning, and gauging pig train.

restriction, the gauging pigs should be run again to verify that the repairs are done properly and the line is indeed free of obstructions.

A typical pipeline flooding, cleaning, and gauging pig train is shown in Figure 14.2. Two cleaning pigs and one gauging pig are shown. Depending upon the individual cases, more cleaning and gauging pigs can be used. For hydrotesting and precommissioning the Yacheng pipeline, four cleaning pigs and two gauging pigs were used (Karklis et al., 1996).

There are two key issues associated with the flooding, cleaning, and gauging operations. One is the control of the pig train velocity at the downhill section. The other is the cleanliness of the pipeline. For the pig train, there are recommended traveling velocities by the manufacturers, normally 3−6 miles per hour. But at the downhill sections, due to the gravity effect, the pig train will travel at higher than the recommended velocity. To help control the velocity, if the pipeline is not very long, it is possible to pressurize the whole pipeline with air. But if the pipeline is too long, this option can be quite expensive. The biggest concern of the flooding, cleaning, and gauging operations is that the pig train may get stuck because of debris. To mitigate this risk, it is very critical to have stringent controls of pipe cleanliness at every stage, from pipe manufacture to installation.

14.2.2 Pipeline Hydrotesting and Leak Testing

The hydrotests are conducted mainly to verify the mechanical strength of pipeline or pipeline sections. Hydrotesting is carried out by pressurizing the system to a specified internal pressure and holding it for a certain period of time to check whether or not there is a pressure drop. If the pressure drops within the hold period, it is assumed that a leak or leaks may exist somewhere in the system. After the holding period is over, the pressure is released and a complete visual inspection follows. Standard codes, such as ASME B31.4, ASME B31.8, and API RP 1110, provide guidance on how to perform pressure tests of gas and liquid pipelines.

Before conducting the tests, it is necessary to establish the specifications for the test procedures and equipment. The procedure specifications should include the following:

- A description of the pipe sections to be tested (lengths, elevation, tic-ins, etc.)
- Test medium (seawater is used for subsea pipelines)
- Chemicals to be mixed with the test medium (biocide and corrosion inhibitors)
- Mixing process of the chemicals with the test medium including the chemical concentrations
- Test pressures
- Test holding time
- Process of pressurizing
- Description of all testing equipment
- Description of testing instrument
- Monitoring and recording of test pressure
- Procedure for depressurizing and discharge of test medium.

The test pressure is recommended to be set at no less than 1.25 times the internal design pressure for both hydrostatic testing and leak testing. The holding time is recommended to be at least 4 h for hydrostatic testing and at least 1 h for leak testing.

A complete description of the test equipment and instruments is very important for the success of the tests. The following is a partial list of the devices needed for the tests:

- A high-volume pump that can fill the line at high enough velocity to remove the debris and to push the pigs.
- A filter that would remove all particles larger than a certain size to ensure the test fluid is clean.
- A meter to measure the quantity of water filled.
- Injection pumps for chemical injections.
- A variable speed, positive displacement pump to pressurize the line. The pump should have a known volume per stroke and should have a stroke counter.
- A pressure recorder that would continuously record the test pressure for the whole test period. The pressure measuring equipment should have an accuracy and repeatability of $\pm 0.1\%$.
- A temperature measurement device that is properly calibrated and should continuously measure the fluid temperature. The device should be able to read in increments of no less than $0.1°F$ ($0.05°C$).
- A temperature device to measure the ambient temperature.
- Pressure relief valves.
- Pig transmitter device or remote pig signaling system.

When planning the hydrotesting and leak testing, a few issues must be taken into account. The tests should be planned so that nowhere in the test segment does the pressure level produce hoop stress near or above the specified minimum yield stress. This will require the test pressure to be determined by taking into full account the effect of the pipeline profile and external conditions. If the test pressure is relatively high because of the high design pressure, the pressure relief valves have to be properly checked and set at the right pressure to protect the pipeline and the involved personnel. When launching a suite of pigs from a test-head launcher and receiving into a similar receiver, there is always concern that not all the launched pigs have been successfully launched or received into the terminal. It is necessary to install a pig transmitter device or some kind of remote pig signaling system on the final pig to confirm the pig launch and reception.

How the test results will be reported and what will be reported should be defined before performing the tests. The test records should include the details of the test operations and details of any failures. The failure report should include the exact location of each failure, the type of failure, the causes for the failure, and the recommended repair methods. When the tests are finished, all waste should be disposed of in the correct manner which should be defined in the company's waste management and disposal policy.

14.2.3 Pipeline Dewatering, Drying, and Purging

For offshore gas transmission pipelines, after successful hydrotesting and leak testing and before introducing gas, the pipelines have to be dewatered, dried, and purged. For offshore liquid pipeline, the water is usually displaced by diesel or dead oil and it is not necessary to dry the pipeline. The process of dewatering, drying, and purging can be quite complicated, depending upon the application. This is because after the tests, the pipeline is full of seawater and the water has to be sufficiently displaced from the pipeline. Otherwise hydrate may form inside the pipeline when the hydrocarbon is introduced. Another reason for displacing the water is that many products react with water to form acids and other corrosive compounds which would corrode the pipeline. This is especially true if the gases contain carbon dioxide. Thus, displacing the water from the pipeline, which is also called dewatering, is a necessary step for pipeline precommissioning. If the pipeline is very short, it may be economical to just inject enough methanol or glycol to treat the water to mitigate hydrate without the need for displacing the water.

The primary function of a dewatering pig train is to displace water efficiently, leaving behind a minimum quantity of fluids for subsequent drying. A typical dewatering system involves a displacement fluid supply, a dewatering pig train, and a valve for water flowrate control. For relatively short pipelines, the pig train may consist of a small number of mechanical pigs which can be driven with nitrogen or air. Sometimes produced gas may be used to push the pig train. Nitrogen is also used for dewatering the

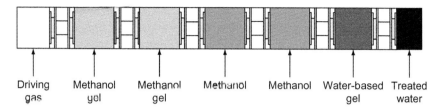

Driving   Methanol   Methanol   Methanol   Methanol   Water-based Treated
  gas   gel   gel   gel   gel   water

FIGURE 14.3   Typical dewatering pig train.

stainless steel pipelines. For long pipelines, a typical dewatering pig train
may include a number of pigs and fluid slugs. The fluid slugs serve different
functions, like providing lubrication of the pig seals and preventing the forward slippage of the driving gas. The speed of the pig train is controlled by
adjusting the water flowrate at the outlet while the pressure is controlled at
the inlet by the gas. A typical dewatering pig train is shown in Figure 14.3.

The dewatering train for the Zeepipe system (Falk et al., 1994) consisted of
10 mechanical pigs, which were separated by slugs of various liquids. Two slugs
of water-based gel were at the front of the train. The purpose of the gel slugs
was to lubricate the first pigs to decrease wear. Behind the gel slugs were four
batches of methanol which were used to coat the pipe wall to inhibit the water
that was left behind. Three batches of methanol gel were at the rear of the train
to prevent gas bypassing forward into the train due to imperfect sealing.

The performance of the gel slugs in the pig train will impact the efficiency
of the dewatering operation. There are a couple of gel systems available
(Schreure et al., 1994). One is the water-based gel system, which is a mixture
of polymers and crosslinkers. The other is the methanol or hydrocarbon gel
system. The gel slugs should satisfy the following function requirements
(Schreure et al., 1994):

- The gel slugs should minimize the fluid bypass across pig seals. The slugs
  should prevent water from backward bypassing and prevent gas from
  forward bypassing.
- The fluids in the gel slugs should be compatible with the pigs and the
  pipeline materials.
- Gels should be strong enough to sustain any shearing and dilution, thereby
  preserving their rheological properties through the whole operation.
- Gels should be chemically stable at the operating conditions for the whole
  operation, which can last for weeks.

After dewatering operations, the pipeline may still have enough water to
cause problems during start-up, and a drying operation is required to further
remove the residual amount of water in the pipeline. Of course, if the pipeline is going to transport water-wet gas, there is no need to dry the pipeline.
If the pipeline is stainless steel, the dewatering operations are normally
carried out with nitrogen and thus no need to dry the pipeline.

There are two common methods for drying pipeline: air drying and vacuum drying. Air drying techniques have been extensively discussed in literature. The main advantages of air drying are as follows:

● All free water can be removed from the pipeline.
● Very low dew points can be achieved down to as low as −90°F.
● The drying process is relatively short.

Unfortunately, air drying techniques are not well suited for offshore pipelines because the equipment requires a large area.

Vacuum drying is based upon the fact that the water will boil at low temperatures if the pipeline pressure is reduced to the saturated vapor pressure for the ambient temperature. Thus, by reducing the system pressure, it is possible to cause the water to boil and be removed from the pipeline as a gas with a vacuum pump. A typical vacuum drying pressure curve is shown in Figure 14.4.

The vacuum drying process can be divided into three stages. The first stage is the evacuation phase in which the pipeline pressure is drawn down from the atmospheric to the saturated vapor pressure by removing the air from the pipeline. The saturated vapor pressure is a function of the pipeline temperature. This phase is also called the drawdown phase.

As the pressure approaches the saturated vapor pressure, water starts to evaporate and the pressure is more or less maintained as a constant. As the pressure tries to fall, more water evaporates and thus the pressure stays constant. This vapor is sucked out of the pipeline by the vacuum pump. This process continues until all free water in the pipeline has evaporated. This phase is also called boiling phase.

When all free water in the pipeline has evaporated, the pressure in the pipeline will start to fall because there is no more water to evaporate and

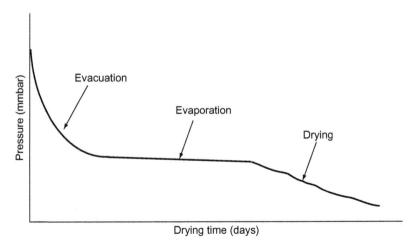

FIGURE 14.4 Typical vacuum drying pressure plot.

maintain the equilibrium. All the air in the pipeline has been evacuated, and the pipeline pressure can be directly correlated to the dew point. This phase is called the final drying phase.

Water evaporation requires heat input. In subsea pipeline, the heat has to come from the surrounding water. If the pipeline is insulated, the heat transfer process from the surrounding water to the pipeline can be quite slow. Thus, the vacuum pump must be properly sized so that the water will not be forced to evaporate faster than the pipeline can absorb the heat of evaporation from the surroundings. Otherwise, ice can form inside the pipeline.

The main advantages of the vacuum drying method are as follows:

- All free water can be removed from the pipeline.
- Very low dew points can be achieved.
- No large space is needed for equipment.
- No significant amount of waste will be produced.

The main disadvantages of vacuum drying are that the drying process can be quite long and the method is not suitable for very long or small bore pipelines.

After drying and before gas up, the pipeline may need to be purged using nitrogen for the following reasons:

- to further verify the line dryness,
- to provide a barrier before the introduction of hydrocarbons.

Even though hydrocarbon can be directly introduced after vacuum drying without the risk of achieving an explosive mixture, the nitrogen purging operations provide an extra safety margin.

After the precommissioning, the pipeline is ready for commissioning and start-up.

REFERENCES

Falk, C., Maribu, J., Eide, L.O., 1994. Commissioning the Zeepipe system sets new standards. Pipeline Gas J 221 (8), 24−33.

Karklis, P., et al., 1996. 1995—The year of the pig: hydrotesting and precommissioning the Yacheng pipeline. Proceedings of the 1996 Pipeline Pigging Conference, Jakarta, Indonesia.

Schreure, G., Burman, P., Hamid, S., Falck, C., Maribu, J., Ashwell, C., 1994. Development of gel system for pipeline dewatering and drying applications. Presented at the 26th Offshore Technology Conference, 2−5 May 1994, Houston, TX, OTC Paper 7577.

FURTHER READING

API RP 1110, 1997. Pressure Testing of Liquid Petroleum Pipelines, 4th Ed. American Petroleum Institute Recommended Practices.

ASME B31.4, 1989. Liquid Transportation System for Hydrocarbons, Liquid Petroleum Gas, Anhydrous Ammonia, and Alcohols (1989 Edition). ASME Standards.

ASME B31.8, 1992. Gas Transmission and Distribution Piping Systems (1992 Edition). ASME Standards.

Chapter 15

Flow Assurance

15.1 INTRODUCTION

The most severe operational hazards of offshore pipelines are the risks associated with the transportation of multiphase fluids. When water, oil, and gas are flowing simultaneously inside the pipeline, there are quite a few potential problems that can occur: water and hydrocarbon fluids can form hydrate and block the pipeline; wax and asphaltene can deposit on the wall and may eventually block the pipeline; with high enough water cut, corrosion may occur; with pressure and temperature changes along the pipeline and/or with incompatible water mixing, scales may form and deposit inside the pipeline and restrict the flow; and severe slugging may form inside the pipeline and cause operational problems to downstream processing facilities. The challenge that engineers will face is, thus, how to design the pipeline and subsea system to assure that multiphase fluids will be safely and economically transported from the bottom of the wells all the way to the downstream processing plant. The practice of identifying, quantifying, and mitigating of all the flow risks associated with offshore pipelines and subsea systems is called flow assurance.

Flow assurance is critical for deepwater pipeline and system operations. In deepwater, the seawater temperature is usually much colder than the surface air temperature. When pipeline is submersed in the deepwater, and if there is no thermal insulation layer surrounding the pipe wall, the fluid heat can be quickly lost to the water. This is especially true if the water current around the pipeline is strong. With an uninsulated pipeline, the heat transfer coefficient at the outer pipe wall can be significant due to the forced convection by the seawater movement (current). If the fluid temperature inside the pipeline becomes too low due to the heat loss, water and hydrocarbon (oil and gas) may form hydrate and block the flow. Furthermore, if the fluid temperature is low enough, wax may start to precipitate and deposit on the pipe wall. Thus, effective preservation of fluid heat is one of the most important design parameters for offshore pipeline.

In deepwater, the pipeline is normally followed by a production riser which goes from the bottom of the sea to the surface processing facilities (topsides). The deeper the water is, the longer the production riser is. With a long riser, the pipeline operating pressure will be higher due to the

Offshore Pipelines. DOI: http://dx.doi.org/10.1016/B978-0-12-397949-0.00015-7

hydrostatic head in the riser. For the same fluid temperature, with higher operating pressure, it is easier for the fluids to form hydrate. With pipeline and riser production system, if the flow conditions are such that severe slugging occurs, the slugs will be proportional to the riser length. The longer the riser, the longer the severe slugs.

How to optimize the pipeline and subsea system design to mitigate the flow assurance issues is a challenge. Flow assurance risks can be managed through robust system design, like heavy thermal insulation, high grade materials, and sophisticated mitigation systems, which would normally drive up the capital cost (CAPEX). On the other hand, flow assurance risks can also be managed through operations, like extensive chemical inhibition, extensive pigging, and flow monitoring, which will drive up the operating costs (OPEX). To balance the CAPEX and OPEX costs, the economics and system uptime are the key parameters.

Flow assurance, as a discipline, is still relatively new. There are many fundamental flow assurance phenomena which are not well understood. The main objectives of this chapter are to describe the fundamental flow assurance concepts and to summarize the mitigation practices used in industry for the flow assurance risks. In the subsequent sections, the major flow assurance issues and some critical parameters that would impact the identifying, quantifying, and mitigating of the flow assurance risks associated with subsea pipeline operations will be covered.

15.2 FLUID SAMPLING AND CHARACTERIZATIONS

One of the most critical steps in identifying and quantifying flow assurance risks is fluid sampling. Whether or not there will be any flow assurance risks in subsea pipeline must be determined from the fluid sample analysis: What is the fluid composition? Is there a potential for wax deposition? Is the potential for asphaltene deposition high, medium, or low? Will the fluid gel when the temperature is low enough after system shutdown? How much energy will be required to remobilize the fluid once it is gelled? All these questions can be answered only by lab or flowloop measurements of the fluid samples. Thus, it is very important to sample the representative fluid that will be transported by pipeline. No matter how accurate the lab measurements and interpretations are, if the fluids do not represent the real production fluids, wrong conclusions may be drawn. Any flow assurance mitigation strategies based upon wrong conclusions will work improperly and the pipeline and subsea system will encounter severe operational risks.

Water samples are also very critical in establishing flow assurance risks. These include scaling, hydrate formation tendencies, corrosivity, compatibility with other water (injected water or water from different production zones), material metallurgy, and design of the water handling equipment. One special challenge associated with flow assurance risk assessment is that there is no water available for sampling because the exploration wells may never reach the aquifer zones. Without water samples, it is very difficult to make accurate flow assurance risk assessments. A lot of times, water samples from nearby fields have to be used, resulting in high levels of uncertainty in the development of the flow assurance mitigation strategies.

15.2.1 Fluid Sampling

There are a lot of discussions in literature on fluid sampling and handling (API RP 44, 1966; API RP 45, 1968; Chancey, 1987; Ostrof, 1979). Fluid samples can be obtained from downhole and/or from surface separator. The downhole samples are the primary samples for power−volume−temperature (PVT) measurements, and surface separator samples are usually treated as backup and can be used as bulk samples for process or reservoir design. It is a good practice to collect at least two downhole samples with one serving as a backup and collect at least three 1-gallon samples from the separator. A certain amount of stock tank oil samples are needed for other crude oil analyses (geochemical and crude assay).

For fluid sampling plans, it is important to know the pros and cons of all available fluid sampling tools. Will drillstem testers (DST) or wireline testers be used? For wireline testers, will RCI (reservoir characterization instrument) or MDT (modular formation dynamics tester) be used? The key is how we can obtain the most representative fluid samples from downhole and transport it unaltered to the surface and to the lab for measurements and analysis. To achieve this, it is important to obtain fluid samples from the main production zone.

Once zone or zones to be sampled are determined, the next question is how to make sure the formation fluids will be sampled with the least mud contamination. During the drilling, with over-balanced drilling, the drilling fluids will penetrate into the formation to form a damaged zone just outside the wellbore. The fluid sampling tool needs to be able to penetrate through the damaged zone to get to the virgin formation fluids. Since it is very difficult to completely avoid mud filtrate contamination during the sampling, it is important that the tool be able to monitor the mud contamination level and thus whether or not the samples are acceptable can be determined. When the fluids are flowing into the tool, the pressure drop between the formation pressure and the pressure in the sample chamber should be kept low so that the fluids will not change phases during the sampling. Gas can come out of oil when the fluid pressure is below the bubble point. Gas may leak out from the tool during the transportation. It is also important to make sure there is no solid, like asphaltene, deposition that may stick onto the chamber wall and not be completely recovered. Otherwise the sampled fluids may not accurately represent those in the formation.

When the sampled fluids are transported from downhole to surface, the pressure of the fluids may drop due to the temperature drop. Whether or not this pressure drop would cause the pressure to be below the bubble point will need to be checked out.

15.2.2 PVT Measurements

Once the fluid samples are in the lab, numerous tests will be performed to measure the fluid properties. Compositional analysis of the downhole sample would be performed through at least C36 + , including density and molecular weight of the Heptanes plus. Pressure—volume relations are determined at reservoir temperature by constant mass expansion. This measurement provides oil compressibility, saturation pressure, single-phase oil density, and phase volumes. The compositions and gas formation volume factors for the equilibrium reservoir gas during primary depletion can be obtained by performing differential vaporization at reservoir temperature. Gas viscosities are then calculated from the composition. Undersaturated and depleted oil viscosity at reservoir temperature can be measured by using capillary tube viscometry.

The following parameters will normally be measured for black oil:

- Stock tank oil density (API gravity)
- Bubble point pressure
- Flash Gas Oil Ratio (GOR)
- Live oil compressibility
- Fluid density at bubble point
- Reservoir oil viscosity
- Flash gas composition
- Flash gas specific gravity
- Reservoir fluid composition.

For gas condensate:

- Condensate density (API gravity)
- Dew point pressure
- Flash GOR
- Flash gas specific gravity
- Flash gas composition
- Stock tank oil composition
- Reservoir fluid composition
- Z factor at dew point.

15.2.3 Specific Flow Assurance Analysis

Other than the PVT measurements, the fluid samples are also used for specific flow assurance measurements. For wax deposition evaluation, the compositional analysis through C70 + will be performed. Measurements, such as wax appearance temperature (WAT) for the dead oils, shear rate, pour point, molecular weight, and total acid number (TAN) would normally be done. For asphaltene analysis, asphaltene titration would be done to determine the stability of asphaltenes. Titration of stock tank oil is normally done with n-heptane or n-pentane while monitoring the percent of asphaltene precipitates to determine stability. If light oil and heavy oil mixed together during transportation, tests would be required to determine the tendency toward asphaltene precipitation of the mixed oil. Even though the hydrate curves of reservoir fluids are usually modeled by software, it is also desirable to confirm the models by performing lab measurements.

Crude oils also have to be tested for foaming tendency and emulsion forming tendency. It is also necessary to evaluate how the water oil emulsion stability can be affected by shearing resulting from pumping and lifting mechanisms. It is also very desirable to measure the live oil water emulsion viscosity at both operating and seabed conditions with water cut ranging between 0% and 100%. The existing public emulsion viscosity models are not universal, and different oil most likely will form emulsions with quite different rheology behavior. Thus, it is important to measure the emulsion viscosity in the lab. The measurement of live oil water emulsion viscosity is quite expensive, and only a few labs are available to do the tests.

There are also chemical screening tests with water samples for corrosion and scale analysis.

15.2.4 Fluid Characterizations

Applications of equation of state and fluid characterizations have been discussed extensively over the last few decades, and excellent papers are available in literature for reference (Huron and Vidal, 1979; Katz and Firoozabadi, 1978; Mathias and Copeman, 1983; Pedersen et al., 1985, 1989, 1992, 2001; Peneloux et al., 1982; Peng and Robinson, 1976, 1978; Reid et al., 1977; Riazi and Daubert, 1980; Soave, 1972; Sorensen et al., 2002; Tsonopoulos and Heidman, 1986).

No matter how many tests we do, the measured parameters will not be able to cover all the application ranges we need. Thus, fluid models (equation of state) that can predict the fluid PVT behavior at different pressure and temperature conditions will be needed in pipeline design. Normally cubic equation of state models are used, like the SRK (Soave−Redlich−Kwong) (Soave, 1972), PR (Peng−Robinson) (Peng and Robinson, 1976), and modified PR (Peng and Robinson, 1978) models. The preferred models would be able to accurately predict the fluid behavior at conditions that cover the whole pressure range of reservoir and topsides processing pressures and the whole temperature range of reservoir and seabed temperatures.

To develop a model to predict the PVT behavior of oil and gas condensate mixture using a cubic equation of state, the critical temperature, the critical pressure, and the acentric factor must be known for each component of the mixture. Unfortunately, oil or gas condensate mixtures may contain thousands of different components. It is thus not practical to develop a model that would cover all the individual components. Some of the components must be grouped together and represented as pseudo-components. A common approach is to characterize the fluids using C7 + , which consists of representing the hydrocarbons with seven or more carbon atoms as a reasonable number of pseudo-components. For each pseudo-component, the parameters of critical pressure, critical temperature, and acentric factor have to be determined (Pedersen et al., 1992). The characterized models are then fine-tuned using the PVT parameters obtained from lab measurements.

It is difficult to tune the model that will match all the lab-measured PVT parameters accurately. One or a few parameters can be tuned to match the lab data well, and the rest of the parameters may not match the lab data well enough. Judgment may be needed to decide what are the critical PVT parameters for the applications. Effort should be made to try to tune the critical parameters to match the lab data well.

15.3 IMPACTS OF PRODUCED WATER ON FLOW ASSURANCE

In offshore production pipeline, there usually exists water together with oil and gas. Water is produced from the reservoir and because water is an excellent solvent, it has dissolved plenty of chemical compounds and gases inside the formation. Water also contains suspended solids and impurities. Inside the reservoir formation, water and the chemical compounds are usually in equilibrium. As water is produced from the formation into the pipeline, because of the changes of pressure and temperature, the equilibrium is destroyed and some compounds would become insoluble and start to precipitate from water and form all kinds of scales. When free water directly contacts the pipeline wall of carbon steel, water would dissolve metal and cause corrosion problems to the pipeline. When water and gases flow together in the pipeline, at certain pressure and temperature conditions, they would form hydrate which can potentially block the pipeline. Produced water presents major flow assurance problems for deep subsea pipelines.

Proper water sampling, handling, and analysis are very critical for flow assurance risk assessment. Many of the water properties, like dissolved gases, suspended solids, and pH values would change with time and would change with pressure and temperature. Both lab and on-site analysis are necessary to get accurate water analysis (API RP 45, 1968, Ostroff, 1979). The main ions in water that are of importance for flow assurance are listed below.

The main negative charged ions (anions) in water are:

Chloride Cl^-
Sulfide HS^-
Sulfate SO_4^{-2}
Bromide Br^-
Bicarbonate HCO_3
Carbonate CO_3^{-2}.

And the main positive charged ions (cations) in water are:

Sodium Na^+
Potassium K^+
Calcium Ca^{+2}
Magnesium Mg^{+2}
Strontium Sr^{+2}
Barium Ba^{+2}
Iron Fe^{+2} and Fe^{+3}
Aluminum Al^{+3}.

Cations and anions can combine and form different substances. When pressure and temperature change, the solubility of each ion will change. The excessive ions will precipitate from water and form solids, like scales. For example, calcium and carbonate would form calcium carbonate scale:

$$Ca^{+2} + CO_3^{-2} \rightarrow CaCO_3 \downarrow$$

Similarly, barium and sulfate can form barium sulfate scale:

$$Ba^{+2} + SO_4^{-2} \rightarrow BaSO_4 \downarrow$$

Water with dissolved salts is also an excellent electrolyte that is required for corrosion to occur. When free water is high enough to wet the inner pipe wall, corrosion may occur. The more salts or ions in the water, the more conductive the water is and the severer the corrosion will be.

It is well known that when free gas and water are mixed together at a certain pressure and temperature, hydrate will form. Hydrates are solids that are similar to ice. Hydrocarbon and free water often coexist inside the offshore pipelines. When the pipeline pressure is high enough and/or fluid temperature is low enough, hydrates will form. If hydrates form inside the offshore pipeline, the pipeline flow can be blocked by the hydrates. Once the pipeline is blocked by hydrates, it can take weeks or months to dissociate the hydrates. Hydrate plugging is one of the major flow assurance risks in the deepwater production system.

Water can significantly change the multiphase flow characteristics inside the pipeline and cause severe slugs to occur. For example, for the same total liquid flowrate and the same gas–oil ratio, the total amount of gas inside the pipeline will be much less with water cut of 90% than with water cut of 0%. With less gas flow, the liquid inventory inside the pipeline will be higher and it is harder for the gas to carry the liquid out of the riser due to less gas energy. Thus, it is easier to form severe slugs.

More detailed descriptions on scales, corrosion, hydrates, and severe slugging will be given in the following sections. But based upon the above brief discussions, it is obvious that produced water has significant impacts on flow assurance risks. The most effective way to mitigate flow assurance risks in production pipelines is to dispose of the water subsea and make sure no water will get into the pipeline. Unfortunately, the most effective way may not be the most economical way nor the most practical way. Currently, the most common ways to mitigate flow assurance risks in offshore pipelines are thermal insulation and chemical inhibitions. But if the amount of water flowing inside the pipeline can be reduced (downhole separation and/or seafloor processing), the amount of chemicals needed for inhibition will also be less, resulting in less operation costs.

15.4 GAS HYDRATES

Gas hydrates are crystalline compounds that occur when small gas molecules contact with water at certain temperatures and pressures. Hydrates are formed when the gas molecules get into the hydrogen-bonded water cages. The physical properties of hydrates are similar to those of ice (Sloan, 1998). But hydrates can form at temperatures well above 32°F in pressurized systems. Commonly found hydrates are composed of water and light gas molecules, like methane, ethane, propane, carbon dioxide, and hydrogen sulfide.

Three hydrate crystal structures have been identified (Sloan, 1998). They are called Structures I, II, and H. The properties of Structures I and II hydrates are well defined. Structure H hydrates are relatively new, and their properties are less well defined. All hydrates contain a lot of gas. A lot of research is being conducted to study hydrates as a potential energy resource (Makogon, 1997).

At certain pressure and temperature conditions, when water molecules form structures that consist of cavities, small gas molecules would get into the cavities to form hydrates. But how the gas molecules get into the cavities and how hydrates form are not well understood (Sloan, 1998). However it is believed that the formation of the hydrate nuclei usually happens at the gas—water interface. The crystals then grow by surface sorption of gas and water molecules (Makogon, 1997). How quickly the hydrate would form and grow depends upon the diffusive flux of the gas and water molecules. If hydrate is forming at the gas—water interface and the water and gas molecules are abundant, the hydrate would grow the highest (Makogon, 1997). That is why hydrate blocks usually occur during re-start-up of pipeline flow where the flow turbulence and agitations would enhance the flux of gas and water molecules.

Gas hydrates are like solids, and their physical properties are similar to those of ice. When hydrates form inside the pipeline, the flow can be blocked by hydrate plugs. Once a hydrate plug is formed, it can take up to weeks and months to dissociate the plugs. It is thus very important to design and operate an offshore pipeline system free of hydrate risks. Hydrates can very easily form downstream of choke where fluid temperature can drop into the hydrate formation region due to Joule–Thompson cooling effects.

15.4.1 Gas Hydrate Formation Curve

Figure 15.1 shows a typical gas hydrate curve. On the left side of the curve is the hydrate formation region. When pressure and temperature are in this region, water and gas will start to form hydrate. On the right side of the curve is the non-hydrate formation region. When pressure and temperature are in this region, water and gas will not form hydrate. Quite a few factors impact the hydrate curve. Fluid compositions, water compositions, and water salinity all affect the hydrate curve. Increasing salinity would shift the curve left and reduce the hydrate formation region.

Figure 15.1 demonstrates that with the initial system in the non-hydrate region, if system pressure is increased while keeping the system temperature constant, hydrate would eventually form. The same is true by reducing the system temperature at a constant system pressure. Hydrate curve is very useful for subsea pipeline design and operations. It provides pressure and temperature conditions that the system should maintain to avoid hydrate formation. Hydrate curves can be calculated by using PVT software. But the key for accurate hydrate curve calculations is to have accurate fluid and water compositions. Again, fluid sampling and analysis are critical for flow assurance risk assessment. If the hydrate temperature is over-conservatively

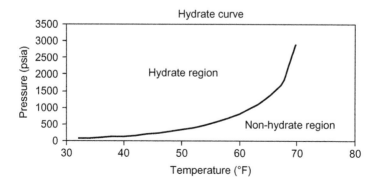

FIGURE 15.1 A typical gas hydrate curve.

calculated by a few degrees, millions of dollars may be wasted in the thermal insulation design.

15.4.2 Hydrate Inhibitors

15.4.2.1 Thermodynamic Inhibitors

As shown in Figure 15.1, no hydrate would form in operating conditions that are on the right-hand side of the hydrate curve. It would therefore help to shift the hydrate curve left so that the non-hydrate region would be larger, and thus the risks for hydrate formation would be smaller. But, for a given pipeline design, the fluid and water compositions are normally specified and the hydrate curve is thus fixed. To shift the hydrate curve left, thermodynamic inhibitors can be used. Thermodynamic inhibitors would not affect the nucleation of hydrate crystals and the growth of the crystals into blockages. They only change the pressure and temperature conditions of hydrate formation. With inhibition, the hydrate formation temperature will be lower or the hydrate formation pressure will be higher. Thus, by applying thermodynamic inhibition, the operating conditions can be shifted out of the stable hydrate region.

Two kinds of thermodynamic inhibitors are commonly used: methanol and monoethylene glycol (MEG). For oil systems, methanol is used the most. Figure 15.2 shows how the hydrate curve shifts with different amounts of methanol inhibition. For system pressure of 1000 psia, the hydrate formation temperature for freshwater is about 62°F. By adding 10 wt% methanol into the freshwater, the hydrate formation temperature is reduced to 54°F. With 20 wt% methanol, the hydrate formation temperature is further reduced to about 44°F. It is obvious that methanol is very effective for hydrate inhibition.

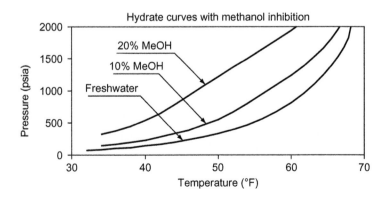

FIGURE 15.2   Gas hydrate curve with different amount of methanol inhibition.

We know that thermodynamic inhibitors can be used to reduce the hydrate formation temperature. But for a given condition, how much inhibitor will be needed? If it is known how much temperature needs to be reduced, the amount of inhibitor needed in the free water can be estimated using the following equation (Hammerschmidt):

$$W_i = \frac{100 M_i \Delta T_h}{(C_i + M_i \Delta T_h)} \tag{15.1}$$

where

W_i = weight percent of the inhibitor in liquid
C_i = constant, 2335 for methanol and 2000 for MEG
M_i = molecule weight of methanol or MEG
ΔT_h = hydrate subcooling which is the temperature needs to be reduced by inhibitor.

For example, for a pipeline system, if the hydrate formation temperature without inhibition is 65°F, then the system operating temperature is 50°F. The hydrate subcooling is the difference of the hydrate formation temperature and the system operating temperature and equals 15°F for this case. The above equation can only calculate the required methanol and MEG in the free water phase. Some methanol or MEG will get lost in gas phase and hydrocarbon liquid phase. The amount of methanol or MEG injected will need to be higher than that calculated by Eq. (15.1). More details on how to estimate the amount of methanol or MEG lost in vapor and condensate can be found in literature (Sloan, 1998).

By adding methanol in the liquid phase, the water concentration in the liquid phase is reduced and the hydrate formation temperature is lowered. The more methanol is added, the more the hydrate formation temperature is lowered. On the other hand, to have the same subcooling, the more water that is added, the more methanol is required. The need for large amounts of methanol may cause problems in storage and handling because of its flammability and toxicity and will result in high chemical OPEX. Furthermore, if a large amount of methanol is carried over into the export lines, it will cause problems for downstream processing. Compared to methanol, MEG is less flammable but is more expensive and less available.

Salt can also affect the hydrate formation conditions. By adding more salt into the water, the hydrate formation curve will shift to the left, as shown in Figure 15.3. The impact of salt on hydrate curve can be significant. By adding 2 mole% NaCl into the freshwater, the hydrate formation temperature will be 4−5°F lower. If the salt concentration is increased to 8 mole%, the hydrate formation temperature will be more than 25°F lower. However, even though salt solution can be used for hydrate inhibition, too much salt can cause salt deposition and scale deposits in the process facilities. Salt solution is also corrosive and can cause corrosion problems to equipment.

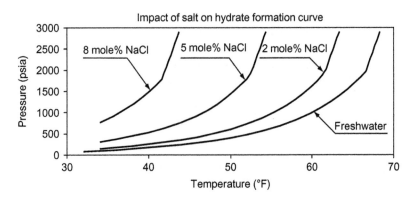

FIGURE 15.3   Gas hydrate curves with different salt concentrations.

15.4.2.2 Low-Dosage Hydrate Inhibitors

As discussed above, high water flow will require large amounts of methanol or MEG for hydrate inhibition, resulting in high OPEX. To mitigate the high-dosage problems, more effective hydrate inhibitors than methanol and MEG are needed for high water flow. The inhibition mechanisms of the new chemicals must be different from the traditional thermodynamic inhibitors to be more effective at low dosage. The chemicals that would effectively inhibit hydrate at low dose rates are called low-dosage hydrate inhibitors (LDHI). Two kinds of LDHI are most popular in the oil industry: one is kinetic hydrate inhibitor and the other is anti-agglomerate (Balson et al., 2002; Kelland et al., 1995; Mehta et al., 2003).

Kinetic hydrate inhibitors tend to slow down the hydrate nucleation process and delay the formation and growth of hydrate crystals for an extended period of time (Balson et al., 2002). But kinetic inhibitors can only delay the kinetics and cannot completely prevent the nucleation process. Thus, kinetic inhibitors can only prevent hydrate from forming for a finite time. Once this period of time has lapsed, there is a very rapid conversion of the remaining water into large hydrate and blockage may occur (Mehta et al., 2003). Thus, when designing the hydrate mitigation strategies for subsea system, it is very critical to make sure the fluid residence time in the pipeline during steady-state flow is less than the "hold-time," which is the time before hydrates start to form rapidly. The "hold-time" of some kinetic inhibitors can be about 24−48 h.

Another important parameter for kinetic inhibitors is the subcooling. It is reported that kinetic inhibitors can only work up to 15−23°F subcooling (Balson et al., 2002). For deepwater applications, the subcooling is normally larger than 25°F and the applications of kinetic inhibitors are severely limited.

Anti-agglomerates are polymers and surfactants that tend to prevent the formation and accumulation of large hydrate crystals into a hydrate blockage so that a transportable slurry can be maintained. It will not delay the nucleation of hydrate crystals but will keep the crystals in hydrocarbon phase by reducing the growth rate of the crystals. The hydrate crystals will be transported with hydrocarbon as slurry flow. Anti-agglomerate has no subcooling limitation as do kinetic inhibitors and can be effective at subcoolings of greater than 40°F (Mehta et al., 2003). Since the crystals have to be carried out of flowline, a liquid hydrocarbon phase is required to suspend the crystals.

15.4.3 Hydrate Mitigation Strategies

As mentioned above, the most effective way to mitigate hydrate is to get rid of water. If there is no water flowing in the pipeline, there is no hydrate risk. But, in the real world, getting rid of water may not be the most practical or most economical way. Some other methods have to be utilized.

15.4.3.1 Thermal Insulation

Based upon the hydrate formation curve, as long as the fluid temperature is above the hydrate formation temperature, no hydrate will be formed. Thus, a very good way to mitigate the hydrate risk is to maintain the fluid temperature inside the pipeline above the hydrate formation temperature. However, for pipelines in deepwater, the water temperature is normally very low and can be below 40°F depending upon the water depth, and steel pipe is not a very good thermal insulator. Thus, it is necessary to put thermal insulation material around the pipeline to prevent the heat from being lost to the surroundings.

There are a few different insulation methods available in the industry. One is to directly cast insulation materials onto the outer surface of the pipeline (cast-in-place). The insulation materials for this application may be a layer of homogeneous material or may consist of multiple layers, with each layer being a different material. Single layer insulation is most used for cases where the insulation thickness is not excessive. For the large insulation thickness cases, multiple layer insulation is required due to mechanical and installation reasons.

Another popular insulation method is pipe-in-pipe, where the hydrocarbon pipeline is put into another concentric pipeline. The annulus between the two pipelines is either completely or partially filled with insulation material. Pipe-in-pipe thermal insulations normally provide better insulation than cast-in-place methods. But pipe-in-pipe methods are also normally more expensive.

Bundles are also used for thermal insulation. Hydrocarbon pipelines and some other pipelines that flow hot water are bundled together. The heat is transferred from the hot water to the hydrocarbon fluids. Thus, the fluid temperature is kept above the hydrate formation temperature.

A couple of industry deepwater pipelines were intentionally buried under the seabed to use soil as a thermal insulation material. But due to the burial process, excessive water will exist in the covering soil and convection may be significant. Burying the pipeline alone will not be enough for thermal insulation. Some extra insulation will be needed.

Thermal insulation is not very effective in mitigating the hydrate risks of gas pipelines. Because the density of gas is much smaller compared to liquid, the thermal mass (density times the heat capacity) of gas is also much smaller than that of liquid. Thus, it is not very effective to thermally insulate the gas pipeline.

Thermal insulation is a very good hydrate mitigation strategy for oil pipeline, especially when the pipeline is in operation. By using insulation, it is easy to keep fluid flowing temperature everywhere along the pipeline above the hydrate temperature. But no matter how much insulation is put on the pipeline, after a long pipeline shutdown, the fluid temperature will fall below the hydrate formation temperature and eventually cool down to the seawater temperature. Thus, thermal insulation itself is not enough for hydrate mitigation for long pipeline shutdowns. Other mitigation strategies, like pipeline depressurization or dead oil displacement, will be needed. But thermal insulations do provide a certain period of cooldown time so that no other mitigation operations are needed. Cooldown time is defined as the time, after pipeline shutdown, before the fluid temperature drops down to the hydrate formation temperature for a given pipeline shutdown pressure. Thus, operations, like pipeline depressurization or dead oil displacement, would have to finish within the cooldown time. Otherwise, hydrate will form inside the shutdown pipeline.

There is another important parameter called "no-touch" time or "hand-free" time. "No-touch" time is defined as the time within which no action needs to take place after a pipeline shutdown. That is why it is also called "hand-free" time. "No-touch" time is always shorter than the cooldown time. It is the difference between the cooldown time and the time needed to perform the operations, like pipeline depressurization. This "no-touch" time provides a very valuable period for the operator to diagnose the problems that cause pipeline shutdown. If the problems are found and fixed within the "no-touch" time, the pipeline can be restarted with no need to use other hydrate mitigation operations. If the problems cannot be fixed within the "no-touch" time period, operators will need to perform operations to mitigate hydrate. The longer the "no-touch" time, the longer the time available for the operator to fix the problems and the less need to perform extra operations. But the longer the "no-touch" time, the thicker the needed insulation layer.

15.4.3.2 Chemical Inhibitions

Thermodynamic inhibitors, like methanol and MEG, and LDHIs, like kinetic inhibitors and anti-agglomerates, are often used for hydrate mitigations.

Chemical inhibitors are not normally used continuously for oil systems; instead they are used after shutdown or during re-start-up. Thermodynamic inhibitors are usually used continuously for gas pipelines because gas pipelines are normally not insulated.

After the "no-touch" time, methanol and MEG are used to inhibit the fluids in the subsea system, like trees, well jumpers, and manifolds. But it is difficult to estimate the amount of water in the system after shutdown, and thus it is difficult to know how much methanol is needed to completely inhibit the fluids. Thus, in practice, methanol or MEG is used to completely displace the fluids in the subsea system. A certain amount of methanol is bullheaded into the well (usually above the surface controlled subsurface valve) to protect the upper portion of the wellbore from forming hydrate. Fluids in pipelines are usually not displaced by methanol because the pipelines are usually too long and too much methanol will be required.

15.4.3.3 Electric Heating

Recently, more research has been conducted on hydrate mitigation using electric heating (Lervik et al., 1997). Electric heating can be divided into two categories: direct and indirect electric heating. With direct electric heating, electric current flows axially through the pipe wall and directly heats the flowline, while with indirect heating, electric current flows through a heating element on the pipe surface and the flowline is then heated through thermal conduction.

Electric heating can be used as a mitigation method for pipeline hydrate problems. After shutdown, electric heating can be used to keep the pipeline fluid temperature above the hydrate formation temperature and no hydrate will form. Electric heating can also be used as an intervention/remediation strategy for hydrate problems. Once a hydrate plug is formed, electric heating can be used to melt the hydrate. In this way, the hydrate will be melted much faster than using pipeline depressurization. Shell's Na Kika project in the Gulf of Mexico used electric heating as a hydrate remediation method (March et al., 2003).

15.4.3.4 Hot-Oil Circulation

Hot-oil circulation is a popular strategy for hydrate mitigation during system re-start-up. After a long shutdown, the fluid in the pipeline will become cold (seawater temperature). If the pipeline is re-started up with cold fluid in it, hydrate risk is very high. To reduce the hydrate risk, hot oil is first circulated through the pipelines (looped pipelines are required) to displace the cold fluid and also warm up the pipelines. The time required to warm up the pipelines depends upon the hot-oil discharge temperature, hot-oil circulation flowrates, and pipeline length. Usually it would take up to 5−10 h to warm up the subsea pipeline.

15.4.3.5 System Depressurization

Pipeline depressurization is used to mitigate hydrate plug after a long shutdown. From the hydrate formation curve, for a given temperature, non-hydrate region can be reached by reducing the pressure. After a long shutdown, the fluid temperature will eventually reach the seawater temperature. Based upon the hydrate curve, the hydrate formation pressure at the seawater temperature can be determined. Thus, the pipelines can be depressurized (also called pipeline blowdown) below the hydrate formation pressure. Once the pipeline pressure is below the hydrate formation pressure, no hydrate will form and the system can be continued to be shut down.

System depressurization is also often used to melt a hydrate plug formed in a pipeline. When the system pressure is below the hydrate formation pressure, the hydrate plug will start to dissociate. The hydrate plug dissociation process is fairly slow. It can take up to weeks or even months to completely melt a long hydrate plug. That is why it is very important to design and operate subsea pipeline out of hydrate region. For safety reasons, it is always better to be able to depressurize the pipeline from both sides of the hydrate plug.

15.5 WAX DEPOSITIONS

Crude oil is a complex mixture of hydrocarbons which consists of aromatics, paraffins, naphthenics, resins, asphaltenes, diamondoids, mercaptans, etc. When the temperature of crude oil is reduced, the heavy components of oil, like paraffin/wax ($C18-C60$), will precipitate and deposit on the pipe wall. The pipe internal diameter will be reduced with wax deposition, resulting in higher pressure drop. Wax deposition problems can become so severe that the whole pipeline can be completely blocked. It would cost millions of dollars to remediate an offshore pipeline that is blocked by wax.

15.5.1 Fundamental Concepts

15.5.1.1 Crude Cloud Point or WAT

Wax solubility in aromatic and naphthenic is low, and it decreases drastically with decreasing temperatures. Thus, it is easy for wax to precipitate at low temperature. The highest temperature below which the paraffins start to precipitate as wax crystals is defined as crude cloud point or WAT. Since light ends can stabilize the paraffin molecules (Meray et al., 1993), the cloud point of live oil with pressure below the bubble point is usually lower than the cloud point of the dead oil or stock tank oil. The cloud point of live oil is more difficult to measure than that of dead oil. Thus, the cloud point of dead oil samples is often used in offshore pipeline thermal insulation design.

This approach is conservative and can practically result in millions of dollars of extra cost in thermal insulation.

When measuring the cloud point, the key is to preheat the oil sample to a high enough temperature to solubilize all the preexisting wax crystals. There are quite a few techniques available for cloud point measurement: viscometry, cold finger, differential scanning calorimetry, cross polarization microscopy, filter plugging, and Fourier transform infrared energy scattering, etc. The cloud points measured using different techniques may differ by more than 10° (Hammami and Raines, 1999; Monger-McClure et al., 1999).

15.5.1.2 Crude Pour Point

When the waxy crude is cooled, paraffins or waxes will precipitate out of the crude oil once the temperature is below the cloud point. The precipitated wax may deposit on the pipe wall in the form of a wax-oil gel (Venkatesan et al., 2002). The gel deposit consists of wax crystals that trap some amount of oil. As the temperature gets cooler, more wax will precipitate and the thickness of the wax gel will increase, causing gradual solidification of the crude. When the wax precipitates so much and forms wax gel, the oil will eventually stop moving. The temperature at which oil sample movement stops is defined as the crude pour point. When crude gets so cold that it stops moving inside the offshore pipeline after shutdown, depending upon the characteristics of the gel, crude oil may not be able to be remobilized during re-start-up. Thus, it is very important to check the re-start-up pressure of the crude by cooling the crude down to below the pour point. Because the seawater temperature can be below the pour point of the crude, wax gel may form after long pipeline shutdown. It is critical to make sure the pipeline will be able to be re-started up after long shutdown.

15.5.2 Wax Deposition Mechanisms

Extensive research has been conducted to try to understand and model the wax deposition process which is a complex problem involving fluid dynamics, mass and heat transfers, and thermodynamics (Brown et al., 1993; Burger et al., 1981; Creek et al., 1999; Hsu et al., 1999; Singh et al., 1999). It is widely accepted that molecular diffusion of paraffin is one of the dominant deposition mechanisms. Whether or not Brownian motion, gravity settling, and shear dispersion play significant roles in wax deposition is still a research topic.

15.5.2.1 Molecular Diffusion

When waxy crude is flowing in offshore pipeline, the temperature at the center of the pipeline is the hottest while the temperature at the pipe wall is the coldest, resulting in a radial temperature profile. Since the wax solubility in the oil is a decreasing function of temperature, when the temperature is lower than the cloud point, wax crystals will come out of solution. Thus, the radial temperature gradient will produce a concentration gradient of wax in oil with the wax concentration in the oil lowest close to the pipe wall. The concentration gradient would thus result in mass transfer of wax from the center of the pipe to the pipe wall by molecular diffusion. Wax mass transfer can be described by the Fick's law as

$$\frac{dm_w}{dt_w} = \rho_w D_w A_d \frac{dC_w}{dr}$$

(15.2)

where

m_w = mass of the deposited wax on the pipe wall
t_w = time
ρ_w = density of the solid wax
D_w = diffusion coefficient of liquid wax
A_d = deposition area
C_w = volume fraction concentration of wax in liquid oil
r = radial coordinate.

The diffusion coefficient is expressed by Burger et al. (1981) as a function of oil viscosity:

$$D_w = \frac{K_w}{\nu}$$

(15.3)

where

K_w = constant
ν = oil dynamic viscosity.

The constant in Eq. (15.3) is often adjusted to match modeled deposition rates with experimental ones.

15.5.2.2 Other Proposed Mechanisms

There are a few wax deposition mechanisms that are not widely accepted, like Brownian diffusion, shear dispersion, and gravity settling. Once the temperature is below the cloud point, wax crystals will precipitate out of solution and be suspended in the oil. The suspended wax crystals will collide with each other and with oil molecules due to Brownian motion. Because of the wax concentration gradient, it is possible that the net effect of Brownian motions is to transport the wax crystals in the direction of decreasing concentration. It is thus suggested that wax deposition can occur due to the Brownian diffusion of wax crystals. But quite a few existing wax deposition models do not take into account the Brownian diffusion.

Gravity settling as one of the possible wax deposition mechanisms is based upon the argument that the wax crystals tend to be denser than the oil and would thus settle in a gravity field and deposit on the bottom of the pipelines. But experiments with horizontal and vertical flows showed that there was no difference in the amount of wax deposited for the two flow configurations. Thus, it is not yet clear how significant a role gravity would play for wax deposition.

Burger et al. (1981) and Weingarten and Euchner (1986) reported possible wax deposition by shear dispersion. They claimed that shear dispersion played a role in wax deposition mainly in laminar flow and proposed the following equation for the deposition rate:

$$\frac{dm_s}{dt} = k_w C_s A_d \gamma \tag{15.4}$$

where

m_s = mass of the deposited wax due to shear dispersion
k_w = constant
C_s = the concentration of solid wax at the pipe wall
A_d = deposition area
γ = shear rate.

But Brown et al. (1993) performed experiments with zero heat flux across the pipe wall (thus no molecular diffusion) and showed no wax deposition due to shear dispersion. Brown et al. concluded that shear dispersion does not contribute to wax deposition.

15.5.3 Wax Mitigation Strategies

15.5.3.1 Thermal Insulation

For subsea production pipeline, the most widely used wax mitigation method is to include enough thermal insulation to maintain the fluid temperature everywhere along the pipeline above the WAT during normal or "steady-state" operations. For transient operations, like shutdown, the fluid temperature inside the pipeline will cool down with time and eventually will reach the seawater temperature within a transient time that is about 12−36 h depending upon the insulation design. Once the pipeline cools down to seawater temperature, there is no temperature gradient between the bulk fluid and the pipe wall and no wax will deposit. During the cooldown transient time, the fluid temperature can be lower than the WAT and some wax will deposit onto the pipe wall. Since the transient time is relatively short, the amount of wax deposited will be very small, because wax deposition is a slow process. Furthermore, the wax deposited during shutdown will be remelted once the pipeline reaches normal operation again.

To be conservative for the insulation design, the WAT of dead oil is normally used. But for subsea pipeline insulation design, the most important drivers are hydrate mitigation and system cooldown time. If the subsea pipeline insulation design satisfies the hydrate mitigation and cooldown time requirements, it will normally also satisfy the requirement that fluid temperature be above the WAT during steady-state flow. Details on subsea thermal insulation are provided in a different chapter of this book.

15.5.3.2 Pigging

Another popular wax mitigation method is to regularly pig the pipeline to remove deposited wax from pipe walls. For some subsea pipelines, especially export lines where hydrate is not a concern, pigging would normally be the main wax management strategy. There are numerous types of pigs, like simple spheres, foam pigs, and smart pigs.

The pig is sent down into the pipeline from a pig launcher and is pushed by the production crude or any other fluids, like dead oil or gas. The pig mechanically scrapes the wax from the pipe wall and redeposits it back into the crude in front of the pig. A regularly scheduled pigging program is very critical for the success of pigging operations. If the pigging operation is not scheduled frequently enough, too much wax may have deposited onto the pipe wall. During the pigging operation, a pig may get stuck inside the pipe due to the excess amount of wax in front of it. The pigging schedule program will be developed based upon wax deposition modeling and will be fine-tuned as more on the wax deposition rate is understood in field operations.

15.5.3.3 Chemical Inhibition/Remediation

Wax chemical inhibitors can be divided into two different types: one is to prevent/delay the formation of wax crystals and thus reduce the WAT and also prevent the wax from depositing onto the pipe wall; the other is to decrease the wax pour point and thus delay the waxy crude solidification when the crude cools down.

The mechanisms whereby chemicals inhibit wax formation and deposition are not well understood. It seems to be accepted that with absorption of polymers and additives onto the surface of wax crystals, it is possible to keep them from agglomerating and to keep the wax crystals dispersed, thus reducing the wax deposition rate. Groffe et al. (2001) performed lab and field studies on wax chemical inhibition. They developed novel chemicals that would have an ability to interfere with the wax crystal growth mechanism and were capable of keeping or making the crystals smaller so that they may cause the settling process much slower. The chemical, if possible, would also be able to absorb onto the metal surfaces and make them oleophobic. They claimed, based upon their lab work, that the chemical they developed was able to lower the WAT of a waxy crude by almost 10°C. It was also noted that the chemical has antisticking properties and was able to reduce the amount of wax deposited onto adhering metal surface.

Wang et al. (2003a,b) tested eight different commercial wax inhibitors and found that all the inhibitors reduced the total amount of wax deposited. They noticed that the inhibitors that depress the WAT the most also are most effective in reducing wax deposition. But the inhibitors could only effectively decrease the deposition of low molecular weight wax (C34 and below) and had little effect on the deposition of high molecular weight wax (C35–C44). They claimed that although the total amount of wax formed was reduced, the absolute amount of high molecular wax was increased. Thus, the net effect of many of the commercial wax inhibitors is to form harder wax which will be more difficult to remove.

With waxy crudes, when the temperature is lowered, the wax crystallizes gradually in the form of needles and thin plates. With further crystallization, these needles and thin plates turn into three-dimensional networks and cause solidification of crude (Groffe et al., 2001). Chemicals of specific polymers and surfactants can prevent formation of these networks by retarding the growth of waxy crystals, resulting in many tiny crystals. Thus, by changing the crystal structure, the ability of wax crystals to intergrow and interlock is reduced, making the pour point of the crude lower.

15.6 ASPHALTENE DEPOSITIONS

Asphaltenes are defined as the compounds in oil that are insoluble in n-pentane or n-hexane, but soluble in toluene or benzene. That is, asphaltene solids would precipitate when excess n-pentane or n-hexane is added to the crude oil. Asphaltene solids are dark brown or black in color and, unlike waxes, will not melt. But like waxes, with changes in pressure, temperature, and composition, asphaltenes tend to flocculate and deposit inside reservoir formation, well tubing, and production flowlines. Mixing reservoir fluids with a different gas (injected gas or gas-lift gas) or mixing two oil streams can also induce asphaltene precipitations (Wang et al., 2003a,b).

15.6.1 Asphaltene Precipitation

The saturation of asphaltenes in crude oil is a key parameter to determine whether or not asphaltene would cause any problems. If asphaltenes are always undersaturated in crude oil, i.e., asphaltenes are stable then no precipitation will occur. On the other hand, asphaltene precipitation would occur if they are highly supersaturated in crude oil. The saturation of asphaltenes in crude oil can change from undersaturated to supersaturated if the pressure, temperature, and composition change. During oil production, temperature and pressure changes between reservoir and production pipeline can be quite significant. Similarly, fluid composition can also change significantly during production: gas can separate from the oil when the oil pressure drops below the bubble point or gas-lift gas can be injected into the oil stream. Thus, during oil production and transportation, asphaltenes precipitation inside the production system is a potential risk.

A parameter that is closely related to asphaltene stability in oil is solubility. Solubility parameters of oil and asphaltenes are key input data for most of the thermodynamic models for asphaltene phase behavior. The solubility parameter is defined as

$$\delta_s^2 = \frac{\Delta u^v}{v_m} \tag{15.5}$$

where

δ_s = solubility parameter
Δu = cohesive energy per mole (the energy change upon isothermal vaporization of one mole of liquid to the ideal gas state)
v_m = molar volume.

Solubility parameter is a measure of the cohesive energy density or the internal pressure that is exerted by molecules within a solution. When two liquids with quite different molecules are mixed together, the liquid with higher solubility parameter will tend to "squeeze" the molecules of the liquid with a lower solubility parameter out of the solution matrix, resulting in two immiscible phases. On the other hand, if two liquids with similar solubility parameters are mixed, it is more likely for the two liquids to be miscible to one another (Burke et al., 1990).

The solubility parameter of asphaltene and of crude oil will strongly affect how much asphaltene will be soluble in the crude oil. If the solubility parameter of the crude oil is similar to the solubility parameter of asphaltene, more asphaltene will be soluble in the crude. Solubility parameter is a function of temperature (Barton, 1991). Increasing the temperature tends to decrease the asphaltene solubility parameter (Buckley et al., 1998; Hirschberg et al., 1984).

The pressure effect on asphaltene solubility depends upon the pressure being above the bubble point or below the bubble point. de Boer et al. (1995) and Hirschberg et al. (1984) studied the pressure dependence of asphaltene solubility and presented similar plots of asphaltene soluble in oil as a function of pressure, as shown in Figure 15.4. When the pressure is above the bubble point, the fluid composition is constant, but with decreasing pressure, the density of crude decreases due to oil expansion, and so does asphaltene solubility as shown in Figure 15.4. The asphaltene solubility reduces to a minimum when pressure is approaching the bubble point. Below the bubble point, gases start to evolve from the live oil and the oil density increases. The asphaltene solubility also increases with decreasing pressure. The loss of light ends improves the asphaltene stability in crude oil.

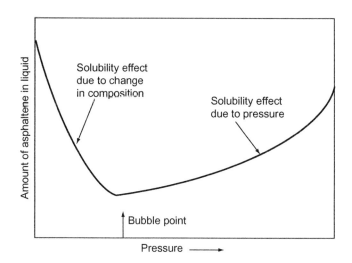

FIGURE 15.4   Pressure dependence of asphaltene solubility in crude oil.

The solubility parameter of a mixed system, like crude oil that consists of many components, can be calculated based upon the solubility parameter of the individual component (de Boer et al., 1995):

$$\delta_m = \sum_{i=1}^{n_c} f_i \delta_i \qquad (15.6)$$

where

δ_m = solubility parameter of mixed system
f_i = the volume fraction of ith component
δ_i = solubility parameter of ith component
n_c = total number of components in the system.

During production, fluid composition will change as a function of pressure and temperature. Especially when the pressure is below the bubble point, gas starts to come out of solution. Similarly, gases like CO_2 can be added to crude for enhanced oil recovery, and gases can also be added to crude through gas-lift operations. All those operations can change crude oil compositions and the crude oil solubility parameters, and thus may potentially induce asphaltene precipitation.

Assuming the asphaltene and crude oil are in equilibrium (no asphaltene precipitation), the maximum volume fraction of asphaltenes soluble in the crude is given by the Flory–Huggins theory (Burke et al., 1990; Hirschberg et al., 1984) as the following:

$$(\phi_a)_{max} = \exp\left\{\frac{V_a}{V_L}\left[1 - \frac{V_L}{V_a} - \frac{V_L}{RT}(\delta_a - \delta_L)^2\right]\right\} \qquad (15.7)$$

where

ϕ_a = volume fraction of asphaltenes in oil
V_a, V_L = molar volume of asphaltenes and liquid oil phase, respectively
δ_a, δ_L = solubility parameter of asphaltenes and liquid oil, respectively
T = temperature
R = ideal gas constant.

The properties of liquid oil (molar volume and solubility parameter) can be calculated from proper equation of state, while the properties of asphaltenes have to be estimated from experimental data.

15.6.2 Onset of Asphaltene Precipitation

Asphaltene solubility parameter can also be affected by other components in the oil, like resins (Hirschberg et al., 1984). Asphaltenes and resins are heterocompounds and form the most polar fraction of crude oil. Resins have a strong tendency to associate with asphaltenes, and they help reduce the asphaltene aggregation. On the other hand, if normal alkane (pentane or heptane) liquids are added to crude oil, the crude oil becomes lighter and resin molecules desorb from the surface of asphaltenes in an attempt to reestablish the thermodynamic equilibrium that existed in the oil before the addition of normal alkane liquids. Because of the desorption of resins, asphaltene micelles start to agglomerate to reduce overall surface free energy (Hammami et al., 2000). If sufficient quantities of normal alkane are added to the oil, the asphaltene molecules aggregate to such an extent that the particles would overcome the Brownian forces of suspension and begin to precipitate.

Hammami et al. (2000) performed experimental studies on the onset of asphaltene precipitation using a solids detection system (SDS), which consists of a visual PVT cell and fiber-optic light transmission probes. The sample oil is first charged into the PVT cell and the pressure of the cell is then lowered isothermally at programmable rates. At each equilibrium pressure, the sample volume is measured and the corresponding density is calculated. The power of the transmitted light is continuously measured. The power of transmitted light is inversely proportional to oil density. When the pressure is above the bubble point, decreasing pressure will result in reduced oil density and thus increased power of transmitted light. The power of transmitted light is also inversely proportional to the particle sizes. If particle sizes increase, as asphaltenes flocculate, the power of transmitted light will decrease. If sufficient gas bubbles exist in the oil, the power of transmitted light will decrease dramatically. Thus, the experiments would stop at the bubble point.

If the crude oil has no asphaltene precipitation and deposition problems, Hammami et al. claimed that the power of transmitted light would more or

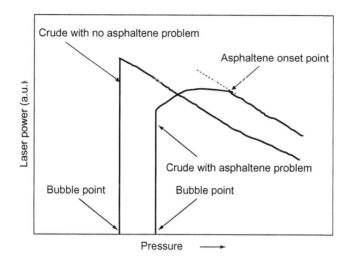

FIGURE 15.5   Power of transmitted light as a function of pressure during pressure depletion.

less linearly increase as the pressure is decreased isothermally from above the bubble point. This is due to the decrease in fluid density with decreasing pressure when the pressure is above the bubble point. When the bubble point is approached, the power of transmitted light would drop dramatically to noise level, as shown in Figure 15.5. If the crude oil has asphaltene precipitation and deposition problems, the trend of the power of the transmitted light is quite different from that of oils without asphaltene deposition problems as pressure is reduced. As the pressure is reduced from above the bubble point, the power of transmitted light would initially increase more or less linearly and then it would reach a plateau with further pressure reduction. After the plateau, the power of transmitted light would gradually decrease and eventually drop to the noise level when the bubble point is reached, as shown in Figure 15.5. The pressure at which the power of transmitted light is deviated from the straight line is defined as the onset point of asphaltene precipitation.

Improved prediction of the onset of asphaltene precipitation may be achieved using refractive index (RI) to characterize crude oils and their mixtures with precipitates and solvents (Buckley et al., 1998; Wang et al., 2003a, b). The RI is calculated based upon the measurement of the total internal reflection angle (critical angle) as shown in Figure 15.6 and is expressed as

$$RI = \frac{1}{\sin \theta_c} \tag{15.8}$$

RI is a function of fluid composition and density. For different fluids, the RI will be different. Based upon experimental studies, Buckley et al. (1998) noticed that the onset of asphaltene precipitation occurred at a characteristic

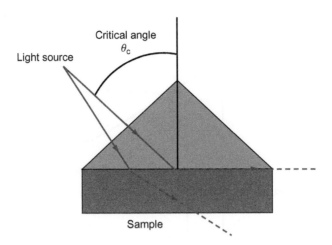

FIGURE 15.6   Schematic diagram of refractometer.

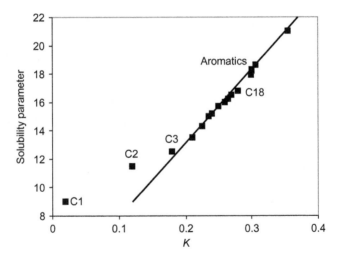

FIGURE 15.7   Relationship between solubility parameter and RI ($K = RI^2 - 1/RI^2 + 2$) for n-alkanes and aromatics *(from Buckley et al., 1998)*.

RI, and the RI can be correlated with the solubility parameter as shown in Figure 15.7. Based upon Figure 15.7, for most of the normal alkanes and some aromatics, the relationship between solubility parameters and ($RI^2 - 1/RI^2 + 2$) is more or less linear.

For a mixture of crude oil and precipitant (like n-alkane), the RI of the mixture is between the RI of the crude oil and the RI of the precipitant, and can be determined based on the RIs and the volume fractions of the crude oil and the precipitant. The precipitation of asphaltenes will occur only when the mixture RI is below a critical RI called P_{RI}. By studying the mixtures of crude oil and n-heptane for 10 different oils, Buckley *et al.* found that the mixture RI ranged between 1.47 and 1.57, while the P_{RI} was more or less a constant of about 1.44. The asphaltene content in those sample oils ranged between 1.2 and 10.9 wt% and no correlation between the asphaltene content and either RI or P_{RI} was found.

With the asphaltene onset RI (P_{RI}) known, whether or not a crude oil would have asphaltene precipitation problems at a given pressure and temperature condition can be determined by measuring the corresponding RI. If the measured RI is larger than P_{RI}, there will be no precipitation risk. On the other hand, if the measured RI is smaller than P_{RI}, asphaltene precipitation is possible. Unfortunately, the RI of live oil under pressure and temperature cannot be easily measured.

The RI of live crude oils under pressure can be estimated from the RI of stock tank oil and gas–oil ratio (Buckley et al., 1998) as

$$\left(\frac{RI^2 - 1}{RI^2 + 2}\right)(p) = \frac{1}{B_o}\left(\frac{RI^2 - 1}{RI^2 + 2}\right)_{STO} + 7.52 \times 10^{-6}\frac{R_s}{B_o}\sum_{i=1}^{m} x_i R_i \qquad (15.9)$$

where

STO = stock tank oil
B_o = formation volume factor
R_s = gas–oil ratio
x_i = mole fraction of ith component in the gas
R_i = molar refraction of ith component in the gas.

Figure 15.8 shows how the RI of live oil may change with pressure. For pressure above the bubble point, the RI of live oil decreases with pressure. The RI of live oil will reach a minimum around the bubble point. Below the bubble point, RI increases with decreasing pressure. When the RI is below the P_{RI}, asphaltenes become unstable and flocculation may occur.

15.6.3 Screening of Crude Oils for Asphaltene Precipitation—de Boer Plot

The above-mentioned RI measurement and transmitted light power measurements are sound experimental methods for detecting the onset of asphaltenes precipitation. But these tests take time. In 1992, de Boer et al. published a simple method for screening crude oils on their tendency for asphaltene precipitation (de Boer et al., 1995). By correlating crude properties, like solubility parameter, molar volume, and asphaltene solubility in oil, with the density of the crude at *in situ* conditions, de Boer et al. was able to develop an asphaltene supersaturation plot, called de Boer plot which has the difference

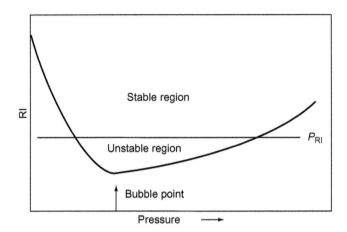

FIGURE 15.8   Live oil RI changes as a function of pressure.

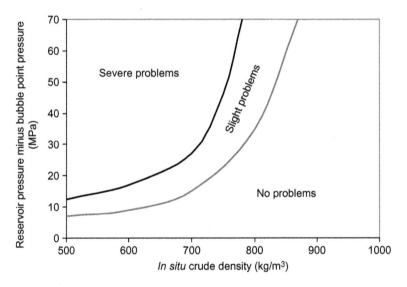

FIGURE 15.9   de Boer crude oil supersaturation plot (based upon de Boer et al., 1995).

of reservoir pressure and bubble point pressure as the *y*-axis and the *in situ* crude density as the *x*-axis. A simplified de Boer plot is shown in Figure 15.9. For given crude and reservoir conditions, the difference of reservoir pressure and bubble point pressure and the crude densities at reservoir conditions can be calculated. Then, Figure 15.9 can be used to quickly assess the risk level of asphaltene precipitation during production.

Based upon field experiences, de Boer et al. (1995) concluded that asphaltene deposition problems are encountered with light crude oils that are high in C1−C3 (>37 mole%) and have a relatively low C7+ content (<46 mole%), with high bubble point pressure (>10 MPa) and a large difference between reservoir and bubble point pressures. The asphaltene content in those light oils is low (<0.5 wt%). The heavy crude oils that have high asphaltene content tend to have fewer problems with asphaltene deposition. This is especially true if the reservoir pressure is close to the bubble point pressure.

15.6.4 Asphaltene Prevention and Remediation

Two kinds of methods are currently being used for asphaltene remediation. One is a mechanical method, which includes pigging, coiled tubing operations, and wireline cutting. The other is to use chemical solvents to dissolve deposited asphaltenes. Chemical inhibitors that are used to prevent asphaltenes from deposition in a production system include pipelines and wellbores.

Pigging can be used to remove asphaltenes inside manifolds and pipelines, provided the manifold and pipeline system can handle pigs. Compared to waxes, asphaltenes are brittle and hard, and thus special pigs are required. Pigs with disks and cups should be used; spheres and foam pigs will not be efficient for removing asphaltene solids. For a successful pigging operation, pigging frequency is important. If the time between pigging operations is too long, too much asphaltene depositions can occur inside the pipeline. Excessive asphaltene deposition can cause pigs to get stuck. On the other hand, pigging operations often require production shutdown and unnecessary pigging operations will result in production loss. Since there is no reliable way to predict how much asphaltenes will deposit inside a pipeline with time, pigging frequency can be optimized only by learning system behavior. It is safest to start at high pigging frequency and monitor the amount of asphaltene solids removed. Once the system behavior is better understood, the pigging frequency can be optimized.

Wireline cutting can be used to remove asphaltene solids inside the wellbore, provided the wellbore can be easily accessed. Coiled tubing systems can be used to remove asphaltene solids inside the wellbore and pipelines. The limitation with coiled tubing is that if the solid deposition is too far away from the coiled tubing deployment point, coiled tubing cannot be used.

Even though asphaltenes are not soluble in alkanes, they are very soluble in aromatic solvents, like benzene. Products of aromatics and alcohol mixtures are available from chemical companies that can be used to remove asphaltene deposits. It is always critical to test the chemicals first to assess their effectiveness for a specific deposit. Chemical solvents are often squeezed into formation to fight asphaltene deposit problems near the wellbore region, which cannot be easily accessed by mechanical means.

Chemicals, like blends of aromatics, surfactants, and oil and water soluble alcohols, are developed to inhibit asphaltene precipitation. Some of the chemicals would increase the surface tension of crude oils and thus help keep asphaltenes from precipitating in the treated hydrocarbon. Some of the chemicals would help supply resins in the oil to stabilize the asphaltene molecules. Again, whether or not an inhibitor is effective for a specific asphaltene problem can be determined only by testing. It is often true that a product effective for dissolving deposits may not be a good inhibitor.

15.7 INORGANIC PRECIPITATES—SCALES

Waxes and asphaltenes are precipitates from crude oils. In this section, potential precipitates from water (inorganic precipitates—called scales) will be discussed. Like wax and asphaltene depositions, scales can potentially cause serious flow assurance concerns by plugging production facilities and topsides processing devices, jamming control valves, and restricting flows in tubing and pipelines. Scales can also form inside the formation and can potentially reduce productivity by plugging the formation. Thus, it is important to understand how scales are formed and how to mitigate the scale problems.

15.7.1 Fundamental Concepts

15.7.1.1 Common Scales

The most common scales occurring in the oil industry are calcium carbonate, barium sulfate, strontium sulfate, and calcium sulfate.

Calcium carbonate ($CaCO_3$) is also called calcite scale. Calcite scale is formed when the calcium ion is combined with the carbonate ion:

$$Ca^{+2} + CO_3^{-2} \rightarrow CaCO_3 \downarrow \qquad (15.10)$$

where Ca^{+2} = calcium ion and CO_3^{-2} = carbonate ion.

Calcium carbonate is a solid and can potentially precipitate from solution to form scale. Similarly, when the calcium ion is combined with the bicarbonate ion, calcium carbonate will also be formed:

$$Ca^{+2} + 2(HCO_3^{-1}) \rightarrow CaCO_3 \downarrow + CO_2 + H_2O \qquad (15.11)$$

where HCO_3^{-1} = bicarbonate ion.

Barium sulfate is formed when the barium ion is combined with the sulfate ion:

$$Ba^{+2} + SO_4^{-2} \rightarrow BaSO_4 \downarrow \qquad (15.12)$$

where Ba^{+2} = barium ion and SO_4^{-2} = sulfate ion.

Like calcium carbonate, barium sulfate is a solid and can potentially precipitate from solution to form scale.

Strontium sulfate is formed when the strontium ion is combined with the sulfate ion:

$$Sr^{+2} + SO_4^{-2} \rightarrow SrSO_4 \downarrow \qquad (15.13)$$

where Sr^{+2} = strontium ion.

Calcium sulfate can precipitate from water if the calcium ion is combined with the sulfate ion:

$$Ca^{+2} + SO_4^{-2} \rightarrow CaSO_4 \downarrow \qquad (15.14)$$

Calcium sulfate scales include anhydrite ($CaSO_4$) and gypsum ($CaSO_4 \bullet 2H_2O$).

Carbonate scales tend to form from formation waters with reduced pressure, increased temperature, and/or increased pH value. Sulfate scales tend to form when formation waters mix with seawater because seawaters normally have high sulfate concentrations.

Some less common scales, like calcium fluoride (CaF_2), are reported in the literature (Yuan et al., 2003). Calcium fluoride is extremely insoluble and there are not currently many chemicals available to effectively treat it.

15.7.1.2 Solubility

Solubility is a parameter used to assess how much a substance can stay in a solution without precipitation and is defined as the maximum amount of a solute that can be dissolved in a solvent under given physical conditions (pressure, temperature, pH, etc.). The higher the solubility of a compound, the larger the amount of the compound that can dissolve in a solution. The solubility of a compound can change when pressure, temperature, and/or compositions change. Different compounds have different solubility. It is well known that the solubility of calcium carbonate, barium sulfate, strontium sulfate, or calcium sulfate in water is relatively small. That is why these compounds tend to precipitate from water to form scales.

15.7.1.3 Saturation Ratio

Saturation ratio (SR) is defined as the ratio of the ion product to the ion product at saturation conditions. For example, for calcium carbonate ($CaCO_3$),

$$SR = \frac{C_{Ca^{+2}} \times C_{CO_3^{-2}}}{(C_{Ca^{+2}} \times C_{CO_3^{-2}})_{saturation}} \qquad (15.15)$$

where

$C_{Ca^{+2}}$ = concentration of Ca^{+2} in solution

$C_{CO_3^{-2}}$ = concentration of CO_3^{-2} in solution.

For a given solution:

$SR = 1$, the solution is saturated with $CaCO_3$.
$SR < 1$, the solution is undersaturated with $CaCO_3$ and precipitation will not occur.
$SR > 1$, the solution is supersaturated with $CaCO_3$ and precipitation can potentially occur.

A concept that is used more often than SR is called saturation index (SI) which is defined as

$$SI = \log_{10}(SR) \qquad (15.16)$$

and if

$SI < 0$, the scaling ions are undersaturated in the solution at the given condition and no scale precipitation.
$SI = 0$, the scaling ions are at equilibrium in the solution.
$SI > 0$, the scaling ions are supersaturated in the solution at the given condition and scale precipitation is possible.

15.7.2 Factors Affecting Scale Precipitation

The major factors affecting the scale precipitation from water are pressure, temperature, pH value, and dissolved solids in water. Table 15.1 summarizes the impacts of these factors for the common scales in the oil industry (Carlberg and Matches, 1973; Jacues and Bourland, 1983; Kan et al., 2001; Rosario and Bezerra, 2001).

Even though the main reasons for carbonate scales to form inside wellbore are pressure drop inside tubing (increased pH due to the escape of CO_2) and high downhole temperature, the main cause for sulfate scales to form is the mixing of different waters. Mixing waters from different fields, from different wells at the same field, from different laterals in the same well, and mixing of formation water and seawater can potentially induce scales to form in the production facilities.

Scale is one of the major flow assurance concerns in deepwater production. There are large pressure and temperature changes throughout the production system (from reservoir to topsides). These pressure and temperature changes may induce scales. Fluids with formation water from different formations and wells are normally mixed in the production pipeline system. Different formation waters may have different compositions, and scales may form when these waters are mixed. For fields where seawater is injected for pressure maintenance, scales may become serious when different seawater factions are being produced into the production system.

TABLE 15.1 Summary of Major Factors Impacting Scale Precipitations

Scales	Temperature Effects	Pressure Effects	pH Value Effects	Dissolved Solids Effects
	Less soluble with increased temperature.	Less soluble with reduced pressure.	Less soluble with increased pH value.	Less soluble with reduced total dissolved solids in water.
Calcium carbonate	More likely scale will form with hotter water.	If waters go through the bubble point, CO_2 would evolve from solution and scale likely to form.		Adding salts into distilled water can increase the solubility.
Barium sulfate	For common temperature range, solubility increase with increased temperature.	Less soluble with reduced pressure.	Little impact.	More soluble with increased dissolved salt.
Strontium sulfate	Less soluble with increased temperature.	Less soluble in NaCl brines with reduced pressure.	Little impact.	More soluble with increased NaCl content.
Calcium sulfate	Less soluble with increased temperature for the common reservoir temperature range.	Less soluble with reduced pressure.	Little impact.	More soluble with increased water salinity.

15.7.3 Scale Prevention and Control

The main means of scale control is chemical inhibition, which includes both continuous chemical injection and periodic scale squeeze into formation. Scale inhibitors prevent scale deposition and they do not normally redissolve the deposits that have already formed. So the key function of a scale inhibitor is prevention, not remedy. Scale control strategies can be different at different stages of field life (Jordan et al., 2001). At early life, only connate water or aquifer water breaks through. The most likely scales will be carbonate scales which will be the main focus for scale control strategy. Scale severity will increase with increased water cut. If seawater is injected at later field life, sulfate scales can be formed when the injected seawater breaks through and mixes with formation water. Strategies at this stage would include controlling both carbonate and sulfate scales. With production, the seawater faction in the produced water will increase with time, and the severity of sulfate scales will change accordingly.

When chemical inhibitors are used for scale control, inhibitors will work with one or more of the following three main mechanisms (Graham et al., 1997; Yuan, 2002, Personal communication):

1. Crystal nucleation inhibition
2. Crystal growth retardation
3. Dispersion of small-scale crystals in the flowing fluid.

An inhibitor molecule works against crystal nucleation by interacting directly with the scaling ions in the brine and thus prevents the ions from agglomerating into nuclei. Inhibitor molecules can also retard crystal growth by either adsorbing onto the crystal surface (the growth sites) or fitting into the crystal lattice to replace one of the scaling ions (usually the anion). By doing so, it distorts the crystal lattice or the growth steps thus preventing the crystal from growing rapidly in a regular morphology.

If small-scale crystals have already formed in solution, an inhibitor may also prevent the crystals from adhering to each other and to other surfaces by dispersing them in the fluid. The small crystals are hence carried along with the fluid, and scale deposition is minimized. A particular inhibitor often inhibits scale formation with a primary inhibition mode. Some are better at exhibiting one particular inhibition mechanism than the other.

Testing and selecting the right inhibitor for a given scale problem are very critical for successful scale control (Graham et al., 2002; Jordan et al., 2001; Rosario and Bezerra, 2001; Yuan, 2003). The most important step for screening an inhibitor is water sampling. With representative samples available, water chemistry data which is the most important information needed to diagnose and analyze the scaling potential of produced waters can be obtained. Water chemistry data include concentrations of ions (anions and cations, like calcium, barium, strontium, bicarbonate, and sulfate) and pH. Accurate chemistry data of the produced water under system conditions (*in situ*), along with system information such as production data, temperature, and pressure as well as gas composition are essential for assessing scale risks and for testing inhibitors.

Obtaining representative water samples requires good practices. For a new oil/gas field, original formation water samples should be collected. Water samples must be preserved and stabilized at the time of sampling. Samples without preservation often go through changes including precipitation of scaling ions, evolution of carbon dioxide (CO_2), and pH drift. If a sample is collected without using a pressurized container, pH and bicarbonate should be determined immediately on-site. This is because both will drift rapidly, resulting in CO_2 evolution from the solution. It is also important to determine whether or not the samples have been contaminated by drilling muds and completion fluids before performing analysis. Finally, the water chemistry data should be reviewed by an expert to ensure the quality.

Once water chemistry data is available, the scale prediction can be performed using simulation packages. There are a few commercial simulation packages available (Kan et al., 2001; Rosario and Bezerra, 2001; Yuan et al., 2003). Based upon the simulations, the nature of scale and potential amount of scale that will precipitate can be assessed. And proper scale control technologies/strategies can be eventually developed. A very successful story on how to develop a new scale inhibitor for a specific field problem in the Gulf of Mexico was reported by Yuan et al. (2003).

A successful scale inhibitor has to have the following properties:

- It must inhibit scale formation at threshold inhibitor levels under a range of brine, temperature, and pressure conditions.
- It should have good compatibility with the produced water to avoid the formation of solids and/or suspensions. Some scale inhibitors will react with calcium, magnesium, or barium ions to form insoluble compounds which can precipitate to form scales, thus creating new problems.
- It should have good compatibility with the valves, wellbore, and flowline materials, e.g., low corrosivity on metals. Thus, a corrosivity test is necessary.
- It should be compatible with other chemicals, like corrosion inhibitors, wax inhibitors, and hydrate inhibitors. The scale inhibitor should be physically compatible with other chemicals so no solids will form. The scale inhibitor should also be compatible with other chemicals so their individual performance will not be significantly interfered. It was reported by Kan et al. (2001) that hydrate inhibitors (methanol and glycols) would impact the solubility of sulfate minerals and the effectiveness of scale inhibitors may be impacted.
- It must be thermally stable under the application temperature and within the residence time. This can be challenging for some fields with high formation temperatures.
- Its residuals in produced brine must be detectable for monitoring purposes.

For controlling scales in wellbore and in pipelines, scale inhibitor is required to be injected continuously so it can inhibit the growth of each scale crystal as it precipitates from the water. To have the maximum effectiveness for inhibiting further growth, scale inhibitor must be present in the water upstream of the point where scale precipitation occurs. That is why in a lot of cases, scale inhibitor is injected at the bottom of the wellbore.

If scale is a risk in formation, especially near the wellbore region, it is not practical to continuously inject inhibitor into formation. Scale squeeze operations to bullhead inhibitor into formation are required. Scale squeeze has been used extensively in North Sea fields for quite a long time (Graham et al., 2002) and is relatively new in South Africa and Gulf of Mexico operations. Extensive literature is available on scale squeeze operations (Bourne et al., 2000; Collins et al., 1999; Graham et al., 2002; Jordan et al., 2001; Lynn and Nasr-El-Din, 2003).

If sulfate scales are due to seawater injection, an alternative scale control strategy is to partially remove the sulfate ions from injected seawater. Sulfate removal can be achieved by using a de-sulfation plant (Davis and McElhiney, 2002; Graham et al., 2002; Vu et al., 2000). Sulfate removal plants can reduce sulfate content from the typical level of 2700−3000 ppm to a value in the range of 40−120 ppm. With the much reduced sulfate ions in the seawater, the tendency for sulfate scale formation will be significantly reduced. However, even with the sulfate removal operation, scale squeeze and/or scale control in the production stream may still be required. But the squeeze frequency will be reduced.

15.7.3.1 Scale Removal

Once scales are formed in the production facilities, they can be removed either by mechanical means, such as pigging, or by dissolving using chemicals. When brush or scraper pigs are run through the pipeline, they can mechanically remove some of the scale deposits on the pipe wall. But if the deposits, which may contain scales, waxes, and/or asphaltenes, are too hard, pigging may not be very effective.

Acids can react with scales and dissolve scale deposits on the pipe wall. For calcium carbonate scales, either hydrochloric acid or chelating agents can be used. Calcium sulfate scale is not soluble to hydrochloric acid. Inorganic converters, like ammonium carbonate ($(NH_4)_2CO_3$), can be used to convert it into calcium carbonate which can then be dissolved using hydrochloric acid. Since it is quite possible that hydrocarbons can deposit on the surface of the scales and hydrocarbons can interfere with the acid reaction with the scales, it is necessary to prewash the scales using hydrocarbon solvents. Furthermore, to keep the acid from dissolving the pipe wall, a corrosive inhibitor is also necessary to be added to the acid.

15.8 CORROSION

With gas, oil, and water flowing through offshore pipeline, one of the important flow assurance issues is corrosion. This is because an aqueous phase is almost always present in the oil and gas fluids. When enough water is flowing through the pipeline, water would wet the pipe inner surface and corrosion can occur. The corrosivity of the liquid phase depends upon the concentrations of CO_2 and H_2S, temperature, pressure, flow regime, and flowrates. Corrosion can occur in subsea production systems with different forms: galvanic corrosion, pitting, cavitation, stress corrosion cracking, hydrogen embrittlement, corrosion fatigue, etc. Corrosion can result in the loss of millions of dollars if a subsea pipeline is not properly protected. Pipeline engineers need to understand corrosion fundamentals to design sound strategies that will effectively control corrosion and protect the pipelines.

15.8.1 Corrosion Fundamentals (Cramer and Covino, 1987; Fontana and Greene, 1967)

The phenomena associated with corrosion in gas, oil, and water multiphase flow are very complex, involving the chemistry of the produced fluids, the metallurgy of the pipeline material, and the multiphase flow hydraulics. During their refining process, metals absorb quite a significant amount of extra energy. Because of the extra energy, metals are unstable in aqueous environments. With the right chemical process, metals would corrode and return to their original lower energy, stable states. Different metals have different energy stored and thus have different corrosion tendency. The metals used for subsea pipeline and well tubulars are not homogeneous, and potential differences of these inhomogeneous materials are the primary causes of corrosion.

Corrosion that involves conductive water is called wet corrosion and is an electrochemical process. There are four basic fundamental elements in a corrosion process:

1. An anode
2. A cathode
3. An electrolyte
4. A conducting circuit.

Figure 15.10 shows the corrosion process. When a piece of metal is put in a conductive fluid, like water, due to the potential differences among different materials, a portion of the metal surface is easier to corrode than the rest. This portion of metal is called an anode, where metal dissolves into the conductive fluid. Thus, during corrosion, metal is lost by dissolving into solution. The chemical reaction is described as

$$Fe \rightarrow Fe^{+2} + 2e \tag{15.17}$$

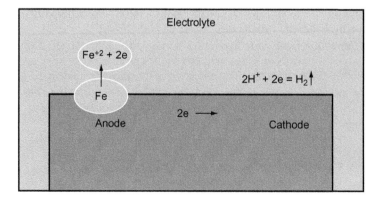

FIGURE 15.10   Schematic of the corrosion process.

where

e = electron
Fe = iron atom
Fe^{+2} = iron ion.

By losing 2 electrons, the iron atom becomes an iron ion that is positively charged. The electrons left behind would travel to another area on the metal surface, which is called the cathode where the electrons are consumed by reaction with ions in the electrolyte. For example, if the electrolyte is conductive water:

$$2H^+ + 2e \rightarrow H_2 \uparrow \qquad (15.18)$$

where

H^+ = hydrogen ion
H_2 = hydrogen gas.

To complete the electric circuit, a conductive solution to conduct current from the anode to the cathode is needed. The conductive solution is called the electrolyte. Water with dissolved solids is a good electrolyte. A path is also needed to conduct the current from the cathode to the anode. The metal itself provides the path and completes the electric circuit. Thus, the anode, the cathode, the electrolyte, and the electron conductor form the essential elements (corrosion cell) for metal corrosion.

The environment for subsea pipeline systems is favorable for formation of corrosion cells. The metals used for the pipeline system can serve as anode, cathode, and the metallic conductor between the two. The water, either produced or injected seawater, provides the electrolyte required to complete the electron circuit. Pipeline consists of dissimilar metals that may have different tendencies to corrode with the higher tendency metal to be the anode and the lower tendency metal to be the cathode. Even with the same metal, due to the inhomogeneity, one local metal area (anode) can be more corrosive than the other (cathode).

The amount of gas dissolved in water greatly impacts its corrosivity. Water with no dissolved gases will normally cause no corrosion problems. On the other hand, if gases, like oxygen, carbon dioxide, and hydrogen sulfide, are dissolved in water, the water would be very corrosive. The corrosion reactions involved with the three gases can be expressed as the following:

For carbon dioxide:
At the anode

$$Fe \rightarrow Fe^{+2} + 2e \qquad (15.19)$$

At the cathode

$$CO^2 + H_2O \rightarrow H^+ + HCO_3^- \rightarrow 2H^+ + CO_3^{-2} \qquad (15.20)$$

Combining the above two equations, we have

$$Fe^{+2} + CO_3^{-2} = FeCO_3 \qquad (15.21)$$

For oxygen:
At the anode

$$Fe \rightarrow Fe^{+2} + 2e \qquad (15.19)$$

At the cathode

$$O_2 + 2H_2O \rightarrow 4OH^- \qquad (15.22)$$

Combining the two equations:

$$4Fe^{+2} + 6H_2O + 3O_2 \rightarrow 4Fe(OH)_3 \rightarrow Fe_2O_3 + 3H_2O \qquad (15.23)$$

For hydrogen sulfide
At the anode

$$Fe \rightarrow Fe^{+2} + 2e \qquad (15.19)$$

At the cathode

$$H^2S + H_2O \rightarrow H^+ + HS^- + H_2O \rightarrow 2H^+ + S^{-2} + OH^- \qquad (15.24)$$

By combining the above two equations, we have

$$Fe^{+2} + S^{-2} \rightarrow FeS \qquad (15.25)$$

From the above discussion, it is clear that there are a few parameters that would control the corrosion reactions: the reactions at the cathode and anode, the flow of electrons from the anode to the cathode, and the conductivity of the electrolyte. These controlling factors are the main parameters dealt with in almost all corrosion prediction models (de Waard and Lotz, 1993; Nesic et al., 1995). If the reactions at both anode and cathode can be reduced, e.g., by using corrosion inhibitors to slow down the ion transport in the electrolyte, the corrosion rate will be slowed down. Similarly, if the electrolyte is less conductive due to the removal of dissolved oxygen, carbon dioxide, or hydrogen sulfide, the corrosion rate will also be reduced. The conductivity of the electrolyte can be reduced by adding chemicals to increase the pH value of the electrolyte. These are the methods that are widely used in the industry for corrosion control (Kolts et al., 1999; Strommen, 2002).

15.8.2 Corrosion Forms

Corrosion can occur in different forms and can be caused by a variety of different reasons.

15.8.2.1 Pitting Corrosion

Pitting corrosion is formed when localized holes or cavities are created in the material due to metal loss. Pitting corrosion can occur if protective film is not uniformly applied to the metal surface. Poorly applied film areas are more easily corroded. Pitting corrosion is very disastrous because it is difficult to detect. One single pit can cause material fatigue, stress corrosion cracking, and may even cause catastrophic failure of subsea pipelines.

15.8.2.2 Galvanic Corrosion

Galvanic corrosion is referred to as the corrosion due to two dissimilar materials coupled in a conductive electrolyte. With galvanic corrosion, one metal which is generally more corrosive becomes the anode, while the less corrosive one becomes the cathode. The anode metal in galvanic corrosion will corrode more rapidly than it would alone, and the cathode metal will corrode more slowly than it would alone. The larger the potential difference between the two metals, the more rapidly the anode will corrode. A very important factor that would impact galvanic corrosion is the ratio of the exposed area of the cathode to the exposed area of the anode. A small anode to cathode area ratio is highly undesirable. Under this condition, current is concentrated in a small anodic area and rapid loss of the dissolving anode will occur. Galvanic corrosion principles can be used favorably to protect the important system by scarifying a dedicated system that will corrode away. This principle is used in the so-called cathodic protection systems where steel is connected to a more corrosive metal, like zinc, and is protected. The steel is the cathode and the zinc is the anode.

15.8.2.3 Cavitation Corrosion

Cavitation occurs when vapor bubbles are repeatedly formed and subsequently collapsed in a liquid on a metal surface. The explosive forces associated with the bubble collapses can damage any protective films and result in increased local corrosion. Cavitation can also cause mechanical damage to system parts, like pump impellers. Cavitation is less likely to occur in offshore pipelines.

15.8.2.4 Hydrogen Attacks

In sour systems, hydrogen can diffuse into metal to fill any voids that may exist in the material. As corrosion continues, hydrogen atoms continue to diffuse into the voids to form hydrogen molecules, increasing the pressure in the voids. Depending upon the hardness of the material, the voids would develop into blisters, which is called hydrogen blistering, or into cracks which is called sulfide stress cracking. Due to stress cracking, materials can fail at stress levels below their yield strength. If materials contain elongated defects that are parallel to the surface, hydrogen can get into the defects and create cracks. Once the cracks on different planes inside the metal are connected, the effective wall thickness is reduced. This kind of hydrogen attack is called hydrogen-induced cracking. Crolet and Adam (2000) reported a form of hydrogen cracking called stress-oriented, hydrogen-induced cracking (SOHIC). SOHIC, which is a hybrid of sulfide stress cracking and hydrogen-induced cracking, is found to be associated with refining in the vicinity of welds that are not stress-relieved. After shutdown, subsea pipelines would experience much colder temperatures compared to normal operation temperatures. The reduced temperature causes thermal contraction and results in increased tensile stress. If the welds were done with imperfections, the welds tend to experience localized corrosion.

15.8.3 Corrosion Control

There are a few methods available for the corrosion control of subsea pipelines: using CRAs (corrosion resistant alloys) instead of carbon steel, applying corrosion inhibitors, isolating the metal from the electrolyte, and using cathodic protection. One or more of these methods may be used together. Cathodic protection and chemical inhibition can both be used to protect a subsea pipeline.

CRA steel is often used to replace carbon steel for corrosive applications. But CRAs are normally more expensive than carbon steel. Thus, whether or not CRAs should be used depends upon overall economics. But in subsea application, the very critical, high-impact components, like trees, jumpers, and manifolds, are often made of CRAs. But pipelines, especially long pipelines, are often made of carbon steel and continuous corrosion inhibitor injection is utilized to protect the pipeline.

Corrosion inhibitors are chemicals that, when added to an environment, would effectively reduce the corrosion rate of a metal that is exposed to that environment. Corrosion inhibitors would react with metal surfaces and adhere to the inner surface of the pipeline and protect the pipe from corrosion. The active compounds in the inhibitor help form a film layer of inhibitor on the surface and prevent the water from touching the pipe wall. A minimum inhibitor concentration is required to provide high inhibition efficiency. But inhibitor concentration that is significantly higher than the minimum required concentration provides little or no additional benefits. Some inhibitors can also slow down the diffusion process of ions to the metal surface and increase the electrical resistance. For example, some specific inhibitors can help slow down the reaction at cathode by forming a deposit layer on the cathode area to increase the resistance and limit the ion diffusion process.

In order for the inhibitor to be distributed evenly around the inner pipe wall perimeter, the fluids inside the pipeline must have a certain high flowing velocity. If the fluid velocity is too low, the inhibitor may not be able to reach the upper portion of the pipe wall and the inhibitor will only form a protective film around the lower portion of pipe wall. On the other hand, if the velocity is too high and causes high near wall shear stress, the protective film may be removed from the pipe wall. For smooth pipeline, the efficiency of corrosion inhibitors can be as high as 85−95%, but can drop if the shear stress increases drastically at locations such as fittings, valves, chokes, bends, and weld beads. These irregular geometries cause flow separation and reattachment of the flow boundary and increase the rates of turbulence.

In gas/condensate pipeline, adding the hydrate inhibitors, like glycol or methanol, can also help reduce the corrosion rate (Strommen, 2002). This is because the hydrate inhibitors absorb free water and make the water phase less corrosive.

The protective layer to isolate the pipe wall from water can also be achieved by using plastic coating and plastic liners. Water injection pipelines and well tubings often use plastic liners to control corrosion problems.

As we discussed in the previous section, one element of the corrosion cell is the current flow. If we stop the current flow from the anode to the cathode, the corrosion is stopped. This is the principle of cathodic protection, which is one of the widely used corrosion control methods in subsea pipeline. The key for the cathodic protection to work is to provide enough current from an external source to overpower the natural current flow. As long as there is no net current flow from the pipeline, there is no pipeline corrosion.

It is well known that different metals have different tendencies for corrosion in seawater. By connecting more corrosive metal to the subsea pipeline, the pipeline metal is forced to be the cathode while the more corrosive metal is the anode, which will corrode away. In this way, pipeline corrosion is significantly reduced. The galvanic anodes used in cathodic protection are usually made of alloys of magnesium, zinc, or aluminum which are much more active in seawater than in steel pipeline.

With proper design, cathodic protection is one of the most reliable corrosion control methods.

15.9 SEVERE SLUGGING

One of the flow assurance issues in subsea pipeline is related to production delivery. From the processing point of view, it is always desirable that the fluids coming from the pipeline are stable in composition and in flow. If the flow arriving topside is not stable, the processing system may experience upsets that often result in shutdown of the whole subsea production system (Song and Peoples, 2003).

A typical subsea production system usually consists of subsea pipeline and the production riser. Depending upon water depth, the riser length ranges from less than one hundred feet, as in shallow water production systems, to a few thousand feet, as in deepwater production systems. With a longer production riser, more energy will be required to deliver stable flow to the processing system. For deepwater production fields, especially at the late field life stage when the reservoir pressure is low and the total production rate is reduced, the gas and liquid velocities in the pipeline are not high enough to continuously carry the fluids out of the riser, resulting in intermittent (unstable) fluid delivery to the processing devices.

When the liquid cannot be continuously produced out of the riser, the liquid will accumulate at the riser base to form a liquid column, called liquid slug. The liquid slug will completely block the gas flow. When the gas pressure behind the slug is high enough, the liquid slug will be pushed out of the riser, resulting in a huge amount of liquid flowing into the processing separator. This phenomenon is called severe slugging. The liquid slug with little or no gas in it would often cause upsets (like high liquid level) in the separator if the separator and its control system are not adequately designed. When liquid slug is being produced, there is little or no gas flowing to the compressors. This can cause compression system problems.

15.9.1 Severe Slugging Description

Typical severe slugging would occur in a pipeline riser configuration as shown in Figure 15.11. The pipeline section coupled with the riser is normally inclined downward. The pipeline sections upstream of the downward inclined section may be upward inclined, horizontal, or downward inclined.

When gas and liquid flowrates are low, the liquid cannot continuously flow out of the riser and start to fall back from the riser and accumulate at the riser base, as shown in Figure 15.11A. This stage is called severe slugging formation or severe slugging generation. During this stage, there is almost no liquid and gas production and no fluid flowing into the separator. While liquid is accumulating at the riser base, gas and liquids are continuously flowing into the riser base from the pipeline inlet. Thus, the liquid column or slug formed at the riser base would continue to grow into the riser and also grow against the flow direction into the pipeline. Depending upon the GOR and other parameters, like system pressure and temperature, the slug inside the pipeline can be a few times longer than the riser height. Since the liquid slug prevents the gas from entering the riser, the pressure behind the slug is building up by the gas flow.

As more and more liquid accumulates at the riser base, the liquid slug would eventually reach the riser top and start to produce the liquid slug, as

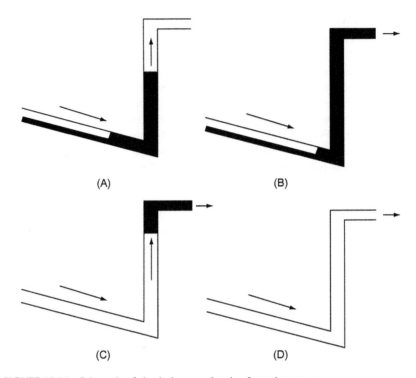

(A)   (B)

(C)   (D)

FIGURE 15.11   Schematic of classical severe slugging formation process.

shown in Figure 15.11B. This stage is called slug production. During this phase, the liquid is producing into the separator at high velocity and little or no gas is being produced. Because of the high-velocity liquid flow and huge amount of liquid associated with the slug, the separator may reach high liquid level and cause trips or upsets.

When the liquid slug is being produced, the gas will eventually enter the riser, as shown in Figure 15.11C. When gas enters the riser, the difference between the separator pressure and the gas pressure behind the slug becomes higher and higher as the liquid slug becomes shorter and shorter inside the riser. The liquid slug is being produced or being pushed by the gas at higher and higher velocity. This stage is called liquid blowout.

After the liquid slug is produced, the huge gas pocket behind the slug will be blown out of the riser and charge into the separator. This stage is called gas blowdown (Figure 15.11D). During this stage, little or no liquid would flow into the separator, and low liquid level may be reached that would eventually cause system upsets and shutdown. The gas blowdown stage in severe slugging can cause as big a problem as the slug production stage.

Once the gas is blown out of the riser, the pipeline system pressure is reduced. The liquid would again fall back from the upper portion of the riser and start to accumulate at the riser base. This is the start of the next severe slugging circle. In summary, severe slugging is undesirable because it would cause a period of no liquid and gas production followed by high liquid and gas flows, resulting in large pressure and flow fluctuations which would often cause processing device shutdown. Once the processing system is shut down, the subsea production system including trees and pipelines will have to be shut down.

15.9.2 Severe Slugging Prediction

Since severe slugging can cause production system shutdown, it is very critical to be able to correctly predict the onset of severe slugging. For a proper design of a subsea pipeline system, the multiphase flow characteristics inside the pipeline must be checked for the whole field life and the processing devices and their control systems must be designed to be able to handle the delivered flow from the pipeline. Whether or not severe slugging is a risk will significantly impact the design philosophy of the processing and control system.

Pots et al. (1987) presented a model to predict the onset of severe slugging:

$$\pi_{ss} = \frac{W_g}{W_l} \frac{ZRT}{M_g g L (1 - H_l)} \leq 1 \tag{15.26}$$

where

π_{ss} = Pots' number, dimensionless
W_g, W_l = gas and liquid mass flowrate, respectively
Z = gas compressibility factor
R = gas universal constant
T = pipeline temperature
M_g = gas molecular weight
L = pipeline length
g = accelerational gravity
H_l = average liquid holdup inside the pipeline.

Severe slugging is expected when the Pots' number is equal to or less than unity. Pots' model can be used to determine the onset of severe slugging, but the model cannot predict how long the severe slugs will be and how fast severe slugs will be produced into the separator. For subsea pipeline design, transient multiphase flow simulators are often used to determine the important flow parameters, like pressure, temperature, flow velocity, flow regime, slug frequency, and slug size.

Figures 15.12–15.14 show typical pipeline inlet pressure, outlet pressure, and outlet gas flowrate as function of time during severe slugging flow (Song and Kouba, 2000). Those are simulation results for a pipeline of 19 in.

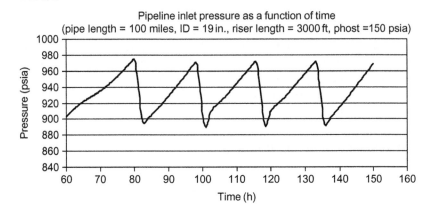

FIGURE 15.12 Pipeline inlet pressure as a function of time for severe slugging flow (GOR = 500 scf/stb, Q_o = 50 mb/d, 19 in. pipe size).

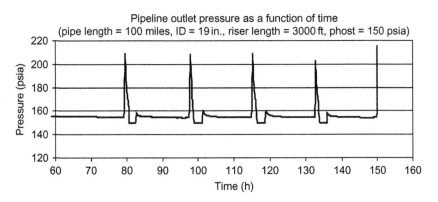

FIGURE 15.13 Pipeline outlet pressure as a function of time for severe slugging flow (GOR = 500 scf/stb, Q_o = 50 mb/d, 19 in. pipe size).

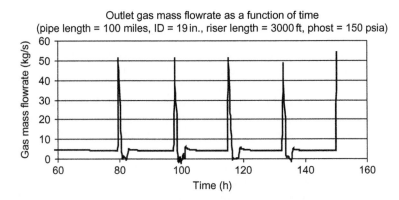

FIGURE 15.14 Outlet gas mass flowrate as a function of time for severe slugging flow (GOR = 500 scf/stb, Q_o = 50 mb/d, 19 in. pipe size).

ID with flowrate of 50 mb/d and GOR of 500 scf/stb. Figure 15.12 shows how the pipeline inlet pressure changes with time during severe slugging. The inlet pressure fluctuates between 890 and 970 psia. The severe slugging occurs once every 20 h. When the severe slugs are being pushed out from the pipeline outlet, the pipeline outlet pressure also increases as shown in Figure 15.13. Before the severe slugs are produced, the pipeline outlet pressure is about 150 psia which equals the platform pressure.

Figure 15.14 shows how the outlet gas mass flowrates change with time during severe slug flow. Before the slug is pushed out of the pipeline, the gas mass flow is a constant and the gas flowrate equals the steady-state flowrate. Once the liquid slug is pushed out, the huge gas pocket behind the liquid slug is produced, resulting in a much higher gas mass flowrate as shown in the figure. Once the gas pocket is produced, for a period of time, no gas is flowing out. The same behavior can be expected for the outlet liquid flow.

15.9.3 Severe Slugging Elimination

There are a few methods that can be used to effectively mitigate severe slugging problems.

15.9.3.1 Favorable Pipeline Bathymetry

A pipeline bathymetry is preferred if the pipeline flow is going upward. In other words, the water depth at the pipeline outlet is preferred to be shallower than that at the inlet. This is because the multiphase slug flow is much less severe with an upwardly inclined pipeline than with a downwardly inclined pipeline. Pipeline A, shown in Figure 15.15, will tend to have more severe slugging problems than Pipeline B. Thus, it is important that, at the pipeline design stage, favorable pipeline routing is chosen, if possible, to eliminate severe slugging risks.

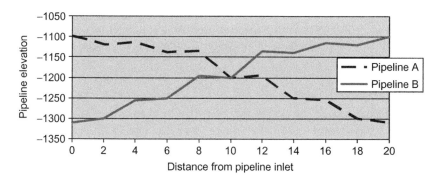

FIGURE 15.15   Upward and downward inclined pipeline profiles.

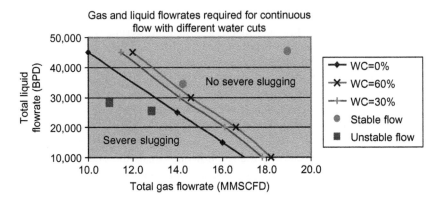

FIGURE 15.16   Required total gas flow for stable flow in flowline for different water cut.

15.9.3.2 Increasing Gas Flow

One of the main reasons that severe slugging occurs is that the gas velocity is too low to carry the liquid out of the riser. If more gas can be introduced into the pipeline riser system, the gas velocity will be increased and the gas will also help lift the liquid out of the riser by reducing the fluid mixture density. Song and Peoples (2003) reported that at a West Africa subsea field when a well is diverted from the production pipeline into the test pipeline for testing, the flow inside the production pipeline would change to severe slugging flow due to the reduced production. But if enough extra gas is added to the flowline before the well is shifted out for well testing, the gas can help stabilize the flow inside the pipeline and no severe slugging will occur. They developed a plot showing the required total liquid and total gas flowrates for continuous stable flow for different water cut, as shown in Figure 15.16. For a given liquid flowrate, if the total gas flowrate is high enough that the flowing condition is above the curve, no severe slugging will occur. Otherwise, severe slugging will occur. The individual points in the figure are field measurements. This information, together with well test data, was used by operations personnel to estimate when severe slug flow was likely. The primary mitigating measure that the operations personnel undertook to avoid slugging was to increase gas flow inside the pipeline.

15.9.3.3 Gas-Lift Riser

If enough gas can be injected at the riser base to change the flow inside the riser to hydrodynamic slug flow, churn flow, or annular flow, the severe slugging problem can be mitigated. With hydrodynamic slug flow or churn flow, the slugs are much shorter than the slugs in the severe slugging flow. Topside separators are normally sized to handle the hydrodynamic slugs and no system shutdown is likely. If sufficient gas is injected at the riser base to change the flow to annular flow, the flow will be even more stable. But to reach annular flow, a significant amount of gas will be needed and may not be practical.

15.9.3.4 Topsides Choking

Severe slugging can be mitigated by choking the flow at the top of the riser. Choking the flow would increase system pressure and make the system "stiffer." With increased system pressure, the gas becomes less compressible. Thus, when the liquid slug formed at the riser base blocks the gas flow, the gas pressure behind the liquid slug would increase more quickly and be able to push the liquid slug out of the riser faster. In this way, the liquid accumulation time is shorter and the liquid slugs will consequently be smaller. So the severe slugs are minimized. But choking would increase the system back pressure and thus reduce the overall production.

15.9.3.5 Subsea Separation

Song and Kouba (2000) performed studies on severe slugging elimination using subsea separation. It is understood that the favorable condition for severe slugs to form is that gas and liquid simultaneously flow through a long riser at low velocities. It is very difficult to change fluid velocity that is controlled by production rates and pipe size. But, it is possible to separate the gas from the liquid and let the gas and the liquid flow through two different pipelines or risers. In this way, the gas and the liquid will not be flowing simultaneously through the riser and severe slugging can, thus, be avoided. Seafloor separation becomes one of the methods that can potentially solve the severe slugging problems associated with deepwater production. Based upon their studies, Song and Kouba concluded:

- Subsea separation can help mitigate severe slugging. By separating the gas from the liquid, it is possible to eliminate severe slugs. This is especially true for riser base separation.
- Subsea separation can boost production by reducing the back pressure assuming single-phase liquid pump is used to boost the liquid.
- There is an optimum separator location for subsea applications. For certain flowrates and pipe sizes, it is more effective to put the separator at the riser base than at the wellhead.
- For the purpose of eliminating severe slugs, the requirement for the subsea separation efficiency is not very high. Based on the study, for the riser base separation, as long as the separation efficiency is higher than 75%, severe slugs can be eliminated.

One disadvantage associated with riser base separation is that two risers will be required.

REFERENCES

API RP 44, 1966. API recommended practice for sampling petroleum reservoir fluids, first ed. Am. Pet. Inst.

API RP 45, 1968. API recommended practice for analysis of oil-field waters, second ed. Am. Pet. Inst. (Dallas).

Balson, T., Craddock, H.A., Dunlop, J., Frampton, G., Payne, G., Reid, P., et al., 2002. The development of advanced kinetic hydrate inhibitors. Royal Soc. Chem. Chem. in the Oil Ind. VII. RSC Publishing, Cambridge, UK.

Barton, A.F.M., 1991. CRC Handbook of Solubility Parameters and Other Cohesive Parameters, second ed. CRC Press, Boca Raton, FL.

Bourne, H.M., Heath, S.M., McKay, S., Fraser, J., Muller, S., 2000. Effective treatment of subsea wells with a solid scale inhibitor system. In: Presented at the International Symposium on Oilfield Scales. 26−27 January 2000, Aberdeen.

Brown, T.S., Niesen, V.G., Erickson, D.D., 1993. Measurement and prediction of the kinetics of paraffin deposition. In: Presented at the SPE Annual Technical Conference and Exhibition, 3−6 October 1993, Houston, Texas, SPE Paper 26548.

Buckley, J.S., Hirasaki, G.J., Liu, Y., Von Drasek, S., Wang, J.X., Gill, B.S., 1998. Asphaltene precipitation and solvent properties of crude oils. Pet. Sci. Technol. 16 (3&4), 251−285.

Burger, E.D., Perkins, T.K., Striegler, J.H., 1981. Studies of wax deposition in the trans Alaska pipeline. J. Pet. Technol. 33, 1075−1086.

Burke, N.E., Hobbs, R.E., Kashou, S., 1990. Measurement and modeling of asphaltene precipitation. J. Pet. Technol. 42 (11), 1440−1446.

Carlberg, B.L., Matches, R.R. 1973. Solubility of calcium sulfate in brine. In: Presented at the Oilfield Chemistry Symposium, 24−25 May 1973, Denver, Colorado, SPE Paper 4353.

Chancey, D.G., 1987. Measuring, sampling, and testing crude oil (Chapter 17). In: Bradley, H.B. (Ed.), Petroleum Engineering Handbook. Society of Petroleum Engineers, Richardson, TX.

Collins, I.R., Cowie, L.G., Nicol, M., Stewart, N.J., 1999. Field application of a scale inhibitor squeeze enhancing additive. SPE Prod. Facil. 14 (1), 21−29.

Cramer, S.D., Covino, B.S., 1987. ASM Handbook Volume 13A: Corrosion: Fundamentals, Testing, and Protection. ASM International, Material Park, OH.

Creek, J.L., Matzain, A., Apte, M.S., Brill, J.P. Volk, M., Delle Case, E. et al., 1999. Mechanisms for wax deposition. In: Presented at the AIChE National Spring Meeting. Houston, TX.

Crolet, J.L., Adam, C., 2000. SOHIC without H_2S. Mater. Perform. 39 (3), 86−90.

Davis, R.A., McElhiney, J.E., 2002. The advancement of sulfate removal from seawater in offshore waterflood operations. In: Presented at the NACE Corrosion Conference, 7-11 April, 2002, Denver, Colorado, NACE International Paper 02314.

de Boer, R.B., Leerlooyer, K., Eigner, M.R.P., van Bergen, A.R.D., 1995. Screening of crude oils for asphalt precipitation: theory, practice, and the selection of inhibitors. SPE Prod. Facil.10 (1), 55−61.

de Waard, C., Lotz, U., 1993. Prediction of CO_2 corrosion of carbon steel, Presented at Corrosion/93, NACE, Houston, Texas, 1993, Paper 69.

Fontana, M.G., Greene, N.D., 1967. Corrosion Engineering. McGraw-Hill, New York, NY.

Graham, G.M., Sorbie, K.S., Jordan, M.M., 1997. How scale inhibitors work and how this affects test methodology. In: Presented at the 3rd International Conference on Advances in Solving Oilfield Scaling. Aberdeen.

Graham, G.M., Mackay, E.J., Dyer, S.J., Bourne, H.M., 2002. The challenges for scale control in deepwater production systems: chemical inhibition and placement. In: Presented at the NACE Corrosion. Denver.

Groffe, D., Groffe, P., Takhar, S., Andersen, S.I., Stenby, E.H., Lindeloff, N., et al., 2001. A wax inhibition solution to problematic fields: a chemical remediation process. Pet. Sci. Technol. 19 (1&2), 205–217.

Hammami, A., Raines, M., 1999. Paraffin deposition from crude oils: comparison of laboratory results to field data. SPE J. 4 (1), 9–18.

Hammami, A., Phelps, C.H., Monger-McClure, T., Little, T.M., 2000. Asphaltene precipitation from live oils: an experimental investigation of the onset conditions and reversibility. Energy Fuels 14 (1), 14–18.

Hirschberg, A., deJong, L.N.J., Schipper, B.A., Meijer, J.G., 1984. Influence of temperature and pressure on asphaltene flocculation. SPE J. 24 (3), 283–293.

Hsu, J.C., Elphingstone, G.M., Greenhill, K.L., 1999. Modeling of multiphase wax deposition. J. Energ. Resour. Technol. Trans. ASME 121 (2), 81–85.

Huron, M.J., Vidal, J., 1979. New mixing rules in simple equations of state for representing vapor-liquid equilibria of strongly non-ideal mixtures. Fluid Phase Equilib. 3 (4), 255–271.

Jacues, D.F., Bourland, B.I., 1983. A study of solubility of strontium sulfate. J. Pet. Technol. 23 (2), 292–300.

Jordan, M.M., Sjuraether, K., Collins, I.R., Feasey, N.D., Emmons, D., 2001. Life cycle management of scale control within subsea fields and its impact on flow assurance, Gulf of Mexico and the north sea basin. In: Proceedings of the SPE Annual Technical Conference and Exhibition. Louisiana, New Orleans, 30 September–3 October 2001, SPE Paper 71557.

Kan, A. T., Fu, G., Tomson, M.B., 2001. Mineral-scale control in subsea completion. In: Presented at the 2001 Offshore Technology Conference. Houston, TX.

Katz, D.L., Firoozabadi, A., 1978. Predicting phase behavior of condensate/crude-oil systems using methane interaction coefficients. J. Pet. Technol. 30 (11), 1649–1655.

Kelland, M.A, Svartaas, T.M., Dybvik, L.A., 1995. Studies on new gas hydrate inhibitors. In: Presented at Offshore Europe, 5–8 September 1995, Aberdeen, United Kingdom, SPE Paper 30420.

Kolts, J., Joosten, M., Salama, M., Danielson, T.J., Humble, P., Belmear, C., et al., 1999. Overview of Britannia subsea corrosion-control philosophy. In: Presented at the 1999 Offshore Technology Conference. 3–6 May 1999, Houston, TX. OTC Paper 11019.

Lervik, J.K., Kulbotten, H., Klevjer, G., 1997. Prevention of hydrate formation in pipelines by electrical methods. In: Proceedings of the Seventh International Offshore and Polar Engineering Conference. 25–30 May, 1997, Honolulu.

Lynn, J.D., Nasr-El-Din, H.A., 2003. A novel low-temperature, forced precipitation phosphonate squeeze for water sensitive, non-carbonate bearing formations. In: Presented at the SPE Annual Technology Conference and Exhibition. 5–8 October, 2003 Denver. SPE Paper 84404.

Makogon, Y.F., 1997. Hydrates of Hydrocarbons. PennWell Books, Tulsa, OK.

March, D.M., Bass, R.M., Phillips, D.K., 2003. Robust technology implementation process applied to a first deepwater electrical heating ready system. In: Presented at the 2003 Offshore Technology Conference. 5–8 May 2003. Houston, TX. OTC Paper 15145.

Mathias, P.M., Copeman, T.W., 1983. Extension of the Peng–Robinson equation of state to complex mixtures: evaluation of the various forms of the local composition concept. Fluid Phase Equilib. 13, 91–108.

Mehta, A.P., Hebert, P.B., Cadena, E.R., Weatherman, J.P., 2003. Fulfilling the promise of low-dosage hydrate inhibitors; journey from academic curiosity to successful field implementation. SPE Prod. Facil. 18 (1), 73−79.

Meray, V.R., Volle, J.L., Schranz, C.J.P., Marechal, P.L., Behar, E., 1993. Influence of light ends on the onset crystallization temperature of waxy crudes within the frame of multiphase transport. In: Presented at the SPE Annual Technical Conference and Exhibition, 3−6 October 1993, Houston, Texas, SPE Paper 26549.

Monger-McClure, T.G., Tackett, J.E., Merrill, L.S., 1999. Comparisons of cloud point measurement and paraffin prediction methods. SPE Prod. Facil. 14 (1), 4−16.

Nesic, S., Postlethwaite, J., Olsen, S., 1995. An electrochemical model for prediction of CO_2 corrosion, Presented at Corrosion/95, NACE Paper 131.

Ostrof, A.G., 1979. Introduction to Oilfield Water Technology, second ed. National Association of Corrosion Engineers, Houston, TX.

Pedersen, K.S., Thomassen, P., Fredenslund, A., 1985. Thermodynamics of petroleum mixtures containing heavy hydrocarbons. 3. Efficient flash calculation procedures using the SRK equation of state. Ind. Eng. Chem. Process Des. Dev. 24 (4), 948−954.

Pedersen, K.S., Fredenslund, A., Thomassen, P., 1989. Properties of Oils and Natural Gases. Gulf Publishing Inc, Houston, TX.

Pedersen, K.S., Blilie, A., Meisingset, K.K., 1992. PVT calculations of petroleum reservoir fluids using measured and estimated compositional data for the plus fraction. Ind. Eng. Chem. Res. 31 (5), 1378−1384.

Pedersen, K.S., Milter, J., Rasmussen, C.P., 2001. Mutual solubility of water and reservoir fluids at high temperatures and pressures, experimental and simulated phase equilibrium data. Fluid Phase Equilib. 189 (1−2), 85−97.

Peneloux, A., Rauzy, E., Fréze, R., 1982. A consistent correlation for Redlich−Kwong−Soave volumes. Fluid Phase Equilib. 8 (1), 7−23.

Peng, D.-Y., Robinson, D.B., 1976. A new two-constant equation of state. Ind. Eng. Chem. Fundam. 15 (1), 59−64.

Peng, D.-Y., Robinson, D.B., 1978. The characterization of the heptanes and heavier fractions for the GPA Peng−Robinson programs, GPA Research Report RR-28.

Pots, B.F.M, Bromilov, I.G., Konijn, M.J.W.F., 1987. Severe slug flow in offshore flowline/riser system. SPE Prod. Eng. 2 (4), 319−324.

Reid, R.C., Prausnitz, J.M., Sherwood, J.K., 1977. The Properties of Gases and Liquids. McGraw-Hill, New York, NY.

Riazi, M.R., Daubert, T.E., 1980. Prediction of the composition of petroleum fractions. Ind. Eng. Chem. Process Des. Dev. 19 (2), 289−294.

Rosario, F.F., Bezerra, M.C., 2001. Scale potential of a deep water field—water characterisation and scaling assessment. In: Presented at the 2001 SPE 3rd International Symposium on Oilfield Scale, 30−31 January 2001, Aberdeen. SPE Paper 68332.

Singh, P., Gogler, H.S., Nagarajan, N., 1999. Prediction of the wax content of the incipient wax-oil gel in a flowloop: an application of the controlled-stress rheometer. J. Rheol. 43 (6), 1437−1459.

Sloan, E.D., 1998. Clathrate Hydrates of Natural Gases, second ed. Marcel Dekker, Inc., New York, NY.

Soave, G., 1972. Equilibrium constants from a modified Redlich−Kwong equation of state. Chem. Eng. Sci. 27 (6), 1197−1203.

Song, S., Kouba, G., 2000. Fluid transport optimization using seabed separation. J. Energy Resources Tech. 122 (3),105−109.

Song, S., Peoples, K., 2003. Impacts of transient analysis on Kuito production operations. In: Presented at the 2003 Offshore Technology Conference. 5–8 May 2003, Houston, TX, OTC Paper 15186.

Sorensen, H., Pedersen, K.S., Christensen, P.L., 2002. Modeling of gas solubility in brine. Org. Geochem. 33 (6), 635–642.

Strommen, R.D., 2002. Seven years experience from subsea, deepwater pipeline internal corrosion monitoring, NACE Corrosion 2002, Paper 2251.

Tsonopoulos, C., Heidman, J.L., 1986. High-pressure vapor-liquid equilibria with cubic equations of state. Fluid Phase Equilib. 29, 391–414.

Venkatesan, R., Singh, P., Fogler, H.S., 2002. Delineating the pour point and gelation temperature of waxy crude oils. SPE J. 7 (4), 349–352.

Vu, V.K., Hurtevent, C., Davis, R.A., 2000. Eliminating the need for scale inhibition treatments for Elf exploration Angola's Girassol field. In: Presented at 2000 International Symposium on Oilfield Scale. 26–27 January 2000. Aberdeen. SPE Paper 60220.

Wang, J.X., Buckley, J.S., Burke, N.A., Creek, J.L., 2003a. Anticipating asphaltene problems offshore—a practical approach. In: Presented at the 2003 Offshore Technology Conference. Houston, TX.

Wang, K.S., Wu, C.H., Creek, J.F., Shuler, P.J., Tang, Y.C., 2003b. Evaluation of effects of selected wax inhibitors on paraffin deposition. Pet. Sci. Technol. 21 (3&4).

Weingarten., J.S., Euchner, J.A., 1986. Methods for predicting wax precipitation and deposition. In: Presented at the SPE Annual Technical Conference and Exhibition.

Yuan, M., Williamson, D.A., Smith, J.K., Lopez, T.H., 2003. Effective control of exotic mineral scales under harsh system conditions. In: Presented at the SPE International Symposium on Oilfield Chemistry. Houston, TX.

FURTHER READING

Vetter, O.J.G., Vandenbroek, I., Nayberg, J., 1983. SrSO$_4$: the basic solubility data. In: Presented at the International Symposium on Oilfield and Geothermal Chemistry. Denver.

Chapter 16

Pigging Operations

16.1 INTRODUCTION

The term pig was originally referred to Go-Devil scrapers driven through the pipeline by the flowing fluid trailing spring-loaded rakes to scrape wax off the internal walls. One of the tales about the origin of the name "pig" is that the rakes made a characteristic loud squealing noise. Pipeline operators now describe any device made to pass through a pipeline for cleaning and other purposes with the word pig. The process of driving the pig through a pipeline by fluid is called a pigging operation.

Although pigs were originally developed to remove deposits, which could obstruct or retard the flow through a pipeline, today pigs are used during all phases in the life of a pipeline for many different reasons. During pipeline construction, pigging is used for debris removing, gauging, cleaning, flooding, and dewatering. During fluid production operations, pigging is utilized for removing wax in oil pipelines, removing liquids in gas pipelines, and meter proving. Pigging is widely employed for pipeline inspection purposes such as wall thickness measurement and detection of spanning and burial. Pigging is also run for coating the inside surface of pipeline with inhibitor and providing pressure resistance during other pipeline maintenance operations. Figure 16.1 shows pipeline deposits displaced by a pig. This chapter describes how to apply different pigging techniques to solve various problems in the pipeline operations.

16.2 PIGGING SYSTEM

A pigging system includes pigs, a launcher, and a receiver. It also includes pumps and compressors, which are not discussed here because they have to be available for transporting the product fluids anyway. Obviously, pigs are the most essential equipment. Although each pipeline has its own set of characteristics that affects how and why pigging is utilized, there are basically three reasons to pig a pipeline: (1) to batch or separate dissimilar products, (2) to displace undesirable materials, and (3) to perform internal inspections. The pigs used to accomplish these tasks fall into three categories:

1. Utility pigs, which are used to perform functions such as cleaning, separating, or dewatering.

Offshore Pipelines. DOI: http://dx.doi.org/10.1016/B978-0-12-397949-0.00016-9

FIGURE 16.1   Pipeline deposits that could obstruct or retard the flow through a pipeline. *(Courtesy of Pigging Products & Services Association)*

2. In-line inspection (ILI) tools, which provide information on the condition of the line, as well as the extent and location of any problems.
3. Gel pigs, which are used in conjunction with conventional pigs to optimize pipeline dewatering, cleaning, and drying tasks.

16.2.1 Utility Pigs

Utility pigs can be divided into two groups based upon their fundamental purpose: (1) cleaning pigs used to remove solid or semisolid deposits or debris from the pipeline and (2) sealing pigs used to provide a good seal in order to either sweep liquids from the line or provide an interface between two dissimilar products within the pipeline. Within these two groups, a further subdivision can be made to differentiate among the various types or forms of pigs: spherical pigs, foam pigs, mandrel pigs, and solid cast pigs.

Spherical pigs, or spheres, are of either a solid composition or inflated to their optimum diameter with glycol and/or water. Figure 16.2 shows some spheres. Spheres have been used for many years as sealing pigs. There are four basic types of spheres: inflatable, solid, foam, and soluble. Soluble spheres are usually used in crude oil pipelines containing microcrystalline wax and paraffin inhibitor. Spheres normally dissolve in a few hours. The dissolving rate depends on fluid temperature, fluid movement, friction, and absorbability of the crude. If the line has never been pigged, it is a good idea to run the soluble pig. If it hangs up in the line, it will not obstruct the flow. Inflatable spheres are manufactured of various elastomers (polyurethane, neoprene, nitrile, and Viton) depending on their applications. An inflatable sphere has a hollow center with filling valves that are used to

FIGURE 16.2 Some spheres used in the pipeline pigging operations. *(Courtesy of Girard Industries, Inc.)*

inflate the sphere with liquid. Spheres are filled with water, or water and glycol, and inflated to the desired size. Spheres should never be inflated with air. Depending on the application and material, the sphere is inflated 1–2% over the pipe inside diameter. As the sphere wears from service, it is resized, extending its life. In small sizes, the sphere can be manufactured solid, eliminating the need to inflate it. The solid sphere does not have the life of an inflatable sphere because it cannot be resized. Spheres can also be manufactured from open-cell polyurethane foam. They can be coated with a polyurethane material to give better wear. For cleaning purposes, they can have wire brushes on the surface. The advantages of the foam sphere are that they are lightweight, economical, and do not need to be inflated. Spheres in general are easy to handle, negotiate short radius 90s, irregular turns, and bends. They go from smaller lateral lines to larger main lines and are easier to automate than other styles of pigs. Spheres are commonly used to remove liquids from wet gas systems, serve to prove fluid meters, control paraffin in crude oil pipelines, flood pipeline to conduct hydrostatic test, and dewater after pipeline rehabilitation or new construction. Special design considerations for the pipeline should be considered when using spheres. They should never be run in lines that do not have special flow tees installed.

Foam pigs, also known as Polly-Pigs, are molded from polyurethane foam with various configurations of solid polyurethane strips and/or abrasive materials permanently bonded to them. Figure 16.3 demonstrates a foam pig and how it works.

Foam pigs are molded from open-cell polyurethane foams of various densities ranging from light density (2 lb/ft^3), medium density (5–8 lb/ft^3), to heavy density (9–10 lb/ft^3). They are normally manufactured in a bullet shape. They can be bare foam or coated with a 90-durometer polyurethane material. Coated pigs may have a spiral coating of polyurethane, various brush

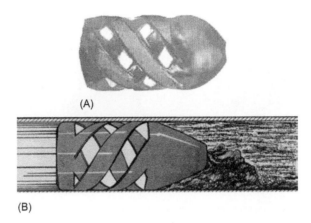

(A)

(B)

FIGURE 16.3 (A) A foam pig and (B) an ideal foam pig cleaning the pipeline. *(Courtesy of Montauk Service, Inc.)*

materials, or silicon carbide coating. If the pig is of bare foam, it will have the base coated. The standard foam pig length is twice the diameter. Foam pigs are compressible, expandable, lightweight, and flexible. They travel through multiple diameter pipelines and go around mitered bends and short radius 90° bends. They make abrupt turns in tees so laterals can be cleaned. They also go through valves with as little as a 65% opening. The disadvantages of foam pigs are that they are one-time use products, shorter length of runs, and high concentrations of some acids will shorten life. Foam pigs are also inexpensive. Foam pigs are used for pipeline proving, drying and wiping, removal of thick soft deposits, condensate removal in wet gas pipelines, and pigging multiple diameter lines. Foam pigs coated with wire brush or silicon carbide are used for scraping and mild abrasion of the pipeline.

A mandrel pig has a central body tube, or mandrel, and various components can be assembled onto the mandrel to configure a pig for a specific duty. Figure 16.4 demonstrates some mandrel pigs.

The pig is equipped with wire brushes or polyurethane blades for cleaning the line. The mandrel pig can be either a cleaning pig, sealing pig, or a combination of both. The seals and brushes can be replaced to make the pig reusable. Cleaning pigs are designed for heavy scraping and can be equipped with wire brushes or polyurethane blades. These pigs are designed for long runs. Bypass holes in the nose of the pig control the speed or act as jet ports to keep debris suspended in front of the pig. The cost of redressing the pig is high, and larger pigs require special handling equipment to load and unload the pig. Occasionally, the wire brush bristles break off and get into instrumentation and other unwanted places. Smaller size mandrel pigs do not negotiate 1.5D bends.

FIGURE 16.4 Some mandrel pigs used in pipeline pigging operations. *(Courtesy of Girard Industries, Inc.)*

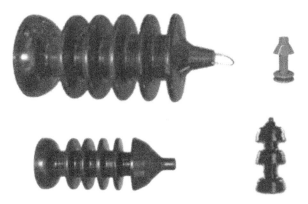

FIGURE 16.5 Some solid cast pigs used in pipeline pigging operations. *(Courtesy of Apache Pipeline Products, Inc.)*

Solid cast pigs are usually molded in one piece, usually from polyure-thane; however, neoprene, nitrile, Viton, and other rubber elastomers are available in smaller size pigs. Figure 16.5 demonstrates some solid cast pigs. Solid cast pigs are considered sealing pigs although some solid cast pigs are available with wraparound brushes and can be used for cleaning purposes. The solid cast pig is available in the cup, disc, or a combination cup/disc design. Most of the pigs are of one-piece construction but several manufac-turers have all urethane pigs with replaceable sealing elements. Because of the cost to redress a mandrel pig, many companies use the solid cast pig up through 14 or 16 in. Some solid cast designs are available in sizes up to 36 in. Solid cast pigs are extremely effective in removing liquids from prod-uct pipelines, removing condensate and water from wet gas systems, and controlling paraffin buildup in crude oil systems.

FIGURE 16.6   An ultrasonic inspection tool. *(Courtesy of Pigging Products and Services Association)*

16.2.2 ILI Tools

ILI tools are used to carry out various types of tasks including

- Measuring pipe diameter/geometry
- Monitoring pipeline curvature
- Determining pipeline profile
- Recording temperature/pressure
- Measuring bend
- Detecting metal loss/corrosion
- Performing photographic inspection
- Detecting crack
- Measuring wax deposition
- Detecting leak
- Taking product samples, and
- Mapping.

A typical ILI tool is an ultrasonic tool shown in Figure 16.6. Ultrasonic ILI tools are used for measuring metal loss and detecting cracks in pipelines. Ultrasonic tools are especially suitable if there are high requirements regarding sensitivity and accuracy, which is especially relevant in offshore pipelines. Ultrasound tools are also well suited with regard to the range of wall thicknesses usually experienced in offshore lines.

16.2.3 Gel Pigs

Gel pigs have been developed for use in pipeline operations, either during initial commissioning or as a part of a continuing maintenance program.

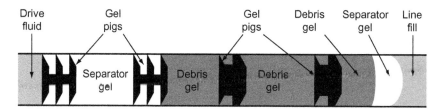

FIGURE 16.7   Application of gel pigs in pipeline pigging operations.

Figure 16.7 shows how gel pigs work. The principal pipeline applications for gel pigs are as follows:

- Product separation
- Debris removal
- Line filling/hydrotesting
- Dewatering and drying
- Condensate removal from gas lines
- Inhibitor and biocide laydown
- Special chemical treatment, and
- Removal of stuck pigs.

Most pipeline gels are water-based, but a range of chemicals, solvents, and even acids can be gelled. Some chemicals can be gelled as the bulk liquid and others only diluted in a carrier. Gelled diesel is commonly used as a carrier of corrosion inhibitor in gas lines. The four main types of gel used in pipeline applications are batching or separator gel, debris pickup gel, hydrocarbon gel, and dehydrating gel. The gel can be pumped through any line accepting liquids. Gel pigs can be used alone (in liquid lines), in place of batching pigs, or in conjunction with various types of conventional pigs. When used with conventional pigs, gelled pigs can improve the overall performance while almost eliminating the risk of sticking a pig. Gel pigs do not wear out in service like conventional pigs. They can, however, be susceptible to dilution and gas cutting. Care must be taken when designing a pig train that incorporates gel pigs to minimize fluid bypass of the pigs and to place a conventional pig at the back of the train when displacing with gas. Specially formulated gels have also been used to seal gate valves during hydrostatic testing. Gels have been developed with a controlled gelation time and a controlled viscosity for temporary pipeline isolation purposes.

16.2.4 Launcher and Receiver

Pigs generally need specially designed launching and receiving vessels (launcher and receiver) to introduce them into the pipeline. The launcher and receiver are installed at the upstream and downstream of the pipeline

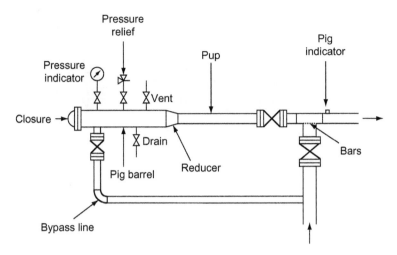

FIGURE 16.8 A typical configuration of a pig launcher for liquid services.

section being pigged, respectively. The distance between the launcher and the receiver depends on the service, location of pump (liquid product) or compressor (gas product) stations, operating procedures, and the materials used in the pig. In crude oil pipeline systems, the distance between launcher and receiver can be as long as 500 miles for spheres and 300 miles for pigs. The amount of sand, wax, and other materials carried along the pig can affect the proper distance. In gas transmission service, the distance between the launcher and the receiver can be as long as 200 miles for spheres and 100 miles for pigs, depending on the amount of lubrication used.

The launcher and receiver consist of a quick opening closure for access, an oversized barrel, a reducer, and a neck pipe for connection to the pipeline. Pigs can be located using fixed signalers along the pipe or electronic tracking systems mounted inside the pig. A typical configuration of a pig launcher for liquid service is illustrated in Figure 16.8. The horizontal barrel holds the pig for loading. Figure 16.9 shows a typical configuration of a pig receiver for liquid service. The horizontal barrel holds the pig for unloading. A barrel diameter of 2 in. larger than the diameter of pipeline has been recommended for both launchers and receivers. The barrel length should be 1.5 times the pig length and long enough to hold 10 or more spheres.

Typical configurations of pig launchers and receivers for gas service are depicted in Figures 16.10 and 16.11, respectively. The inclined barrels should be long enough to hold 10 or more spheres. In large-diameter gas pipelines, the barrel diameter can be 1 in. larger than the pipeline.

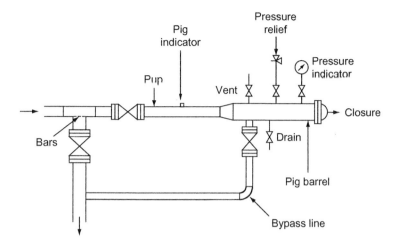

FIGURE 16.9   A typical configuration of a pig receiver for liquid services.

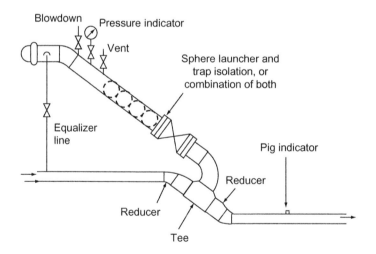

FIGURE 16.10   A typical configuration of a pig launcher for gas services.

16.3 SELECTION OF PIGS

The purpose of operational pigging is to obtain and maintain the efficiency of the pipeline. The pipeline's efficiency depends on two things: first, it must operate continuously and second, the required throughput must be obtained at the lowest operating cost. The type of pig to be used and its optimum configuration for a particular task in a particular pipeline should be determined based upon several criteria including the following:

- Purpose of pigging
  - type, location, and volume of the substance to be removed or displaced
  - type of information to be gathered from an intelligent pig run
  - objectives and goals for the pig run.
- Line contents
  - contents of the line while pigging
  - available versus required driving pressure
  - velocity of the pig.
- Characteristics of the pipeline
  - the minimum and maximum internal line sizes
  - the maximum distance the pig must travel
  - the minimum bend radius and bend angles
  - additional features such as valve types, branch connections, and the elevation profile.

Cleaning pigs: Cleaning pigs are designed to remove solids or accumulated debris in the pipeline. This increases efficiency and lowers operating cost. They have wire brushes to scrape the walls of the pipe to remove solids. Pigs of 14 in. and smaller normally use rotary wire wheel brushes. These brushes are easy to replace and inexpensive. Special rotary brushes are used on some larger pigs. Larger pigs have wear-compensating brushes. These brushes can be individually replaced as needed and are mounted on either leaf springs, cantilever springs, or coil springs. The

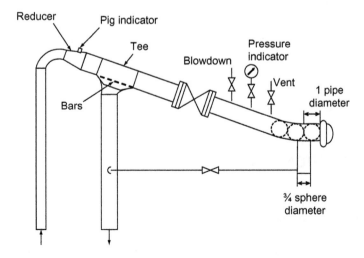

FIGURE 16.11 A typical configuration of a pig receiver for gas services.

springs push the brushes against the pipe wall. As the wire brushes wear, the force of the spring keeps them in contact with the pipe wall, compensating for brush wear. Many different brush materials are available. Standard brushes are made of fine or coarse carbon steel wire. For pipelines with internal coatings, Prostran is the material of choice. Some services require a stainless steel brush. Special brush designs, such as the pit cleaning brush, are also available. When soft deposits of paraffin, mud, etc., need to be removed, the urethane blade is an excellent choice. The blade design is interchangeable with the brushes. Bypass ports are installed in the nose or on the body of the pig. These ports are used to control fluid bypass. If the ports are on the body of the pig, the flow will also pass through the brushes and keep them clean. As the fluid passes through the ports on the nose of the pig, it helps keep the debris in front of the pig stirred up and moving. Plugs are used to regulate the bypass. The sealing elements are either elastomer cups or discs. They are used as a combination cleaning and sealing element to remove soft deposits. Cups are of standard or conical design. Specialty cups are available for some applications. The cup and disc material is normally manufactured from a polyurethane material, which gives outstanding abrasion and tear resistance but is limited in temperature range. Neoprene, nitrile, EPDM, and Viton are available for higher-temperature applications.

The best choices for cleaning applications are normally pigs with discs, conical cups, spring-mounted brushes, and bypass ports. Figure 16.12 shows the details of two pigs of this type. Discs are effective for pushing out solids and providing good support for the pig. Conical cups provide sealing

FIGURE 16.12 Some mandrel pigs used in pipeline pigging operations. *(Courtesy of T. D. Williamson, Inc.)*

characteristics, good support, and long wear. Spring-mounted brushes provide continuous forceful scraping for removal of rust, scale, and other build-ups on the pipe wall. Instead of brushes, polyurethane scraper blades can also be selected for cleaning waxy crude oil lines because the scraper blades are easier to clean than brushes. Bypass ports allow some of the flow to pass through the pig. This can help minimize the solids buildup in front of the pig. For a new pipeline construction, it is a good practice to include a magnetic cleaning assembly in the pig.

Gauging pigs: Gauging pigs are used after constructing the pipeline to determine if there are any obstructions in the pipeline. Obstructions can be caused by partially closed valves, wrinkle bends, ovality caused by overburden, dents caused by rocks underneath the pipe, third-party damage, buckles caused by flooding, earthquakes, etc. Gauging pigs assure that any ovality of the line is within the accepted tolerance. The gauging plate may be mounted on the front or rear of the pig and is made of a mild steel or an aluminum. The plate may be slotted or solid. The outside diameter of the plate is 90−95% of the pipe's inside diameter. Gauging runs are normally done during new construction and prior to running a corrosion inspection pig. The best practice is to choose inspection tools that can provide critical information about the line, such as determining the location (distance), o'clock position, and the severity of a reduction.

Caliper pigs: Caliper pigs are used to measure pipe internal geometry. They have an array of levers mounted in one of the pig cups. The levers are connected to a recording device in the pig body. The body is normally compact, about 60% of the internal diameter, which combined with flexible cups allows the pig to pass constrictions up to 15% of bore. Caliper pigs can be used as gauging pigs. The ability of a caliper pig to pass constrictions means minimal risk of jamming. This is very important for subsea pipelines where it would be very difficult and expensive to locate a stuck pig.

Displacement pigs: Displacement pigs displace one fluid with another based on a sealing mechanism. They can be bidirectional or unidirectional in design. They are used in the testing and commissioning phase of the pipeline, i.e., hydrostatic testing, line fills, and dewatering. Line evacuation and abandonment is another application for the displacement pig. Bidirectional (Figure 16.13) pigs can be sent back to the launch site by reversing flow if they encounter an obstacle. They are also used in the filling and dewatering associated with hydrostatic testing when the water used to fill the line must be pushed back to its source after completion of the test.

The best choice of displacement pigs is normally pigs with multi-lipped conical cups (Figure 16.14). Conical cups can maintain contact with the pipe wall even in out-of-round pipe which is more common in large-diameter pipelines. Conventional cups and discs usually cannot maintain a seal in out-of-round pipe. Multi-lipped cups have numerous, independent sealing lips on each cup, which significantly improves their ability to maintain a seal.

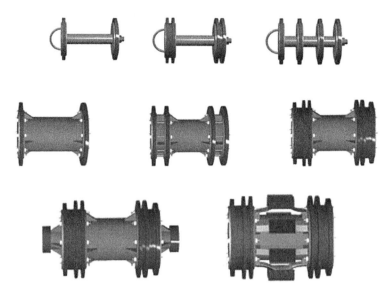

FIGURE 16.13 Some bidirectional pigs used in pipeline pigging operations. *(Courtesy of Apache Pipeline Products, Inc.)*

FIGURE 16.14   Pig with multi-lipped conical cups. *(Courtesy of T. D. Williamson, Inc.)*

Profile pig: A profile pig is a gauging pig with multiple (usually three) gauging plates. One plate is mounted on the front, one in the middle, and one on the rear of the pig. It is normally used before running an ILI tool to assure the tool's passage around bends and through the pipeline.

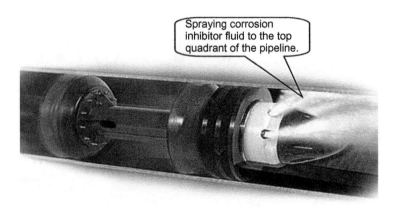

Spraying corrosion
inhibitor fluid to the top
quadrant of the pipeline.

FIGURE 16.15   A special pig for spraying corrosion inhibitor. *(Courtesy of T. D. Williamson, Inc.)*

Transmitter pigs: Occasionally pigs will get stuck in a line. The location of the stuck pig can be found by using a detector pig with a transmitter in its body. The transmitter emits a signal so it can be located with a receiver. Transmitters are normally mounted into a mandrel, solid cast, or Polly-Pig.

Special pigs: Many applications require special pigs. Manufacturers in the pigging industry have made special pigs for many applications. Figure 16.15 illustrates that a special pig can be used for spraying corrosion inhibitor to the upper side of pipe interior. Dual diameter pigs are designed for pigging dual diameter pipelines. They are usually mandrel pigs fitted with solid discs for the smaller line and slotted discs for the larger line. If it is a cleaning pig, the brushes will support it in the line and keep the pig centered. The Polly-Pig is also widely used in this application.

Other special pigs include pinwheel pigs, which use steel pins with hardened tips. They were developed to remove wax and scale from a pipeline. Magnetic cleaning pigs were developed to pick up ferrous debris left in the pipeline.

There are many pig configurations to choose from, but some configurations will not work in some pipelines. It is very important to compare pipeline information to the pig specifications. The best way to stay out of trouble is to provide the pipeline specifications to the pig manufacturer and ask them to recommend a pig.

16.4 MAJOR APPLICATIONS

Major applications of pigging are found during pipeline construction, operation, inspection, and maintenance. Depending on application type and pipeline conditions, different kinds of pigs are chosen to minimize the cost of pigging operations. Tiratsoo (1992) presents a comprehensive description of applications of pigging in the pipeline industry.

16.4.1 Construction

During pipeline construction, it is quite possible for construction debris to get inside the pipeline. The debris could harm downstream equipment such as filters and pumps. The only way to remove possible debris is to run a pig through the pipeline. Typically, debris removal is done section by section as the lay barge moves forward. An air-driven cleaning pig is usually sent through the pipeline section to sweep out the debris. Features of the cleaning pig should be selected based on anticipated pipeline conditions. The most effective way to clear debris is by the use of a magnetic cleaning assembly which can be mounted on conventional pigs. Removal of this type of debris is a must before attempting to run corrosion inspection pigs.

Pipelines subjected to subsea conditions may buckle in certain sections. The place for a buckle to occur during pipe laying is most likely the sag bend just before touchdown on the seabed. To detect the buckle, a gauging pig is pulled along behind the touchdown point. If the pig encounters a buckle, the towing line goes taut, indicating that it is necessary to retrieve and replace the buckled section of pipeline. Features of the gauging pig should be selected based on anticipated pipeline conditions. Caliper pigs can be used as gauging pigs after completion of construction. The ability of caliper pigs to pass constrictions can reduce the possibility of jamming, which is vitally important for subsea pipelines where it would be very difficult and expensive to locate a stuck pig.

Upon completion of construction, the pipeline should be cleaned to remove rust, dirt, and mill scale that contaminate product fluids. These contaminates also reduce the effectiveness of corrosion inhibitor. A typical cleaning operation would consist of sending through a train of displacement pigs with different features suitable to pipeline conditions. Gel slugs are used to pick up debris into suspension, clearing the pipeline more efficiently. Corrosion inhibitor can also be added to the interior of the pipeline in the trip of cleaning pigging.

After cleaning, the pipeline is flooded with water for hydrotesting. Air must be completely removed so that the pipeline can be efficiently pressurized. Pigging with displacement pigs is normally the best solution for flooding a pipeline. Use of bidirectional batching pigs is favorable for the afterward-dewatering operation.

Upon a successful hydrotest, water is usually displaced with air, nitrogen, or the product fluid. Since dewatering is the reverse process of flooding, a bidirectional batching pig used to flood the pipeline, left during the hydrotest, can be used to dewater the pipeline. In cases of gas service pipelines, it is necessary to dry the pipeline to prevent formation of hydrates and waxy solids. For this purpose, methanol or glycol slugs can be sent through the pipeline between batching pigs. An alternative means of drying the pipeline is to vacuum the pipeline with vacuum pumps.

16.4.2 Operation

During fluid production operations, pigging is utilized to maintain efficiency of pipelines by removing wax in oil pipelines and liquids (water and condensate) in gas pipelines. Sometimes pigging operations are for meter proving. Pipeline wax is characterized as long-chain paraffin formed and deposited in pipelines due to changes in pressure and temperature. Accumulation of wax in pipeline reduces the effective pipeline hydraulic diameter and hence efficiency of the pipeline. A variety of cleaning pigs are available to remove wax. Most of them work on the principle of causing a bypass flow through the body of the pig over the brushes or scrapers and out to front. Pigs used for removing wax should be selected to have features inducing the bypass flow. The action of the pig also polishes wax remaining on the pipe wall, leaving a surface for low flow resistance of product fluids. To remove hard-scale deposits, aggressive and progressive pigs are the best choice. They can be used with cleaning fluids that attack the deposits and/or help to keep the deposits in suspension while being pushed out of the line. This is a very special application that would normally be provided by a pipeline cleaning service company. Samples of deposits are usually required for chemical analysis and to determine what cleaning fluids are best suited. Sometimes chemical cleaning is used for removal of specific types of pipe deposits. Chemical cleaning is a process of using pigs in conjunction with environmentally friendly detergent-based cleaning fluids and is almost always done by pipeline cleaning service companies. The detergents help to suspend solid particles and keep them in a slurry, thus allowing the removal of large volumes of solids in one pig run. Samples of material to be removed from the line are required in order to select the best cleaning fluid. The cleaning fluids are captured between batching and cleaning pigs, and normally a slug of the fluid is introduced in front of the first pig.

In gas service pipelines, liquid water and/or gas condensate can form and accumulate on the bottom of the pipeline. The liquid accumulation reduces the flow efficiency of the pipeline. It can also develop slug flow, causing problems with the processing facilities. Different types of displacement and cleaning pigs are available to remove the liquids. Because gas is the drive fluid, the pigs used for removing the liquids in the gas pipeline should be selected to have features of good sealing. Spheres are usually the preferred choice for liquid removal from wet gas systems. Most of these systems are designed to automatically (remotely) launch and receive spheres. A large number of spheres can be loaded into the automatic launcher and launched at predetermined frequencies. At the receiving end of the line is a slug catcher to capture all the liquid pushed in by a sphere. If more liquid is brought in than the slug catcher can handle, the plant normally shuts down. Thus, spheres are launched at a frequency that prevents exceeding the capacity of the slug catcher. Pipeline systems are normally designed for use with spheres or pigs but not both. Pipelines designed for spheres may require modifications of launchers and receivers in order to run conventional pigs.

To clean pipelines with known internal corrosion, special pigs are available equipped with independent scraping wires that will go into a pit to break up and remove deposits that prevent corrosion inhibitors from getting to the corroding area. Brushes on conventional pigs will not extend into a pit. To clean internally coated pipelines, the preferred choice is a pig with discs and cups because these will normally remove deposits from the coating due to the "Teflon-like" characteristics of epoxy coatings. Conventional cleaning pigs with "prostran" brushes or polyurethane blades can also be used on internally coated pipelines.

16.4.3 Inspection

A variety of intelligent pigs have been employed for pipeline inspection purposes, including detection of not only dents and buckles but also corrosion pitting, cracks, spanning and burial, and measurement of wall thickness. The information obtained from the pigging operations is used for assessment of pipeline safety and integrity.

Magnetic flux leakage pigs have been used for detection of dents and buckles and measurement of pipe ovality and wall thickness over the entire pipe surface. The principle of magnetic flux leakage detection relies on measurement of metal loss, and hence the size of defect. Usually, a series of survey that runs over years are required to establish trends. Magnetic flux leakage pigging can be utilized in liquid and gas pipelines.

Ultrasonic intelligent pigs are used to make direct measurement of wall thickness of the entire pipe surface. They are better suited to liquid pipelines and cannot be used in gas pipelines without a liquid couplant.

Pipeline spans have traditionally been found by external inspection using side-scan sonar or ROVs. In recent years, neutron-scatter pigs have been employed to detect spanning and burial in subsea pipelines with lower cost and better accuracy.

16.4.4 Maintenance

Pigging is also run for maintenance of pipelines, for coating the inside surface of pipeline, providing pressure resistance, and installing barrier valves. Traditionally, the internal surfaces of pipe joints are precoated with a smooth epoxy liner, leaving the welds uncoated. Recently, a pigging system has been developed to coat the entire internal surface of pipeline by first cleaning the surface and then pushing through a number of slugs of epoxy paint.

Shutting down offshore, especially deepwater, pipelines for maintenance is very expensive. With advanced technology, it is possible to carry out some maintenance jobs without shutting down the pipeline. In cases where there are not enough isolation valves, a pressure-resisting plug may be pigged into the pipeline to seal off downstream operation.

Corrosion inhibitors are normally injected into the line on a continuous basis and carried through the line with the product flow. Sometimes inhibitors are batched between two pigs, but there is no way to guarantee the effectiveness of this method, especially at the 12 o'clock position. Special pigs have been developed that spray inhibitor onto the top of the pipe as they travel through the pipe. This is done by using a siphoning effect created by bypass flow through an orifice specifically designed to pick up inhibitor from the bottom of the pipe.

16.5 PIGGING PROCEDURE

Pressure and flow rate: Any pigging operation should follow a safe procedure that is suitable to the given pipeline conditions. Operating pigging pressures and fluid flow rates should be carefully controlled. Velocity of driving fluid is usually between 3 and 5 ft/s during pigging. Recommended ranges of operating pressures and flow rates are presented in Table 16.1.

Prerun inspection: The pig must be in good condition if it is to do the job it was selected to do. If the pig has been run before, it should be inspected to assure that it will run again without stopping in the pipeline. Measure the outside diameter of the pig's sealing surface. This diameter must be larger than the inside pipe diameter to maintain a good seal. Inspect the sealing surfaces to assure there are no cuts, tears, punctures, or other damage that will affect the pig's ability to run in the pipeline. The unrestrained diameter of brush pigs should also be measured to assure that the brushes will maintain contact with the pipe wall during the complete run. When using brush-type mandrel cleaning pigs, the brushes should be inspected for corrosion or breakage. Every precaution should be taken to prevent these brushes from breaking in the pipeline. Loose bristles can damage valves, instrumentation, and other pipeline equipment. All components of brush-type mandrel pigs should be checked to be certain that they are tight and in good condition.

Pig launching and receiving: Pig launchers are used to launch the pig into the pipeline, and pig receivers are used to receive the pigs after they have made a successful run. The design of these pig traps will depend on the type of pig to be run and the pipeline design conditions. Provisions in the station design should include the handling equipment for pigs of 20 in. and larger. Caution should be taken for liquid spillage from the pig traps.

The following pig launching procedures can be used as guidelines for developing operating procedures. Since company policies vary regarding

TABLE 16.1 Recommended Pigging Pressures and Flow Rates

Pipe Inner Diameter (in.)	Typical Pigging Pressure (psig)		Liquid Flow Rate (GPM)		Gas Flow Rate (SCFM)	
	Launching	Running	3 FPS	5 FPS	5 FPS	10 FPS
2	100–200	40–100	20	40	30	60
3	100–150	35–85	60	100	70	140
4	75–125	30–80	110	190	120	240
6	50–100	30–75	260	430	270	540
8	30–80	25–70	460	770	440	880
10	30–60	25–50	720	1200	580	1200
12	30–50	20–45	1040	1700	760	1500
14	20–50	15–40	1400	2300	930	1900
16	15–45	10–40	1800	3100	1100	2200
18	15–40	10–30	2300	3900	1200	2400
20	10–25	5–20	2900	4800	1200	2400
24	10–25	5–20	4100	6900	1700	3400
30	10–20	5–15	6500	10,900	2400	4800
36	10–20	5–10	9400	15,700	3200	6400
40	10–20	5–10	11,600	19,400	4000	8000
42	10–20	5–10	12,800	21,400	4400	8800
48	10–20	5–10	16,700	27,900	5800	11,600
54	10–20	5–10	21,200	35,300	7300	14,600
60	10–20	5–10	26,200	43,600	9000	18,000
72	10–20	5–10	37,700	62,900	13,000	26,000

whether the pig launcher is left on stream or isolated from the pipeline after the pig is launched, the operator should verify that the trap is isolated from the pipeline and depressurized before commencing any part of the launch procedure.

To launch pigs, make sure that the isolation valve and the kicker valves are closed. In liquid systems, open the drain valve and allow air to displace the liquid by opening the vent valve. In natural gas systems, open the vent and vent the launcher to atmospheric pressure. When the pig launcher is completely drained (no pressure left), with the vent and drain valves still open, open the trap (closure) door. Install the pig with the nose firmly in contact with the reducer between the barrel and the nominal bore section of the launcher. Clean the closure seal and other sealing surfaces, lubricate if necessary, and close and secure the closure door. Close the drain valve. Slowly fill the trap by gradually opening the kicker valve and venting through the vent valve. When the filling is completed, close the vent valve to allow pressure to equalize across the isolation valve. Open the isolation valve. The pig is ready for launching. Partially close the main line valve. This will increase the flow through the kicker valve and behind the pig. Continue to close the main line valve until the pig leaves the trap into the main line as indicated by the pig signaler. After the pig leaves the trap and enters the main line, fully open the main line valve. Close the isolation valve and the kicker valve. The pig launching is complete.

To receive pigs, make sure the receiver is pressurized. Fully open the bypass valve. Fully open the isolation valve and partially close the main line valve. Monitor the pig signaler for pig arrival. Close the isolation valve and bypass valve. Open the drain valve and the vent valve. Check the pressure gauge on the receiver to assure the trap is completely depressurized. Open the trap closure and remove the pig from the receiver. Clean the closure seal and other sealing surfaces, lubricate if necessary, and close and secure the trap (closure) door. Return the receiver to the original condition.

Freeing a "stuck" pig: The goals of "pigging" a pipeline include not only running pigs to remove a product or to clean the line, but to do the work without sticking the pig. Getting the pig stuck rarely happens in pipeline that is pigged routinely, but can happen when pigging a pipeline that has been neglected or never pigged before. It is a good practice to run a low-density (2 lb/ft^3) foam pig in any "suspect" pipeline and examine the foam pig for wear patterns, tears, gouges, etc. The pigging project should be continued only after feeling comfortable that the line is piggable. If a pig becomes stuck, it is important to identify the cause. Retrieving the pig is the first priority. When bidirectional pigs are used, stuck pigs may be recovered with reverse flow.

Pig tracking is normally done on critical projects and when attempting to locate stuck pigs. A pig tracking system consists of a transmitter mounted on the pig, an antenna, and a receiver that records and stores each pig passage. In addition, the operator can see and hear the signal of the pig passing under the antenna. The antenna and receiver are simply laid on the ground above and in line with the pipe, and the passage of the pig is heard, seen, and recorded. Inexpensive audible pig tracking systems are also available; however, they cannot be used to find a stuck pig because they rely on the noise the pig makes as it travels through the line. Sometimes a pig without a transmitter fails to come into the receiver because it gets stuck somewhere in the line. When this happens, the pig cups usually flip forward and flow continues around the stuck pig. In order to find the stuck pig, another pig with a transmitter is launched and tracked closely at all points that are readily accessible. When the transmitter pig passes one tracking point but never reaches the next point, it is assumed the transmitter pig has reached the stuck pig and they are both stuck. The line is then walked, carrying the antenna and receiver until the transmitter pig is pinpointed. Both the pigs and the debris ahead of the pigs is then removed by cutting the pipe behind and well ahead of the stuck pig.

REFERENCE

Tiratsoo, J.N.H., 1992. Pipeline Pigging Technology, second ed. Gulf Publishing Company, Accrington, UK.

FURTHER READING

Kennedy, J.L., 1993. Oil and Gas Pipeline Fundamentals. PennWell Books, Tulsa, OK.

Mare, R.F.D., 1985. Advances in Offshore Oil & Gas Pipeline Technology. Gulf Publishing Company, Boston, MA.

McAllister, E.W., 2002. Pipeline Rules of Thumb Handbook. Gulf Publishing Company, Boston, MA.

Muhlbauer, W.K., 1992. Pipeline Risk Management Manual. Gulf Publishing Company, Houston, TX.

Part IV

Condition-Based Maintenance

Tian Ran Lin with contributions from Yong Sun

Chapter 17

An Introduction to Condition-Based Maintenance

17.1 INTRODUCTION

Reliability and continuous operation of engineering assets are always priority for asset managers and plant owners. Nevertheless, health conditions of engineering assets deteriorate over time. Maintenance is thus required to ensure the health state and reliability of an engineering asset and to restore the asset back to its functional level in the original design during the asset useful life cycle. Engineering asset maintenance refers to any technical or administrative actions, or combination of both to keep engineering assets in, or bring them back to their required functional states (British Standards Institution, 1984). Maintenance has so far progressed from corrective maintenance to preventive maintenance, then to condition-based maintenance (CBM) (also termed predictive maintenance) and further to proactive maintenance (Barnard, 2009; Moubray, 1997). Some new concepts such as self-maintenance (Lee and Scott, 2006) and self-healing machines (Gao, 2006) have also been proposed to enhance maintenance excellence.

Corrective maintenance is typically reactive maintenance. It is used to fix a broken down engineering asset. Preventive maintenance, CBM, and proactive maintenance are used to avoid severe consequences caused by asset failures. A common feature of these maintenance policies is to conduct maintenance activities before an engineering asset fails. A major difference among them is that preventive maintenance is conducted according to the ages or duties of assets; CBM is carried out based on asset health conditions; and proactive maintenance is applied to eliminate the root causes of failures. While all activities of conventional maintenance are undertaken by humans, self-maintenance requires engineering assets to monitor, diagnose, and maintain themselves.

Compared to other maintenance policies, CBM is often more effective in avoiding over- or under-maintenance and thus has been widely adopted in engineering asset management (EAM). The concept of CBM was first proposed in the late 1940s by the Rio Grande Railway Company (Wiseman, online access 2011). Since then, CBM has spread rapidly among various industries. This is largely due to the development of condition monitoring

Offshore Pipelines. DOI: http://dx.doi.org/10.1016/B978-0-12-397949-0.00017-0

techniques. CBM is typically employed to reduce maintenance costs by avoiding unnecessary maintenance activities based on condition monitoring information. This maintenance policy requires a maintenance action on an engineering asset when the evidence of asset abnormality appears in the condition monitoring data. However, it is difficult for asset managers to plan ahead and optimize their maintenance activities based on condition monitoring data alone due to the typical short time frame between the detection of an abnormality and the failure of an asset. To address this issue, longer term asset health predictions are needed. In recent years, various techniques such as CBM + (Niu et al., 2010) and condition-based prediction approach (Ma, 2007) have been proposed to meet the need.

CBM usually consists of two steps: (1) asset health condition assessment and prediction and (2) maintenance decision optimization. Detailed procedure of CBM is described in the next chapter. As a general introduction, this chapter focuses on the discussion of commonly used asset condition monitoring methods and signal processing techniques, asset health assessment and prediction approaches, as well as maintenance decision optimization methods. When deploying CBM, maintenance personnel need to know the current states and future trends of their asset's health conditions through asset heath condition assessment and prediction. Two critical terms commonly used to describe asset conditions are "fault" and "failure." As these two terms are closely related and can be misused in practice, they should be properly defined. The International Standard Organization (ISO) has provided a clear definition of failures and faults of machines (ISO 17359, 2003a). This definition is extended here to more general engineering assets including machinery and other engineering systems as follows:

An engineering asset has a fault if any of its components or subsystems has a behaviour or condition that is below its designated level, but the asset can still perform its principal functions.

An engineering asset suffers from a failure when it can no longer accomplish one or several of its required principal functions.

Engineering assets here refer to man-made physical structures and machines. They can be classified into (1) nonlinear assets; (2) linear assets; and (3) hybrid systems (Sun et al., 2008). Nonlinear assets are also called discrete assets. This type of asset has a clear physical boundary. They normally form a hierarchy of machines, assemblies, and components, e.g., a car has an engine which includes bearings and other components. Linear assets are also called continuous assets. This type of asset normally is physically presented in the form of lines, i.e., their lengths are much greater than their diameters or widths. Examples of linear assets include cables, pipelines, and roads. Their boundaries are usually defined by field asset operational and/or maintenance practitioners. A linear asset is often divided into segments for maintenance. Hybrid systems are combinations of nonlinear and linear assets.

An offshore platform is an example of hybrid systems. In this chapter, when asset is mentioned, it can be an engineering system, a machine, a piece of equipment, an engineering structure, a segment of an engineering infrastructure, an assembly, or a component, unless specified. In CBM, we often need to take into account asset types when selecting maintenance strategies and/or condition assessment and prediction methods.

Asset health condition assessment and prediction methods can be classified into different categories based on different criteria. According to the timescale, they can be divided into: (1) methods for the next failure time or remaining useful life prediction within one maintenance cycle and (2) techniques for long-term prediction that typically covers multiple maintenance cycles. According to the prediction methodology, they can be divided into (1) trending and degradation curves; (2) covariate models; (3) space state methods; and (4) artificial intelligent (AI) approaches. According to the prediction scope, they can be divided into (1) conventional condition monitoring techniques and (2) condition-based prediction approaches. Condition monitoring typically involves diagnosis to detect faults and prognosis to predict the next failure time. It focuses on failure modes at the component level. On the other hand, the condition-based prediction approaches focus on the prediction of overall health conditions at machine or system levels rather than specific failure modes. They attempt to predict asset degradation processes. The organization of this chapter is based on the prediction scope while considering the prediction methodologies.

17.2 CONDITION MONITORING

Condition monitoring is the primary and one of the most critical steps for CBM in EAM. It provides the firsthand information about the status of the operating condition of an engineering asset to enable an informed maintenance decision making. Condition monitoring is an integrated part of a diagnosis process. It comprises two major processes: (1) data collection and (2) data processing and analysis. Important parameters and indicators extracted from the processing and analysis are also critical for a reliable prognosis process. International standards for data processing in machine condition monitoring and diagnostics have been well defined and documented (ISO 13374-1:2003b(E); ISO/FDIS 13374-2:2006(E)).

17.2.1 Data Collection and Condition Monitoring Techniques

Data collection is a process to acquire information regarding the health status of an engineering asset for fault diagnosis and prognosis in CBM practices. There are many different types of condition monitoring approaches in current industry practices which produce diverse types of condition monitoring data. Each condition monitoring approach has its unique features and applications.

Common current condition monitoring approaches include vibration monitoring, acoustic monitoring, acoustic emission (AE) and ultrasonic monitoring, oil and wear debris monitoring, ferrography monitoring, thermography monitoring, and process parameter monitoring such as pressure, temperature, humidity, and environmental data analysis. Additional to these condition monitoring approaches, due to the rapid development of electronic devices and signal processing technologies, a current trend in condition monitoring is the multisensor approach which combines several condition monitoring techniques in one application to overcome the limitation of a single technique and to provide more reliable and comprehensive information for a detailed health status monitoring of an engineering asset. For instance, the simultaneous measurement of instantaneous crank angular speed, AE, in-cylinder pressure, and vibration of a diesel engine will provide a detailed knowledge of the overall operating status and performance of the engine including power output, pressure torque, incipient faults, and the location of faulty components. The approach will also enable a visualization of mechanical movements inside the cylinders at any time instant. Figure 17.1 provides an example of synchronized time waveforms acquired from a diesel engine using a multisensor approach.

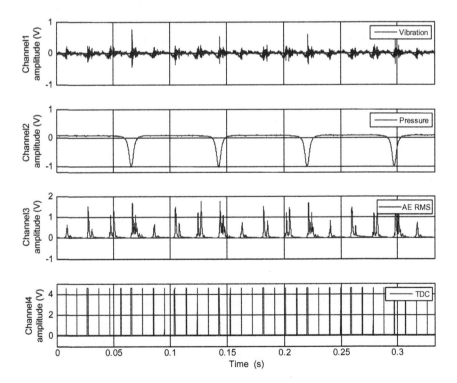

FIGURE 17.1   Synchronized multisensor data of a diesel engine; TDC denotes top dead center.

Several commonly employed condition monitoring techniques are briefly introduced in the subsequent text.

17.2.1.1 Vibration Technique

Vibration technique is one of the most classic techniques in condition monitoring practices. It is often employed to monitor the imbalance, misalignment, looseness, wear, and deflection of mechanical components in a mechanical system. The frequency range for vibration measurement can be as low as in the infrasonic region (below 20 Hz) such as seismic movements and spans across the audible frequency range (between 20 Hz and 20 kHz). The parameters in vibration measurement can be displacement, velocity, or acceleration depending upon the frequency content of the signal. Proximity sensors such as Eddy current probes, magnetic pickups, velocity pickups, and accelerometers are commonly employed in the vibration measurement. In particular, accelerometers are the most used sensors in vibration measurement which come in all shapes and sizes, and have a very large dynamic range and a broad frequency range. Mounting of an accelerometer on a mechanical system being monitored is critical to obtain a reliable condition monitoring data. As a rule of thumb, accelerometers should always be mounted solidly to the vibrating surface of a mechanical system in the designated measuring direction coinciding with the main sensitivity axis of the sensor. An accelerometer can be either permanently mounted (e.g., solidly mounted or stud mounted) or temporally mounted (e.g., using magnetic mount, thin layer of wax, or handheld probe) depending upon individual application and data requirement.

17.2.1.2 Acoustic Technique

Acoustic technique works in a similar way as vibration technique except that one measures airborne noise (sound pressure level) and the other measures the structure-borne sound (vibration). Microphones are the main measurement sensor for the acoustic measurement. Microphones can be generally classified into: (1) condenser microphone; (2) dynamic microphone; and (3) ceramic microphone, according to the sensing element used in the microphone. Condenser microphones have the distinct advantage within the three in a typical industry application due to their insensitivity to vibration and the wide frequency (0.1 Hz to 140 kHz) and high dynamic (up to 140 dB) ranges. Handheld sound level meter is the most commonly seen noise measurement device in industry for condition monitoring applications. Because of the application frequency range of these techniques, vibration and acoustic methods are often considered as low-frequency condition monitoring techniques. This is in comparison with the high-frequency techniques such as ultrasound and AE techniques to be discussed in the subsequent section.

17.2.1.3 Ultrasound Technique

Ultrasound technique operates in the frequency range above the audible frequency (typically between 20 kHz and 100 kHz but can also be extended to the megahertz range). An ultrasound can be either an airborne sound or a structure-borne sound. A heterodyne circuit in an ultrasound device is the main electronic component in the instrument, which converts the ultrasound signal detected by a transducer into an audible signal. The heterodyne signal is enhanced by an amplifier to be heard using standard headphones or is processed through a converter to produce a quantitative output signal in decibels (dB) to be read on a meter/display panel. The signal can also be recorded through a conventional data acquisition system. Abnormalities/defects in a system are usually detected by listening to the characteristics of sound or reading the root mean square (RMS) indicator on the panel in an ultrasonic probe. Nowadays, ultrasound techniques are commonly employed to detect deteriorating or worn off components in a system. It is also used to detect poor lubrication or subsurface fretting of a bearing, compressed air or hydraulic fluid leaks, vacuum leaks, steam trap leaks, and tank leaks caused by compressed gas or fluid forced through a small opening or poorly seated valves. For instance, ultrasound technique is a popular nondestructive testing technique to provide qualitative measurement of pipeline condition with high accuracy and reliability. Ultrasound devices enable inspectors/technicians to detect and diagnose a defect on the spot by discriminating the heterodyne sound among various equipment sounds. A major drawback of ultrasound techniques is that it is difficult to trend the output of ultrasound devices. The output unit is not scalable and is sensitive to the distance between the sensor and the mechanical system being monitored. Another drawback is that ultrasound devices are highly directional and thus, care needs to be taken to ensure the source of ultrasound is effectively monitored. Piezoelectric and fiber-optic elements are the most typical materials used for ultrasound sensors. Piezoelectric-based sensors are more common than fiber-optic sensors in ultrasound devices due to the small sensing element size, and the sensors are less expensive. Another advantage of piezoelectric sensors is that they have a broad operating frequency range from 50 kHz to 2 MHz. On the contrary, applications of fiber-optic ultrasound sensors are limited due to the low sensitivity range and the low frequency range (less than 200 kHz) of the sensors.

17.2.1.4 AE Technique

AE refers to the generation of transient elastic waves caused by the sudden stress redistribution in a material. For example, a sudden release of energy by localized sources in a mechanical system due to an external stimulus (change in pressure, load, or temperature) generates stress waves, which then propagate to the system surface and can be recorded by AE sensors. AE sources range from crack initiation, fractures, dislocations, melting, twinning, phase transformations in materials. Matrix cracking, fiber breakage, and debonding of composite materials, earthquakes, mechanical impacts, and friction also generate AE. AE technique works in a similar fashion as ultrasound techniques except that the formal technique operates in an even higher frequency range (typically above 100 kHz and can be up to several MHz). AE sources are nondirectional point sources which radiate energy in a spherical wave front, and therefore they can be detected by sensors located anywhere in the vicinity of a source. A major benefit of employing AE in condition monitoring of mechanical systems is that it can produce high-quality signals in a noisy machine operating environment due to the high-frequency nature of the technique. AE signals are less likely to be affected by the dominating noise and vibration generated by moving mechanical components of a monitored mechanical system. However, care should also be taken when mounting an AE sensor to minimize the energy loss along the AE propagation path for better signal clarity. AE also comes with inherited problems such as sensor calibration, nonlinearity, data processing, and interpretation. Furthermore, AE techniques can be expensive in mass industrial deployments due to the requirement of highly specialized data acquisition devices.

The sensors used in the AE measurement can generally be grouped into two classes: (1) resonance-type AE sensors and (2) wideband AE sensors. Resonance-type AE sensors operate in a specific frequency band near the resonant frequency of the sensors and have a relatively high sensitivity within the operating frequency band. In contrast, wideband sensors typically operate in a wide frequency band from 100 kHz to 1 MHz. To improve signal clarity, some AE sensors also come with built in preamplifier to overcome the limitation of small AE amplitude and the high-energy decay rate of the high-frequency AE signal.

17.2.1.5 Oil and Wear Debris Analysis and Ferrography Analysis

Oil and wear debris analysis, as well as ferrography analysis are the most frequently used tribology approaches for machine condition monitoring. These tribological analyses are normally undertaken at laboratories using spectrometers and scanning electron microscopes. However, applications of tribological approaches in condition monitoring are limited to lubrication-related or wear-related problems.

17.2.1.6 Thermography Technique

Thermography uses thermo-infrared devices to measure the emission of infrared energy from a machine to determine the machine's operating condition. Thermography is the most appropriate technique to detect problems related to heat transfer and retention of a system. Laser thermography devices or thermo-infrared cameras are the most common instruments used for such measurement. However, care should be taken for thermography measurements since the measurement accuracy of infrared emissions is sensitive to variations of the surrounding environment.

17.2.2 Signal Processing and Analysis

A preliminary step in signal processing is data cleansing to remove the erroneous condition monitoring data recorded by faulty sensors (Jardine et al., 2006). After cleansing the data, the next processing step is to analyze the data and interpret the result to understand the "message" conveyed in condition monitoring signals regarding the health status of a mechanical system. The analysis techniques in signal processing can be generally classified into three main categories: (1) time waveform analysis; (2) frequency domain analysis; and (3) time−frequency analysis. The following sections provide a summary of approaches and parameters used in these data analysis techniques.

17.2.2.1 Time Waveform Analysis

A measured raw time waveform can be analyzed directly in the time domain by calculating its statistical features such as mean, RMS, peak, crest factor, standard deviation, skewness, kurtosis, entropy, and others for fault diagnosis and trending analysis. Equations for calculating these features are given in the following.

The mean in a data analysis is often referred to the statistic arithmetic mean. For a measured time waveform signal $s(i)$ with finite length N, the arithmetic mean and the RMS value can be calculated respectively as

$$\hat{s} = \frac{1}{N} \sum_{i=1}^{N} s(i) \tag{17.1}$$

and

$$\text{RMS} = \sqrt{\frac{\sum_{i=1}^{N} s(i)^2}{N}} \tag{17.2}$$

The standard deviation of the data and the peak value are calculated respectively as

$$\sigma = \sqrt{\frac{1}{N} \sum_{i=1}^{N} (s(i) - \hat{s})^2} \tag{17.3}$$

and

$$\text{Peak} = \max(|s(i)|) \tag{17.4}$$

The crest factor which is defined as the ratio of the peak value and the RMS value of a data series is

$$C_f = \frac{\text{Peak}}{\text{RMS}} \tag{17.5}$$

The skewness is the third statistic moment that measures the degree of asymmetric distribution of a dataset, which is defined as

$$\beta = \frac{\sum_{i=1}^{N} [s(i) - \hat{s}]^3}{N\sigma^3} \tag{17.6}$$

The kurtosis is the fourth statistic moment of a dataset that measures the "peakedness" of the data distribution, which is defined as

$$k = \frac{\sum_{i=1}^{N} [s(i) - \hat{s}]^4}{N\sigma^4} \tag{17.7}$$

Entropy, which was first applied in thermodynamics, is a measure of the uncertainty of a process. Shannon extended the concept of entropy to information theory which quantifies the expected value of the information contained in a message (Shannon and Weaver, 1949). For a set of events Y with probability density function (PDF) of $\{y_i, i = 1,2,\ldots,N\}$ the Shannon entropy is given by (Papoulis, 1991)

$$H(Y) = -\sum_{i=1}^{N} p_i \log_b p_i \tag{17.8}$$

where p_i are the probabilities calculated from the distribution Y, b is the base of the logarithm which takes the common logarithm base values of either 2, e, or 10 depending on the application. For events with a probability around 0 or 1, the term $p_i \log_b p_i$ converges to zero and the entropy is minimum, and for random signals with uniform PDF such as pure noise, the entropy is maximum.

A popular technique to eliminate or minimize the effect of randomness and amplitude variation in time waveform analysis for a better interpretation and understanding of the inherent characteristics of a data is the time synchronous averaging. The time synchronous averaging process averages the raw data over several periods in order to reduce noise, randomness, and other effects from the data, and to enhance the signal component of interest for a more accurate condition monitoring and fault diagnosis. A mathematical description of this averaging process is given by

$$\bar{s}(t) = \frac{1}{r}\sum_{n=0}^{r-1} s(t + nT), \quad 0 \le t < T \tag{17.9}$$

where $s(t)$ denotes the raw time waveform, T is the data length in an averaging period, and r is the number of periods used in the synchronous averaging.

For a quasi-periodic signal, the averaging process is normally triggered by major events of a mechanical system such as the event driven synchronous averaging technique described by Lin and Tan (2011). It was shown that the technique can effectively remove the noise and randomness in the signal while preserving the most useful information relating to the major events of a diesel engine. Figure 17.2 shows an example which compares the raw and synchronous averaging signals of a diesel engine under test.

Time series models such as autoregressive model (Baillie and Mathew, 1996) and autoregressive moving average model (Huzii, 2007) which extracts useful features by fitting time waveform data to a parametric time series are also employed in time waveform analysis. However, applications of these time series modeling techniques are difficult to implement due to the complexity in modeling, especially the need to determine the order of the model (Jardine et al., 2006).

Although time domain analysis can capture transient signals in a system well and provide quick evaluations of the health condition of a mechanical system using statistical trending features, a major drawback of time domain analysis is that it cannot provide details and accurate information to pinpoint the failing mechanical components in a multifaceted system. This is particularly so when weak defect signals (e.g., bearing defect frequencies) are hidden below large dominant signals generated by moving mechanical components. Detailed information about discrete frequency components of a condition monitoring signal is thus required to identify the failing components of a complex system without the need to implement harmful intrusive maintenance measures. Discrete frequency components can be analyzed and linked to the defect mechanical components of a physical system via frequency domain analysis, which is elaborated in the next section.

17.2.2.2 Frequency Domain Analysis

Spectrum analysis is one of the most popular frequency domain analysis techniques, which transforms the time waveform data into discrete frequency components by taking a fast Fourier transform (FFT) analysis. For a digital time waveform $s(n)$ of length N, the discrete Fourier transform of the signal can be expressed by

$$S_k(f) = \sum_{n=0}^{N-1} s(n)e^{-j2\pi\frac{nk}{N}} \qquad (17.10)$$

where k is the length of the frequency spectrum after the transform and j is the imaginary number. By manipulating Eq. (17.10) numerically and employing a non-rectangular windowing function to prevent leakages in the transform, the computational effort of the transform can be reduced substantially and the discrete Fourier transform becomes a FFT (Rangarao and Mallik, 2005). The best advantage of frequency spectrum analysis compared to the time

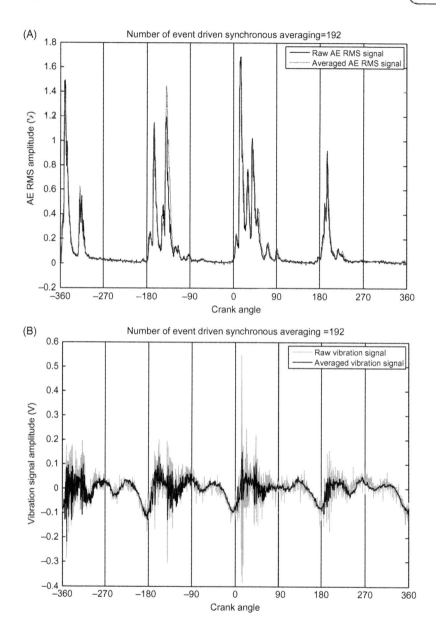

FIGURE 17.2   Comparison of the raw and synchronous averaged signals at the unload condition of a diesel engine. (A) AE RMS signal and (B) vibration signal.

waveform analysis is that discrete frequency components of interest can be identified and related to the corresponding mechanical components of a system being monitored. Advanced signal processing techniques such as envelope analysis (also termed as high-frequency resonance technique), Cepstrum analysis, and other techniques have also been developed for fault detection and diagnosis of rotating machinery in complementing to the more general spectrum analysis. These advanced signal processing techniques provide the important and necessary tools for the detection of localized bearing defects of rotating machinery. It is also desirable that an analysis technique can disclose a bearing defect without any human interventions for an automatic diagnosis. Therefore, critical features need to be extracted from a spectrum to provide a direct indication when a bearing fault is presented. To this end, Shiroishi et al. (1997) proposed a peak ratio (PR), a ratio between the sum of the peak values of a defect frequency and its harmonics and the averaged value of the overall spectrum, as an indicator to identify the presence of bearing faults in a spectrum. To improve the effectiveness and reliability of this technique, Kim et al. (2008) modified the definition of PR slightly to take the form

$$mPR_I = 20 \log_{10} \frac{\sum_{i=1}^{m}(Ps_{ij} - A_s) + \sum_{j=1}^{n}(P_j - A_s)}{A_s} \tag{17.11}$$

where the subscript I denotes that the calculated mPR is for the bearing inner race defect and

$$A_s = \frac{\sum_{k=f_a}^{f_b} S_k}{(f_b - f_a)} \tag{17.12}$$

In Eqs (17.11) and (17.12), A_s is the average spectrum amplitude in a frequency band between f_a and f_b, S_k is the spectrum amplitude at the kth frequency. P_j is the peak amplitude of a bearing defect frequency or its harmonics, Ps_{ij} denotes the spectrum amplitude of the sideband near the jth harmonic of the defect frequency.

Order analysis and waterfall (i.e., order tracking) analysis are two other useful tools in condition monitoring of rotating machinery. In the order analysis, the frequency axis of the spectrum is normalized with a particular fundamental frequency such as the shaft rotating frequency of a machine. The mechanical components of the system having an interrelationship with the rotating shaft then generate respective order harmonics at multiples of the shaft rotating frequency. The order information obtained from such analysis can then be used to reveal the generation mechanism of the order components of interest such as defect components, critical speeds, excited resonances, instabilities, or varying loads of a mechanical system. Figure 17.3 shows an example of the order analysis from the measured instantaneous angular speed of a four-stroke four-cylinder diesel engine at the unload condition. The order spectrum shows that it has a large component at twice the shaft rotating frequency. This order component is proportional to the engine loading condition and performance as shown in Figure 17.4. By comparing the power spectrum amplitude of this order component with that of the

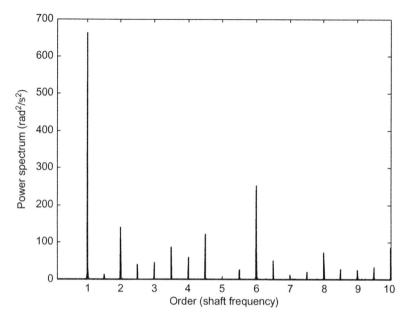

FIGURE 17.3 The order analysis of the measured instantaneous angular speed of a four-cylinder diesel engine at the unload condition.

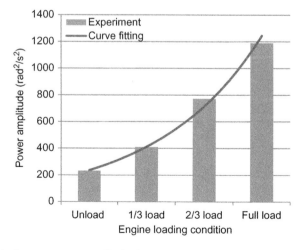

FIGURE 17.4 Power spectrum amplitude of the order component corresponding to the engine firing frequency.

healthy engine, it was found that this order component can be used to detect the incipient exhaust valve fault of the diesel engine.

Figure 17.4 shows that the power spectrum amplitude of the order component corresponding to the engine firing frequency (at twice the shaft rotating speed) of the instantaneous crank angular speed of a diesel engine increases exponentially as the engine loading increases. This observed relationship was found consistent for all three engine operating conditions under investigation (Lin et al., 2011): (1) normal running engine condition; (2) engine operating with one faulty injector; and (3) engine operating with a leaking exhaust valve.

A major limitation of spectrum analysis is that it is mainly useful for stationary signals. Presence of transient or nonstationary signals in the data (a frequently encountered phenomenon in the condition monitoring signal of rotating machinery) would not be captured by invoking the FFT transform. To overcome this limitation, time–frequency analysis technique has been developed. Time–frequency analysis enables a better understanding of how the spectral properties of a signal change with time.

17.2.2.3 Time–Frequency Analysis

In time–frequency analysis, waveform signals are analyzed in both time and frequency domains to capture the progressive change of spectrum components with time (e.g., a transient or nonstationary signal such as music). Energy of the analyzed waveform is then displayed in both time and frequency domains to reveal the signal characteristics for a more accurate fault diagnosis. Short time Fourier transform (STFT) and wavelet analysis are the two commonly employed time–frequency analysis techniques. In STFT, the continuing time waveform signal $s(t)$ is multiplied by a finite analysis time window $\gamma^*(t - \tau)$ and a Fourier transform is then computed on the windowed signal as

$$F(\tau, \omega) = \int_{-\infty}^{\infty} s(t)\gamma^*(t - \tau)e^{-j\omega t}\, dt \qquad (17.13)$$

to produce a local spectrum within a time span defined by the analysis time window $\gamma^*(t - \tau)$. The asterisk sign (*) in the time window indicates a complex conjugate. The analysis window can also be regarded as the impulse response of a low pass filter. The underlying principle of STFT is that if a signal can be considered stationary over the length of the analysis time window, a satisfactory time–frequency analysis can then be obtained by sliding the window along the time axis and performing a Fourier transform for each new position of the window. A main feature of STFT analysis is that it has a uniform resolution in the time–frequency plane which indicates the time–frequency window used in the transform is independent of τ and ω. In practical applications, it is also desirable to have a smaller time–frequency window for better time and frequency resolutions in an STFT. However, the uncertainty principle implies that the area of a time−frequency window cannot be made arbitrarily small. The interdependent relationship between time and frequency resolutions indicates that a small time window will yield a good time resolution but a poor frequency resolution in the transform and vice versa.

Similar to the FFT, the discrete STFT of a discrete time signal $s(n)$ can be written as

$$S(m,k) = \sum_{n=0}^{L-1} s(n)\gamma^*(n-m)e^{-j\frac{2\pi nk}{L}} \tag{17.14}$$

where m and k are respectively the spans of the time and frequency domains, L is the number of samples in the analysis window function $\gamma^*(t-\tau)$. An implicit assumption in obtaining Eq. (17.14) is that the sampling rate of the signal is higher than the rate used in the STFT calculation.

Analogous to the power spectrum in the FFT analysis, spectrogram, which is the square magnitude of an STFT transform, is often used in the display of the transformation result for further data processing and interpretation. Because of the limitation of time−frequency resolution of an STFT, applications of the technique are limited to non-stationary signals or signals with slow changing dynamics. To overcome this limitation, another time−frequency analysis technique known as wavelet transform has been developed. The beauty of this later transformation is that it has good frequency resolution at low frequencies and good time resolution at high frequencies. Thus, the technique is particularly useful for signals with long duration low frequencies and short duration high frequencies. The technique is also well employed to reduce noise (denoising) in raw signals.

Wavelet transform was first introduced by Morlet et al. (1982) in the early 1980s in the analysis of seismic signals. Many wavelet transforms have since been developed for a great range of applications such as imaging compression, denoising, and pattern recognition. Wavelet transform is in fact a timescale representation of a signal. It is the inner product of a signal and the translated and scaled family of a mother wavelet function $\psi(t)$. For a continuing signal $s(t)$, the continuing wavelet transform can be written as

$$X(a,b) = \frac{1}{\sqrt{a}} \int_{-\infty}^{\infty} s(t)\psi^*\left(\frac{t-b}{a}\right) dt \tag{17.15}$$

where a is the scale parameter and b is the time translation. The factor $(1/\sqrt{a})$ is introduced in the equation to ensure that all scale functions have the same energy. If one considers the wavelet function $\psi(t)$ as a bandpass impulse response, then the wavelet transform, Eq. (17.15), is simply a bandpass analysis. Morlet, Haar, Daubechies, Shannon, Gaussian, and Mexican

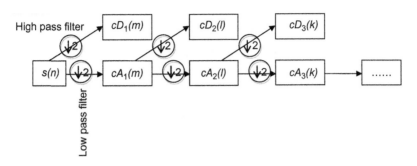

FIGURE 17.5   A diagram of wavelet decomposition in DWT analysis.

hat are some of the commonly used wavelets. Care should also be taken when using a wavelet transform for a better understanding of the properties of a signal. One should ensure that the signal can be perfectly reconstructed from its wavelet representation. Thus, a wavelet transform has to meet the criterion which is also known as the admissibility condition:

$$C_\psi = \int_{-\infty}^{\infty} \frac{|\Psi(\omega)|^2}{|\omega|} d\omega < \infty \qquad (17.16)$$

where $\Psi(\omega)$ is the Fourier transform of the mother wavelet $\psi(t)$. Like the spectrogram in an STFT analysis, scalogram is the most typical display representation of a wavelet transform, which is the squared magnitude of a wavelet transform, $|X(a,b)|^2$.

Although continuing wavelet transform has been widely employed in data analysis of scientific research, the discrete version of the transform is more popular in practical applications due to the small computational cost and excellent signal compaction properties. A discrete wavelet transform (DWT) decomposes a signal into different frequency bands by passing it through a series of filters. Figure 17.5 shows a diagram of the multistage filtering process in a DWT in decomposing a discrete time signal $s(n)$. In each stage of decomposition, the signal is passed through a pair of (quadrature mirror) low and high pass filters simultaneously as well as down sampling.

A more general and an extension of wavelet decomposition is the so-called wavelet packet decomposition which offers a richer range of possibilities for signal analysis. In the orthogonal wavelet decomposition (see Figure 17.5), a signal is split into two parts, one contains a vector of approximation coefficients and the other contains a vector of detail coefficients in each stage of decomposition. The process is repeated by itself and there are $n + 1$ possible ways to decompose the signal for an n-level decomposition. While in the wavelet packet decomposition, both the details and the approximations can be split. This offers a boarder range of possibilities to decode a signal than ordinary wavelet analysis. A graphical illustration of a three-level

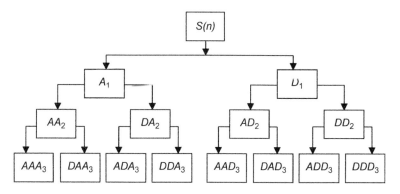

FIGURE 17.6   A three-level wavelet packet decomposition tree. *Note: A* represents the low-frequency part and *D* represents the high-frequency part in each level of decomposition.

wavelet packet decomposition is shown in Figure 17.6. Wavelet transform has been successfully employed for data analysis and machine fault detections in the past three decades since its invention. Peng and Chu (2004) provided a detailed review of applications of wavelet transform for signal processing in condition monitoring and machine fault detections.

Another popular time–frequency analysis is the Wigner–Ville (WV) transform which is a nonlinear time–frequency representation typical for nonstationary signals. WV distribution is functionally similar to a spectrogram but produces better temporal and frequency resolutions than spectrogram. This is because the bilinear WV transform is not based on the signal segmentation and thus not restricted by the uncertainty principle of time–frequency resolution. The continuous and discrete WV transform of a signal $s(t)$ are given respectively by

$$W(t, \omega) = \int_{-\infty}^{\infty} s\left(t + \frac{\tau}{2}\right) s^*\left(t - \frac{\tau}{2}\right) e^{-j\omega\tau}\, d\tau \qquad (17.17)$$

and (Boashash, 1992)

$$W(n, k) = \sum_{m=-M}^{M} s\left(n + \frac{m}{2}\right) s^*\left(n - \frac{m}{2}\right) e^{-j\frac{2\pi mk}{M}} \qquad (17.18)$$

where $M = (N - 1/2)$ and N is the number of independent samples in the data.

To avoid aliasing effects in the WV distribution, care needs to be taken to ensure the bandwidth signal $s(t)$ to be sampled at a rate four times greater than the maximum frequency of the signal instead of the typical twice the Nyquist frequency in frequency analysis due to the bilinear nature of the spectrum. A drawback of WV transform is that it also induces many artifacts and negative values, which need to be properly compensated in the analysis.

17.2.2.4 Event Data Analysis

Data analysis for event data only is well known as reliability analysis, which fits the event data to a time between events probability distribution and uses the fitted distribution for further analysis (Jardine et al., 2006). A popular approach of event data analysis is the failure mode, effects, and criticality analysis (FMECA), an inductive method that performs analysis at either the functional or component level and charts the probability of failure modes against the severity of consequences. The result generated from FMECA highlights failure modes with relatively high probability and severity of consequences, allowing remedial effort to be directed for optimum maintenance decision making and outcomes. The approach is also known as reliability-centered maintenance (RCM). Therefore, the subject is not going to be discussed further in this chapter.

17.2.3 Diagnostics and Prognostics

Diagnostics and prognostics are two of the main approaches in CBM programs. The two techniques differ in their approaches and have different objectives in a CBM program although the two methods cannot be completely cut off from each other. Diagnostics detects and identifies faults in a mechanical system when they occur via the analysis of condition monitoring data of the system. The approach is thus a passive posterior analysis technique. On the other hand, prognostics predicts an impending fault in a system before it materializes and estimates the remaining useful life or the possibility of a system failure based on the current health condition (i.e., the condition monitoring and event data) and past operating profile of the system. The technique is an active forward approach which aims to provide reliability predictions to assist the maintenance decision making in determining inspection intervals, minimizing the maintenance cost, and reducing the unexpected system failure. Prognostics is considered a more effective approach than diagnostics in preventing a system from unexpected failures. Nevertheless, prognostics also comes with its own limitations. For instance, it cannot completely eliminate unexpected faults and failures in a system. Furthermore, predictions based on statistical prognostics models are not hundred percent reliable and can fail to provide early warnings of an incoming system failure. To this regard, diagnostics is thus a necessary tool for providing maintenance decision support. In addition, diagnostics is also useful in providing accurate machine information to establish a more reliable prognostics model. A shortcoming of diagnostics and prognostics approaches is that they generally do not yield adequate information for the long-term, especially in the case of asset life cycle decisions (Ma, 2007). Long-term asset operational and maintenance planning often needs long-term prediction of overall degradation information, rather than specific failure modes of engineering assets. Condition-based prediction approach (CBPA) (Ma, 2007) can be employed to meet this need.

17.3 CONDITION-BASED ASSET HEALTH PREDICTION MODELS AND METHODS

CBPA aims to predict the overall health conditions of engineering assets (often degradation processes) using all possible data/information including condition monitoring measurements, diagnosis and prognosis results, human inspection records and assessments, asset performance indicators, asset maintenance and operational information, asset working environmental data, as well as expert's knowledge and experiences. Although the concept of CBPA has been proposed only for a couple of years, the methods/techniques used in this approach are not completely new. Existing CBPAs largely include state-space-based approaches, covariate models, and AI techniques. Some well-established or recently developed condition-based prediction methods are discussed in the following sections.

17.3.1 Markov Chain

Markov chain is a type of state-space-based modeling technique. It was developed by Markov to model discrete random processes whose future states only rely on their current states and are independent of their past states. If a random process described by a sequence of random variables $X_1, X_2, X_3, \ldots$, has Markov property, then

$$\Pr(X_k = x_k | X_{k-1} = x_{k-1}, X_k = x_{k-2}, \ldots, X_2 = x_2, X_1 = x_1)$$
$$= \Pr(X_k = x_k | X_{k-1} = x_{k-1}) \tag{17.19}$$

where x_k, x_{k-1}, x_{k-2}, $\ldots$, x_2, x_1 are some values (often called states) in the state space which consists of a set of states $\{S_1, S_2, \ldots, S_n\}$. The random process moves among these states step by step with certain probabilities. After a step, the process stays in a state, e.g., S_i ($i = 1, 2, \ldots, n$) and then in the next step, it moves to another state, e.g., S_j ($j = 1, 2, \ldots, n$) with a probability p_{ij} (of course, when $j = i$, we usually say it stays in its current state). The probability p_{ij} is termed the transition probability from state S_i to state S_j. According to probability theory,

$$\sum_{j=1}^{n} p_{ij} = 1 \tag{17.20}$$

Markov chains are often represented using directed graphs. Figure 17.7 shows an example of a Markov chain with four states S_1, S_2, S_3, and S_4. In this figure, circles represent states and arrow lines represent the possible

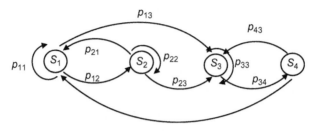

FIGURE 17.7   An example of Markov chain.

movements from one state to another, i.e., the transition probabilities of these movements are not zero.

For a Markov chain with a finite state space, its n^2 transition probabilities can form an $n \times n$ transition matrix in which the ith ($i = 1, 2, \ldots, n$) row contains all transition probabilities from state S_i to all states:

$$\vec{P} = \begin{bmatrix} p_{11} & \cdots & p_{1n} \\ \vdots & \ddots & \vdots \\ p_{n1} & \cdots & p_{nn} \end{bmatrix} \tag{17.21}$$

where $\vec{P}$ represents a transition matrix. If the $\vec{P}$ of a Markov chain does not change over time, the Markov chain is time-homogeneous. According to Markovian theory, the probability distribution that a Markov chain stays in the states of its state space after m steps purely depends on its transition probabilities and its initial probability distribution in these states. For a time-homogeneous Markov chain,

$$\vec{X}_m = \vec{X}_0 \vec{P}^m \tag{17.22}$$

where $\vec{X}_0$ and $\vec{X}_m$ are the probability vectors that a Markov chain stays in the states of its state space initially and after m steps, respectively.

To model the changes of an asset degradation process over time, it is necessary to link moving steps with time explicitly. To do so, we assume that the degradation process moves one step after a certain time (e.g., 1 year). This assumption changes a continuously changing degradation process into a process that jumps between states at discrete time points. With this assumption, Markov chain can be applied to engineering asset health predictions. The procedure is as follows:

Step 1: Check if the asset degradation process can be approximated by a Markov chain. In this step, the following major assumptions need to be justified:

The degradation process can be discretized into finite states. The asset's condition will be in one of these states at a particular time point and only changes once between two discrete time points (i.e., within one time interval). After each change, the asset's condition either remains in its current state or deteriorates to the next state with a certain transition probability. The transition probabilities are time independent. The resident time of the asset's condition at its current state does not affect its transition probabilities moving from the current state to the next state. The state of the asset's condition in the next time point is only determined by its current state and not affected by its historical conditions.

Step 2: Discretize the entire degradation process of assets into finite condition states. The number of states is determined by the characteristics of the degradation process, condition assessment, or monitoring methods, and the availability of historical condition monitoring data. Those key points that depict the characteristics of the degradation process have to be chosen as states. Under the conditions that the states can sufficiently describe the major characteristics of the process and satisfy the assumptions for Markov chains, try to minimize the number of states. More states can describe a process more precisely but also mean that more historical condition observation data are needed to estimate transition probabilities. In engineering practice, an engineering asset degradation process can usually be divided into a state referred to as good, a state referred to as failed, and one or two other states between these two states.

Step 3: Determine time intervals for calculating the changes of asset condition status. The time intervals are mainly determined by the changing speed of the degradation process, inspection, or condition data acquisition intervals and the condition states defined in *Step 2*. Time intervals cannot be too small or too large. If time intervals are too small, the calculated transition probabilities could be very small, while too large intervals may violate the assumption that the asset's condition only changes once during one time interval. In practice, one should try to use equal time intervals and make the intervals as large as possible. In addition, the time interval should be a common factor of the inspection intervals so that all observation data of asset health conditions can be used for the estimation of transition probabilities.

Step 4: Draw the Markov chain of the degradation process. This step is to help understand the changes between states and identify the transition probabilities to be estimated. An example is shown in Figure 17.8. In this

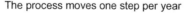

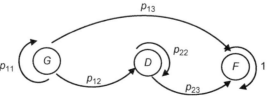

FIGURE 17.8   An example of the Markov chain of a degradation process.

example, a degradation process is divided into three states $\{S_1 = \text{good},$ $S_2 = \text{deteriorated}, S_3 = \text{failed}\}$. The time interval for the calculation is 1 year. An asset currently in a good state can still be good, or deteriorated, or failed in the next year, but a failed asset will be still in failed state in the next year when maintenance is not considered. Theoretically, for a process divided into n states, there are n^2 transition probabilities to be estimated. However, in asset management practice, most of these transition probabilities can be identified from knowledge of asset degradations. In this example, only five transition probabilities $\{p_{11}, p_{12}, p_{13}, p_{22}, p_{23}\}$ need to be estimated as $p_{33} = 1$ and the others are equal to zero.

Step 5: Estimate the transition probabilities. This is the most important step in the Markov-chain-based asset health prediction approach because as mentioned above, the asset health status at any time can be predicted once the transition probabilities and asset initial status are known. Hoskins et al. (1999) presented a least squares approach for the estimation of transition probabilities using historical health inspection and assessment data. This approach aims to minimize the total errors between the expected number and the observed number of assets in each state. Assume there is a population of identical assets and their degradation process can be divided into n states $\{S_1, S_2, ..., S_n\}$. M health condition assessments have been made on these assets at M inspection times $\{t_1, t_2, ..., t_M\}$ in terms of the condition states. The condition assessments can be given by human or condition monitoring systems and recorded using Table 17.1.

Then, these historical condition assessment data can be used to estimate the transition probabilities.

Let $e(\vec{P})$ be the total errors. Here $\vec{P}$ is the transition matrix comprising all transition probabilities to be estimated.

Let N_k^t be the total number of identical assets that were inspected at inspection time t_k ($k = 1, 2, ..., M$); N_{ki}^o the total number of identical assets that were observed in state S_i ($i = 1, 2, ..., n$) at inspection time t_k; N_{ki}^e the

---

TABLE 17.1 Assessment Data of Asset (Asset Name and/or ID)

Inspection Time (age)	Total Inspected Assets	The Number of Assets in S_1	The Number of Assets in S_2	...	The Number of Assets in S_n
t_1				...	
t_2				...	
...				...	
t_M				...	

total number of identical assets that are expected in state S_i ($i = 1,2,...,n$) at inspection time t_k; $P(S_i|t_k)$ the probability that an asset stays in state S_i ($i = 1,2,...,n$) at inspection time t_k. This probability can be calculated using the transition probabilities based on Markov theory.

Then, the expected number of assets in state S_i ($i = 1,2,...,n$) at inspection time t_k and the total errors are given by

$$N_{ki}^e = N_k^t P(S_i|t_k) \tag{17.23}$$

$$e(\vec{P}) = \sum_{1 \le i \le n \quad 1 \le k \le M} (N_{ki}^e - N_{ki}^o)^2 \tag{17.24}$$

The probabilities $\hat{p}_{ij}(i,j = 1,2,...,n)$ that make $\sum_{1 \le i \le n \quad 1 \le k \le M} (N_{ki}^e - N_{ki}^o)^2$ minimal are the estimated transition probabilities.

Step 6: Predict the health conditions of an asset using the estimated transition probabilities, and its initial probability vector which describes its start health status at start time t_0. The start time is the time point for measuring inspection times $\{t_1,t_2,...,t_M\}$. For engineering assets, the asset commission time is usually used as the start time. As an engineering asset is normally at a good condition at the commission time, its initial probability vector corresponding to the example shown in Figure 17.8 is $\{1,0,0\}$.

The Markov-chain-based health prediction method is simple but requires a large population of the identical assets. Therefore, this method is more suitable for linear assets which can be divided into a large number of identical segments. This method has been extended through combining it with Bayesian method (Sun et al., 2011). A limitation of Markov chains is that their transition probabilities are independent of their resident times in the current states. Therefore, the Markov-chain-based method is not suitable for the scenario where the asset condition is more likely to move from the current state to the next state if it has resided in the current state for a longer time. To model this scenario, a semi-Markov approach can be considered (Black et al., 2005). Another limitation of the Markov-chain-based method is that it requires direct health assessment information (often termed direct indicators) for estimating transition probabilities. In reality, direct indicators could be unavailable. In this case, other state space models that enable health prediction using indirect indicators, which represent asset health conditions indirectly, can be applied. The hidden Markov model (HMM) (Jardine et al., 2006) and the Gamma-based state space model (Zhou et al., 2011) are two of such models. HMM has been well established and reported in the literature, while the Gamma-based state space model appeared recently. Therefore, in the following section, we only introduce the Gamma-based state space model.

17.3.2 Gamma-Based State Space Model

The Gamma-based state space model was recently proposed for asset health condition predictions (Zhou et al., 2011). The basic concept of this model is that the degradation state process is not observable or only partially observable, but it can be revealed by some indirect indicators such as vibration. The indirect indicators can be observed or measured. Their measurements are called observations which often include the information of the indirect indicators and some measurement noise (see Figure 17.9). Then the observations can be used for assessment and prediction of the asset degradation process, although the relationship between the observations and the degradation process may have some uncertainties.

This model is superior to other state space models as it no longer requires the discretization of continuous time and continuous degradation processes. The commonly adopted assumptions that the increments of degradation processes follow the *Gaussian* distribution and the observations have linear relationships with degradation processes are also relaxed. Unlike the Markov chain which models asset degradation processes only, the Gamma-based state space model considers both the degradation processes and the indirect health condition indicators of assets. Most condition monitoring parameters are indirect indicators of asset health conditions. The Gamma-based state space model assumes that the increments of an asset degradation process

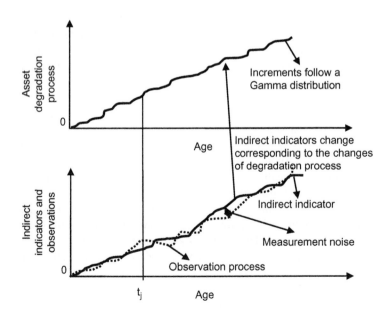

FIGURE 17.9   Concept of the Gamma-based state space model.

follow a Gamma distribution and the indirect indicators are correlated with the degradation process and can reveal the degradation indirectly.

Mathematically, the degradation process of an engineering asset is described by the following state equation:

$$S(t + \Delta t) - S(t) \sim \text{Gamma}[\Delta \xi(t), \gamma] \qquad (17.25)$$

where t is the age of the asset and $S(t)$ is the degradation process of the asset. Gamma$[\Delta \xi(t), \gamma]$ represents a Gamma distribution function with a time-dependent shape parameter and a constant scale parameter. The function $\Delta \xi(t)$ is commonly calculated using the following equation:

$$\Delta \xi(t) = \xi(t + \Delta t) - \xi(t) = \alpha(t + \Delta t)^\beta - \alpha t^\beta \qquad (17.26)$$

where α and β are two constants.

The indirect indicators are given by the following observation equation:

$$X(t) = \varnothing[S(t)] + \varepsilon \qquad (17.27)$$

where $X(t)$ is the observation process. Parameter ε includes both the measurement noise and the uncertain relationships between the degradation process and the indirect indicators. $\varnothing[S(t)]$ represents the changing process of the indirect indicators.

In practice, the relationship between the indirect indicators and the corresponding degradation process is often formulated using a power function, and the noise is often assumed to be independent and identically Gaussian distributed with zero mean, i.e.,

$$X(t) = \zeta \cdot [S(t)]^\eta \qquad (17.28)$$

and

$$\varepsilon \sim N(0, \sigma^2) \qquad (17.29)$$

where ζ and η are two constants, and σ is the standard deviation of the noise.

To use the Gamma-based state space model, an optional parameter setting is $\{\alpha, \beta, \gamma, \zeta, \eta, \sigma\}$. An expectation-maximization (EM) algorithm has been developed to estimate these parameters when partial direct indicators are available based on the following scenario (Zhou et al., 2011):

1. An engineering asset has been inspected M times at time t_k ($k = 1, 2, \ldots, M$).
2. Over the M inspections, only $l(1 \leq l < M)$ asset's health assessments (i.e., the data of direct indicators), state s_i ($i = 1, 2, \ldots, l$), have been obtained. Note that at least one direct indicator data should be available.
3. Only one indirect indicator is monitored. Over the M inspections, M indirect indicator measurements x_k ($k = 1, 2, \ldots, M$) have been obtained. These indirect indicator data can be divided into two categories based on the availability of direct indictor data: one category includes those indirect

indicator measurements that have corresponding direct indicator data. The indirect indicator measurements in this category are denoted by x_{kA} ($kA = 1,2,\ldots,l$). The other indirect indicators are in the second category and denoted by x_{kU} ($kU = 1,2,\ldots,M - l$).

The procedure of the proposed EM algorithm is shown in Figure 17.10. Here, we only describe these steps briefly. For more details, refer to Zhou et al. (2011).

Step 1: If some asset health assessment data (or direct indicators) are available and their corresponding indirect indicators have also been obtained, a complete likelihood function can be developed:

$$l(\vec{\theta}|s_i, x_i, i = 1{:}l) = \ln[f(s_i, x_i|\vec{\theta}, i = 1{:}l)] \qquad (17.30)$$

where $l(\cdot)$ and $f(\cdot)$ are log-likelihood function and PDF, respectively. The initial parameters can be estimated by maximizing the likelihood function. When the observations of direct indicators are not sufficient, the method of moments can be used to obtain the initial parameters.

Step 2 (E-step): The complete likelihood function is divided into two parts: the first part is only associated with the degradation process and

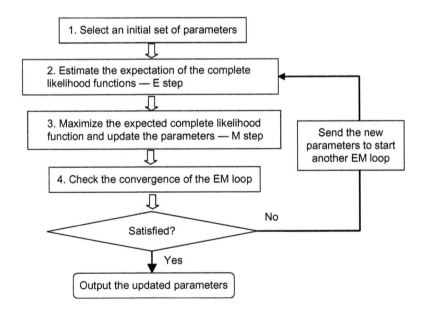

FIGURE 17.10   The process of the EM algorithm.

the first three parameters $\vec{\theta}_1 = \{\alpha, \beta, \gamma\}$ and the second part is only associated with the observations and the last three parameters $\vec{\theta}_2 = \{\zeta, \eta, \sigma\}$:

$$\ln[f(s_k, x_k|\vec{\theta}, k = 1{:}M)] = \ln[f(s_k|\vec{\theta}_1, k = 1{:}M)] + \ln[f(x_k|\vec{\theta}_2, s_k, k = 1{:}M)]$$

$$(17.31)$$

The expectation of the two parts is estimated using a particle smoothing algorithm:

$$E[\ln[f(s_k|\vec{\theta}_1, k = 1{:}M)]] = E\left[\ln\left[\prod_{k=2}^{M} \text{Gamma}(s_k - s_{k-1};\ \alpha t_k^{\beta} - \alpha t_{k-1}^{\beta}, \gamma)\right]\right]$$

$$(17.32)$$

$$E\left[\ln\left[f(x_k|\vec{\theta}_2, s_k, k = 1{:}M)\right]\right]$$
$$= -M\ln(\sigma) - \frac{M}{2}\ln(2\pi) - \frac{1}{2\sigma^2}\sum_{k=1}^{M}(x_k^2 - 2x_k\zeta E[s_k^{\eta}] + \zeta^2 E[s_k^{2\eta}]) \qquad (17.33)$$

where $E[\,\cdot\,]$ stands for the expectation of $[\,\cdot\,]$.

This step also involves the identification of the state process of the Gamma-based state space model.

Step 3 (M-step): Generate a new set of parameters through maximizing the expected complete likelihood function that is established in E-step.

Step 4: The convergence check is to ensure both the accuracy of the estimated parameters and computational efficiency. A commonly used criterion for the convenience of an EM algorithm is the increment of the marginal likelihood function. However, due to recursive Monte Carlo sampling, this criterion is not suitable for the Gamma-based state space model. The computation could be time consuming. Therefore, to increase computational efficiency, one should avoid using the increment of the marginal likelihood function as the convenience indicator for the EM algorithm of the Gamma-based state space model. A good alternative is the increment of the relative likelihood function because calculating this function does not involve additional Monte Carlo sampling.

After the required six parameters have been estimated, the Gamma-based state space model can be used to predict the failure time or remaining useful time of an engineering asset. As the Gamma-based state space model describes the degradation processes of engineering assets, the failure mentioned here is a type of "soft" failure, i.e., the failure happens when the degradation process of an asset reaches a predefined threshold. Therefore, predicting a failure is basically predicting when a degradation process will cross its predefined threshold. As "soft" failures are not real physical break-down of assets, they may or may not be detectable. If the failure of an engineering asset is detectable and the current inspection found that it has not failed, its reliability function of the asset $R(t)$ is given by

$$R(t) \quad = \Pr(T > t | x_k, s_c < s_h, k = 1{:}c)$$
$$\qquad = \int_0^{s_h} \Pr(T > t | s_c) f(s_c | x_k, s_c < s_h, k = 1{:}c) ds_c \tag{17.34}$$

where c stands for the current inspection, s_c is the current direct indicator's result, s_h is the predefined failure threshold, and T is the time to the first failure.

If the failure of an engineering asset is detectable,

$$R(t) = \Pr(T > t | x_k, k = 1{:}c) \tag{17.35}$$

The reliability function, Eq. (17.35), can be expressed as a function of the results of particle smoothing so that Monte Carlo simulations can be used to predict the reliability of the asset. The major advantage of the state space models is that it can well depict asset degradation processes using condition monitoring data. A drawback of this modeling approach is that it is normally complex and often involves a large number of simulations. In addition, when modeling an asset degradation process using a specific state space model, one should check if the real degradation process can be simplified as a particular stochastic process required by the model. From the above description, it can be seen that the Gamma-based space state model formulates the relationship between the health degradation process of an asset and its indirect condition indicators, and updates the asset health assessment and predictions using the recent measurements of these indicators. This observation is also true to other state space models such as HMM. However, in asset management practice, maintenance personnel may also want to see the influences of environmental factors of an asset such as working load and maintenance activities to the health conditions of the asset. In this case, one should use other models such as additive hazard model (AHM).

17.3.3 Additive Hazard Model

AHM is also called additive risk model (ARM) (Ma et al., 2006; McKeague and Sasieni, 1994; Seetharaman, 2004). It is a member of the covariate model family. Unlike the state space models which express the health conditions of engineering assets to various states, covariate models describe asset health condition using hazard functions. The hazard of an asset is its instantaneous failure rate and often denoted by $h(t)$. Mathematically, hazard is defined as

$$h(t) = \lim_{\Delta t \to 0} \frac{F(t + \Delta t) - F(t)}{\Delta t \, R(t)} \tag{17.36}$$

where $F(t)$ is the cumulative distribution function (CDF). Hazard functions have an important property:

$$\text{Their values can never be negative, i.e., } h(t) \geq 0 \qquad (17.37)$$

For more details about hazard, CDF and reliability function, see the next chapter. Hazard can be used to describe the health conditions of engineering assets because more deteriorated assets are more likely to fail, i.e., they have a higher hazard. Great efforts have been made to model the relationships between hazard and condition indicators which are often called covariates in reliability engineering (Jardine et al., 2006). Correspondingly, a model that describes the relationships between hazard and covariates is termed a covariate model. Covariates can be classified into internal covariates and external covariates based on their relationships to asset degradation processes. The indirect condition indicators used in the state space models generally belong to internal covariates which reflect the degradation conditions of engineering assets. In contrast to the internal covariates, external covariates are those factors that can influence (accelerate or decelerate) the degradation processes of engineering assets. Examples of external examples include work loading, lubrication conditions, and environmental temperature. As an example, Figure 17.11 illustrates the relationship of asset health conditions, internal covariates, and external covariates.

Various covariate models have been developed. More than 10 such models are reviewed in Gorjian et al. (2009). These models have been developed according to different assumptions about the relationship between hazard and covariates. In particular, AHM assumes that an asset has a baseline hazard and its real hazard equals the sum of its baseline hazard plus the influence of its covariates. The underlying concept of the AHM is that the changes of some factors (covariates) of an asset will cause changes of its hazard. Mathematically, AHM is expressed as

$$h(t) = h_0(t) + \Psi[\vec{Z}(t)] \qquad (17.38)$$

where $h_0(t)$ is the baseline hazard representing the hazard without the influence of the covariates, i.e., the hazard when $\Psi[\vec{Z}(t)] = 0$. Vector $\vec{Z}(t)$ consists of the covariates that affect the hazard and $\Psi[\vec{Z}(t)]$ is a covariate function.

Selection of the forms of the baseline hazard function and the covariate function is a skilful task. Basically, the baseline hazard function of an asset should be determined from the historical failure characteristics of the asset. When this is difficult because of the lack of historical failure data, the baseline hazard function is often assumed to have the form of a Weibull distribution:

$$h_0(t) = \frac{\beta_w}{\alpha_w} \left(\frac{t}{\alpha_w} \right)^{\beta_w - 1} \qquad (17.39)$$

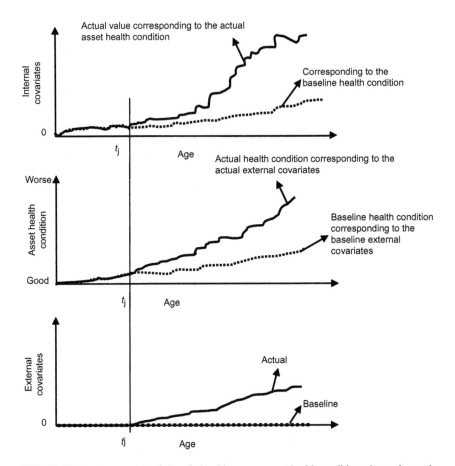

FIGURE 17.11   An example of the relationship among asset health conditions, internal covariates and external covariates.

where α_w and β_w are the scale and shape parameters of the Weibull distribution, respectively.

The covariate function can take the form of an exponential function (Seetharaman, 2004):

$$\Psi[\vec{Z}(t)] = e^{-\vec{\gamma} \cdot \vec{Z}_{(t)}} \tag{17.40}$$

where $\vec{\gamma}$ is a vector consisting of the weighting parameters of individual covariates. The outcome of $\vec{\gamma} \cdot \vec{Z}(t)$ can be interpreted as a composite indicator.

In this case, the AHM has the following form:

$$h(t) = h_0(t) + e^{-\vec{\gamma} \cdot \vec{Z}_{(t)}} \tag{17.41}$$

Note that $e^{-\vec{\gamma}\cdot\vec{Z}(t)}$ can never be zero. As a result, the hazard given in Eq. (17.41) will be always greater than its baseline hazard even though the asset is operating at its baseline conditions. This result is inconsistent with the definition of the baseline hazard.

Another commonly used form of covariate functions is multiplicity (Ma et al., 2006), i.e.,

$$\Psi[\vec{Z}(t)] = \vec{\gamma}\cdot\vec{Z}(t) \qquad (17.42)$$

In this case, the AHM becomes

$$h(t) = h_0(t) + \vec{\gamma}\cdot\vec{Z}(t) \qquad (17.43)$$

Use of this form can avoid the above problem. When $\vec{Z}(t) = 0$, the actual hazard of an asset is equal to its baseline hazard. However, this multiplicity form sometimes can generate negative hazard results, which violates the property of the hazard.

Equation (17.43) has a more general form (McKeague and Sasieni, 1994):

$$h(t) = \alpha_0(t)\vec{Z}_1(t) + \vec{\gamma}\cdot\vec{Z}_2(t) \qquad (17.44)$$

where $\alpha_0(t)$ is a time-dependent regression parameter (Ma et al., 2006). $\vec{Z}_1(t)$ and $\vec{Z}_2(t)$ are two covariate vectors.

The unknown parameters in Eq. (17.38) can be estimated using maximum likelihood estimation method. Rewrite Eq. (17.36) as the following form:

$$f(t) = h(t)\, R(t) \qquad (17.45)$$

where $f(t)$ is PDF, and

$$R(t) = \exp\left[-\int_0^t h(\tau)\mathrm{d}\tau\right] \qquad (17.46)$$

By using Eq. (17.46), Eq. (17.45) becomes

$$f(t) = h(t)\exp\left[-\int_0^t h(\tau)\mathrm{d}\tau\right] \qquad (17.47)$$

Substituting Eq. (17.38) into Eq. (17.47) gives

$$f(t, \vec{Z}(t)) = (h_0(t) + \Psi[\vec{Z}(t)])\exp\left[-\int_0^t (h_0(\tau) + \Psi[\vec{Z}(\tau)])\mathrm{d}\tau\right] \qquad (17.48)$$

Then the corresponding likelihood function is given by

$$L = \prod_{i=1}^{n} f(t_i, \vec{Z}(t_i)|\vec{\theta}) \tag{17.49}$$

where L is the likelihood function and n is the total number of observed assets that have independent, identical distributions. Parameter t_i is the failure time of ith asset and the $\vec{Z}(t_i)$ are the values of covariates at time t_i. Vector $\vec{\theta}$ includes all unknown parameters which need to be estimated.

In additional to AHM, another important member in the covariate model family is proportional hazard model (PHM). PHM follows the same concept as AHM, i.e., the changes of covariates of an engineering asset will alter the hazard of the system. However, it assumes that the real hazard of an asset is equal to the product of its baseline hazard multiplied by a covariate function of the asset. PHM is a commonly used covariate model in engineering asset reliability analysis. Although having been used for both external covariates and internal covariates, PHM is more suitable for modeling the influence of external covariates on hazards. The details of this model will be introduced in the next chapter.

Unlike AHM and PHM, proportional covariate model (PCM) adopts the similar concept used by the Gamma-based state space model. It assumes that the changes of hazard of an asset can be revealed by some internal covariates (condition indicators) of the asset. The relationships between hazards and the internal covariates can be represented by a baseline covariate function. The magnitudes of the internal covariates equal the product of the baseline covariate function multiplied by the hazard of the asset. Obviously, PCM is a covariate model specifically developed for modeling the relationships between hazards and internal covariates. The details of PCM are also described in the next chapter.

So far we have only considered the scenario where the failure times of all the assets have been observed, i.e., we have complete failure data. However, in reality, complete failure data are not always available. The scenario where only part of failure times have been observed, i.e., having incomplete data or censored data are discussed in the following section and the next chapter.

17.3.4 Artificial Neural–Network-Based Prediction Approach

Artificial neural network (ANN) is one of the AI technologies. The methods described so far can be categorized as statistical methods. These methods usually adopt different assumptions about failure distributions and/or the relationships between hazards and covariates. In reality, these assumptions are often indefensible because of the diversity of covariates and the complexity of engineering systems. Compared with these statistical methods, AI methods do not require such assumptions because they can capture the hidden complex relationships from data. Therefore, various AI methods including ANN, expert systems, fuzzy logic, Bayesian networks, multiagent systems, and support vector machine (SVM) have been used for engineering asset condition assessment and predictions. Due to the limitation in the length of chapters, we only introduce ANN-based prediction approaches.

ANNs are computer programs that are coded to simulate the property of biological neural networks in the human brains. An ANN is composed of an input layer, an output layer, and at least one hidden layer between the input layer and the output layer. Each of these layers consists of a number of nodes that mimic biological neurons, but an output layer often has one node only. An ANN can be represented using a directed acyclic graph as shown in Figure 17.12.

ANNs have supervised learning ability, i.e., they can learn from examples and recognize the complex nonlinear relationships between inputs and outputs. This ability of ANNs can be employed to build up the relationships between condition data and asset health conditions, and predict future conditions of assets.

ANN-based approaches have been developed to predict engineering asset reliability (Heng et al., 2009) or hazard (Yu et al., 2010) based on asset failure history and covariates. Figure 17.13 illustrates the major procedure of an ANN-based reliability hazard prediction approach given by Heng et al. (2009).

The most critical step in this procedure is *Step 1*. This step is to generate a target dataset for a selected asset i ($i = 1, 2, \ldots, n$) for training and testing a neural network. The dataset contains a number of the historical reliability values of the asset at successive time points:

$$\vec{R}_i = \left\{ \begin{array}{c} R_i(t_k) \\ R_i(t_{k-1}) \\ \vdots \\ R_i(t_{k-d+1}) \end{array} \right\}, \quad i = 1, 2, \ldots, n \tag{17.50}$$

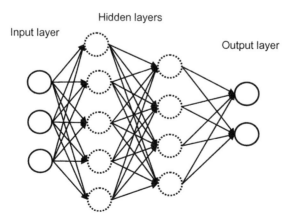

FIGURE 17.12   An example of a two-hidden-layer feedforward ANN.

where $\vec{R}_i$ stands for the target dataset and d is the size of the dataset. $R_i(t_k)$ is the consolidated reliability function of asset i ($i = 1,2,\ldots,n$) at time point t_k. According to Heng et al. (2009), this reliability takes the mean value of the population reliability, $R_{pi}(t)$, and the condition-indicated reliability, $R_{ci}(t_k)$, of asset i. Reliability $R_{pi}(t)$ is estimated based on the historical lifetime data and maintenance history of a population of assets including asset i, whereas reliability $R_{ci}(t_k)$ is estimated using the historical internal covariate data that indicate the degradation of asset i. Consideration of $R_{ci}(t_k)$ is necessary because $R_{pi}(t)$ is the reliability of the entire population of the engineering assets. However, in EAM, the reliabilities of individual assets are more important for making CBM decisions. The health condition indicators (i.e., internal covariates) of an asset certainly provide more accurate information about the reliability condition of the asset.

The collection of the lifetime data can be one of the following three scenarios:

1. The failure of an asset has been observed and the failure time has been recorded. As mentioned in Section 17.3.2, when condition degradation indicators (internal covariates) are employed, "soft" failures are considered. A "soft" failure occurs in an engineering asset when its condition degradation indicators cross its predefined failure threshold which is usually determined by historical data and requirements for the asset's functionalities.

2. The operation of an asset is terminated before it reaches its degradation failure state. The degradation condition just before the termination has not been recorded. An example of this scenario is that an asset has an unexpected sudden failure.

3. The operation of an asset is terminated before it reaches its degradation failure state, but the degradation condition just before the termination has

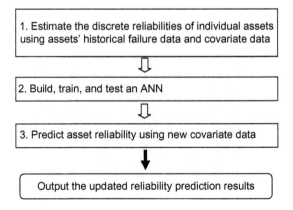

FIGURE 17.13 The process of the ANN-based reliability prediction approach.

been recorded. An example of this scenario is that a preventive replacement is carried out as condition monitoring shows the asset is going to fail soon.

A lifetime dataset of an asset population is considered to have complete failure data if it comprises the failure times of all assets in the population, or otherwise, it is regarded to have censored data (incomplete failure data). For complete failure data, assume that $n_s(t_{j-1})$ identical assets were functional at time t_{j-1} and $n_f(t_j - t_{j-1})$ of these assets failed over the period of $t_{j} - t_{j-1}$ $(j = 1,2,\ldots,k)$. Then the discrete population reliability can be calculated using the following formula:

$$R_{pi}(t_k) = \prod_{j=1}^{k} \left(\frac{n_s(t_{j-1}) - n_f(t_j - t_{j-1})}{n_s(t_{j-1})} \right) \tag{17.51}$$

For censored data, the Kaplan–Meier method (Kaplan and Meier, 1958) can be used to estimate the discrete population reliability (Heng et al., 2009).

The condition-indicated reliability of an asset is defined as the probability that the asset's covariate values have not exceeded its predefined failure threshold:

$$R_{ci}(t_k) = \Pr(T > t_k | \psi_i[\vec{Z}_i(t_k)] < \psi_{ih}) = \int_0^{\psi_{ih}} f(\psi_i(\vec{Z}_i)|t_k)d\psi_i \tag{17.52}$$

where $f(\cdot)$ is the PDF of the covariate values of asset i over time. Function $\psi_i(\vec{Z}_i)$ is the composite health indicator of the asset, and ψ_{ih} is the predefined failure threshold.

It is worth pointing out that the reliability predicted using the ANN-based reliability prediction approach can increase at some time points because of the uncertainties of the condition monitoring data (Heng et al., 2009). These increased results violate the property that reliability decreases monotonically. The ANN-based hazard prediction approach (Yu et al., 2010) can avoid this drawback. The major procedure of the ANN-based hazard prediction approach is the same as that used in the ANN-based reliability prediction approach as shown in Figure 17.13. One only needs to replace the discrete reliabilities with discrete hazards in *Step 1*. There are different definitions of the discrete hazard (Elsayed, 1996; Shaked et al., 1995). In this chapter, in line with the definition of hazard given in Eq. (17.36), discrete hazard at time t_k is defined as the average failure rate of an asset during a specific time interval immediately after time t_k given it has survived until the start point t_k of this interval.

For complete failure data, the historical discrete hazards of an engineering asset population is calculated by (Elsayed, 1996)

$$h_k = \frac{n_f(t_{k+1} - t_k)}{(t_{k+1} - t_k)n_s(t_k)} \tag{17.53}$$

where t_k and t_{k+1} are the start points of the kth and $(k + 1)$th time intervals, respectively. Parameter h_k is the discrete hazard at time t_k, $n_s(t_k)$ is the number of assets that have survived until time t_k, and $n_f(t_{k+1} - t_k)$ is the number of assets that failed in the interval $(t_k,t_{k+1}]$.

After the target dataset is obtained, a neural network can be constructed and then be trained and tested using the target dataset and historical covariate data (i.e., *Step 2* in Figure 17.13). Note that covariate data can also have incomplete problems (Heng et al., 2009; Yu et al., 2011). Various factors can cause covariate data to be missing. Two major causes are the missing of some data records and the failure of condition monitoring system which results in the required data during a certain time period have not been properly collected. However, even though all data have been collected and recorded, we may still encounter incomplete problems when multiple covariates are used in computation. The reason is that different covariates may have different data acquisition frequencies and not be collected at the same time. Some techniques to tackle incomplete covariate data problems include the following:

- Discarding incomplete segments if they only occupy a very small portion of the entire dataset.
- Obtaining the missing part of a covariate dataset using the available data of the same covariate through interpolation methods.
- Calculating the missing part of a covariate using the available data of other covariates that are correlated with it through regression method, Gaussian mixture model, and AI approaches.

Nevertheless, we are not going to discuss these techniques further. Readers who are interested in this topic can refer to other articles such as Yu et al. (2011).

Once a neural network has been trained and tested, it can be utilized for reliability/hazard prediction with the current covariate data, allowing for optimizing maintenance decisions.

17.4 CBM DECISION MAKING

CBM decision making involves the following major tasks:

1. Determine which assets need to be monitored and what type of condition monitoring technologies should be deployed.
2. Determine what data should be collected, how frequently they should be collected, and how long they should be stored.
3. Determine maintenance types and degradation thresholds for carrying out different maintenance activities.

The first task often involves FMECA (ISO17359, 2003a). The second task often involves symptom analysis and failure characteristics analysis.

However, the decisions for these two tasks heavily depend on the decisions for the third task. These are made according to business constraints and objectives. CBM is also objective driven and usually involves cost–benefit analysis. We will further discuss the cost–benefit analysis in the next chapter.

EAM involves various decisions which may have the same or different objectives. These decisions can be divided into four categories: (1) strategic decisions; (2) technical decisions; (3) implementation decisions; and (4) reactive decisions (Sun et al., 2008). Strategic decisions are normally made at the executive level in an organization to decide the disposal or replacement of large capital assets. This type of decision making often needs long-term overall asset health predictions, typically over several years or even decades. Technical decisions are usually made at the managerial level of an organization, focusing on major preventive maintenance and renewal of critical equipment and engineering systems. These decisions often require medium-term (annual, quarterly, or monthly) overall health predictions. Implementation decisions are typically made at the operational level of an organization to decide when to shut down a severely degraded machine and what type of preventive maintenance needs to be carried out. Making of this type of decisions often demands more specific health condition prediction, typically at components level with the information of failure modes, failure locations, as well as asset remaining useful life. Reactive decisions are generally made at operational level to avoid catastrophic failures or severe failure consequences caused by abnormal assets. To make reactive decisions, decision makers need to know which components have been damaged, what are the failure modes, how severe the damage is, and what consequences will be caused if the abnormal assets are not shut down immediately.

CBM can be used to save engineering asset maintenance cost only when decision makers have right asset condition information and other required data such as maintenance cost and business demand. Without the necessary data, it is hard, if not impossible, to make optimal maintenance decisions. Either over- or under-maintaining assets increases cost. In addition, data collection, storage, and analysis are also costly. To understand data requirements, one needs to understand EAM decision-making processes which provide an effective tool for identifying data requirements. Compared with various specific decision-making processes, the generic split asset management decision-making process (Sun et al., 2008) is more useful because of its flexibility and adaptability. A simplified version of this process is shown in Figure 17.14.

From Figure 17.14, it can be seen that the first thing in asset management decision making is the identification of decision objectives; this is critical for making effective decisions. Decision objectives are generally derived from an organization's key performance indicators (KPIs). A most common decision objective is to minimize the total life cycle cost of an asset, which includes acquisition cost, inspection cost, maintenance cost, operational cost, loss due to shutdown of the asset, and disposal cost. Other commonly considered objectives include minimizing asset failure risks, maximizing business profit, maximizing customers' satisfaction, and maximizing environmental benefits. In practice, decisions are often optimized by basing them on a single objective. In this case, other business objectives become constraints. However, sometimes it could be necessary to optimize several objectives simultaneously. In this case, a multigoal optimization method is needed. In addition to the selection of decision objectives and optimization methods, decision makers also need to be aware of the following points:

1. Optimal asset management decisions are not fixed decisions. They could change over time, especially for the long-term maintenance decisions. Nearly all long-term maintenance decisions are preventive maintenance decisions. These decisions often need to be reviewed and updated, based on the up-to-date asset conditions, or otherwise, a decision which is optimal in the past could become a bad decision nowadays because of the changes in asset health condition, business demand, as well as asset operational and maintenance strategies.

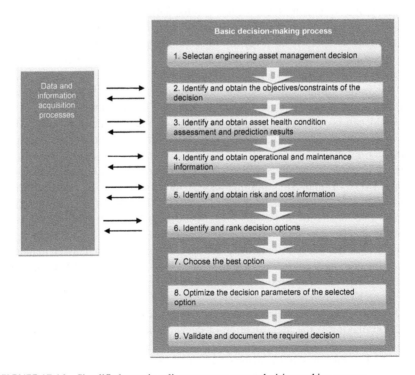

FIGURE 17.14   Simplified generic split asset management decision-making process.

2. Most existing maintenance decision optimization models/methods focus on a single asset or equipment. To achieve global optimization for the whole organization, the interactions among various assets and maintenance activities should be considered. For example, two assets have two different but near optimal renewal times according to their own total cost rate curves and business requirements. However, due to budget and resource limitation, as well as other constraints such as the requirements for service interruptions, these two assets often need to be renewed at the same time using a maintenance grouping strategy (Dekker et al., 1997). In this case, the original locally optimized renewal times of individual assets have to be adjusted to achieve a global optimization. There are various maintenance grouping optimization models. Some of them have been reviewed by Dekker et al. (1997).

More discussion of CBM optimization is presented in the next chapter.

REFERENCES

Baillie, D.C., Mathew, J., 1996. A comparison of autoregressive modelling techniques for fault diagnosis of rolling element bearings. Mech. Sys. Signal Proc. 10, 1–17.

Barnard, I., 2009. Engineering Asset Management: An Insurance Perspective. Fort Myers, FL, <reliabilityweb.com/>.

Black, M., Brint, A.T., Brailsford, J.R., 2005. A semi-Markov approach for modelling asset deterioration. J. Operat. Res. Soc. 56, 1241–1249.

Boashash, B., 1992. Estimating and interpreting the instantaneous frequency of a signal—Part 2: Algorithms and applications. Proc. IEEE 80, 540–568.

British Standards Institution, 1984. BS3811 Glossary of Maintenance Terms in Terotechnology. British Standards Institution, London.

Dekker, R., Wildeman, R.E., Schouten, F., 1997. A review of multi-component maintenance models with economic dependence. Math. Methods Operat. Res. 45, 411–435.

Elsayed, E.A., 1996. Reliability Engineering. Addison Wesley Longman, Inc., Reading, MA.

Gao, J.J., 2006. Fault self recovery of industrial plants—vision and state of the art. In: Mathew, J., et al. (Eds.), Proceedings of the First World Congress on Engineering Asset Management, Gold Coast, Australia, Springer, London.

Gorjian, N., Ma, L., Mittinty, M., Yarlagadda, P., Sun, Y., 2009. A review on reliability models with covariates. In: Kiritsis, D., et al. (Eds.), Proceedings of the Fourth World Congress on Engineering Asset Management, Athens, Greece, Springer, London.

Heng, A., Tan, A.C.C., Mathew, J., Montgomery, N., Banjevic, D., Jardine, A.K.S., 2009. Intelligent condition-based prediction of machinery reliability. Mech. Sys. Signal Proc. 23, 1600–1614.

Hoskins, R.P., Strbac, G., Brint, A.T., 1999. Modelling the degradation of condition indices. In IEEE Proceedings C: Generation, 146. Transmission and Distribution. pp. 386–392.

Huzii, M., 2007. Embedding a Gaussian discrete-time autoregressive moving average process in a Gaussian continuous-time autoregressive moving average process. J. Time Ser. Anal. 28, 498–520.

International Standard, 2003a. Condition Monitoring and Diagnostics of Machines—General Guidelines. ISO 17359.

International Standard, 2003b. Condition Monitoring and Diagnostics of Machines—Data processing, communication and presentation, Part 1: General Guidelines. ISO 13374-1:2003(E).

International Standard, 2006. Condition Monitoring and Diagnostics of Machines—Data processing, communication and presentation, Part 2: Data Processing", ISO/FDIS 13374-2:2006(E).

Jardine, A.K.S., Lin, D., Banjevic, D., 2006. A review on machinery diagnostics and prognostics implementing condition based maintenance. Mech. Sys. Signal Proc. 20, 1483—1510.

Kaplan, E.L., Meier, P., 1958. Nonparametric estimation from incomplete observations. J. Am. Stat. Assoc. 53, 457—481.

Kim, E., Tan, A.C.C., Yang, B.S., 2008. Parameter comparison of acoustic emission signals for condition monitoring of low-speed bearings. Aust. J. Mech. Eng. 6, 1—8.

Lee, J., Scott, L.W., 2006. Zero-breakdown machines and systems: productivity needs for next-generation maintenance. In: Mathew, J., et al. (Eds.), Proceedings of the First World Congress on Engineering Asset Management, Gold Coast, Australia, Springer, London.

Lin, T.R., Tan, A.C.C., 2011. Characterising the signal pattern of a four-cylinder diesel engine using acoustic emission and vibration analysis. In: Shen, G., Wu, Z. (Eds.), Proceedings of the World Conference on Acoustic Emission, 506—515, Chinese Society for Non-destructive Testing, Beijing, China.

Lin, T.R., Tan, A.C.C., Ma, L., Mathew, J., 2011. Estimating the loading condition of a diesel engine using instantaneous angular speed analysis. In: Lee, J., et al. (Eds.), Proceedings of the Sixth Annual World Congress on Engineering Asset Management (WCEAM 2011), Cincinnati, OH, Springer, London.

Ma, L., 2007. Condition monitoring in engineering asset management. In: Proceedings of the 12th Asia pacific Vibration Conference, Sapporo, Japan.

Ma, S., Kosorok, M.R., Fine, J.P., 2006. Additive risk models for survival data with high-dimensional covariates. Biometrics 62, 202—210.

McKeague, I.W., Sasieni, P.D., 1994. A partly parametric additive risk model. Biometrika 81, 501—514.

Morlet, J., Arens, G., Fourgeau, I., Giard, D., 1982. Wave propagation and sampling theory. Geophysics 47, 203—236.

Moubray, J., 1997. Reliability Centred Maintenance. Industrial Press, New York, NY.

Niu, G., Yang, B.S., Pecht, M., 2010. Development of an optimized condition-based maintenance system by data fusion and reliability-centered maintenance. Reliab. Eng. Sys. Saf. 95, 786—796.

Papoulis, A., 1991. Probability, Random Variables and Stochastic Processes. McGraw-Hill, New York, NY.

Peng, Z.K., Chu, F.L., 2004. Application of the wavelet transform in machine condition monitoring and fault diagnostics: a review with bibliography. Mech. Sys. Signal Proc. 18, 199—221.

Rangarao, K.V., Mallik, R.K., 2005. Digital Signal Processing: A Practitioner's Approach. John Wiley & Sons, Chichester, England.

Seetharaman, P.B., 2004. The additive risk model for purchase timing. Mark. Sci. 23, 234—242.

Shaked, M., Shanthikumar, J.G., Valdez-Torres, J.B., 1995. Discrete hazard rate functions. Comput. Operat. Res. 22, 391—402.

Shannon, C.E., Weaver, W., 1949. A Mathematical Theory of Communication. University of Illinois Press, Urbana, IL.

Shiroishi, J., Li, Y., Liang, S., Kurfess, T., Danyluk, S., 1997. Bearing condition diagnostics via vibration and acoustic emission measurements. Mech. Sys. Signal Proc. 11, 693—705.

Sun, Y., Fidge, C., Ma, L., 2008. A generic split process model for asset management decision-making. In: Ni, J., et al. (Eds.), Proceedings of the Third World Congress on Engineering Asset Management and Intelligent Maintenance Systems (WCEAM-IMS 2008), Beijing, China, Springer, London.

Sun, Y., Ma, L., Fidge, C., 2011. A Bayesian Markov process for reliability prediction. In: Singh, M., et al. (Eds.), Proceedings of the 24th International Congress on Condition Monitoring and Diagnostics Engineering Management, COMADEM International, Stavanger, Norway.

Wiseman, M., A history of CBM. <http://www.omdec.com/moxie/Technical/Reliability/a-history-of-cbm.shtml/> (accessed online 2011).

Yu, Y., Ma, L., Sun, Y., Gu, Y.T., 2010. Remaining useful life prediction using elliptical basis function network and Markov chain. Eng. Tech. 71, 494–498.

Yu, Y., Ma, L., Sun, Y., Gu, Y.T., 2011. Handling incomplete data in survival analysis with multiple covariates. In: Mathew, J., et al. (Eds.), Proceedings of the Fifth World Congress on Engineering Asset Management (WCEAM 2010), Brisbane, Australia, Springer, London.

Zhou, Y.F., Sun, Y., Mathew, J., Wolff, R., Ma, L., 2011. Latent degradation indicators estimation and prediction: a Monte Carlo approach. Mech. Sys. Signal Proc. 25, 222–236.

Chapter 18

Pipeline Vibration and Condition Based Maintenance

18.1 INTRODUCTION

Vibration of pipelines has been identified as the main cause of fatigue-induced piping failures (Blevins, 1990), which could lead to catastrophic financial and environment consequences. Two recent incidents in the oil and gas industry, the oil leakage from a ruptured pipeline transporting oil from Canada to America and the explosion of the Deepwater Horizon drilling rig in the Gulf of Mexico in 2010, have highlighted the importance of pipeline vibration mitigation and proactive/preventive pipeline maintenance to prevent similar disastrous occurrences.

Major causes to pipeline vibration can be grouped into two main categories: (1) mechanical-induced vibration and (2) flow-induced vibration. Mechanical-induced pipeline vibration is caused by the vibration transmission from pump, compressor, turbine, control valve, piping supports, and from random impact excitations. For instance, cavitation due to pump (e.g., insufficient net positive suction head, NPSH) or control valve in a pipeline system is one of the most typical phenomena causing pipeline vibration. Cavitation creates high-localized stress and emits shock waves forcing a pipeline to vibrate. The emitting shock waves can cause large pipe movement when the frequency of cavitation coincides with the natural frequencies of the pipe (Miller, 2001). Sudden closure or opening of valves can create "water hammer" effect in the pipeline causing tremendous pipeline vibration. Water hammer is the result of a sudden change in liquid velocity in a pipeline when the flow in the pipeline is quickly started, stopped, or is forced to make a rapid change in direction. The pressure spike (an acoustic wave) created at rapid valve closure can be many times higher than the working pressure of a pipeline. This acoustic pressure wave can cause severe problems from pipeline vibration to pipeline collapse due to fatigue. Miller (2001) discussed some of pipeline vibration involving control valves and presented a design to mitigate such vibration problem by limiting the fluid kinetic energy exiting the throttling point of the valve trim. Lewis et al. (1998) presented a case study to reduce the control valve-induced pipeline vibration in a pulp mill by using variable speed motor to drive the pumping system.

Offshore Pipelines. DOI: http://dx.doi.org/10.1016/B978-0-12-397949-0.00018-2

Flow-induced vibration can be caused by either internal flow or external flow to a pipeline. Structural failure due to flow-induced vibration is a common problem affecting the performance and reliability of pipelines. For internal flow, pipeline vibration is caused by unsteady flow, change of flow direction, change of pipe diameter, sudden closure or opening of control valves, and is also affected by the internal fluid flow velocity. Flow-induced vibration often occurs in the pipelines on offshore platforms when they are used to convey high-pressure oil and gas, and there are many undesirable geometric shapes such as too many bends and large variation sections in pipelines on offshore platforms. Large pipeline vibration would be generated when the frequency of internal flow disturbance in a pipeline coincides with the natural frequencies of the pipeline. Flow-induced vibration causes damage to supporting structures or pipeline ruptures leading to costly shutdown and severe environmental problems. In marine applications, pipelines vibrate in the presence of external cross flow of seawater caused by ocean current and wave. Flow-induced vibration of pipes has been a subject of considerable research in the last few decades.

Zou et al. (2005) employed the Ritz method in deriving a numerical solution to study the effects of internal fluid flow velocity, pressure, and the prestress of the pipeline on the vibration of composite pipelines. Zhong et al. (2007) conducted an experimental study on the effect of slug flow to pipeline vibration and found that pipeline vibration decreases in severe slug flow when gas flow rate increases and the liquid flow rate is kept at constant. Pipeline vibration increases and is up to the maximum amplitude when the liquid flow rate increases, and the gas flow rate is kept at constant. After this stage, the amplitude of pipeline vibration decreases but the vibration frequency increases. Housner (1952) provided an analytical solution to study the effect of internal flowing fluid on pipeline vibration. Laura et al. (1987) provided a brief analysis to correct an algebraic error in the solution of flexural vibration of a Pined-Pined supported fluid carrying pipeline given by Housner and others. The effect of such algebraic error on the result is found to be less than 5%. Lee and Kim (1999) presented a full coupled finite element formulation to study the vibration response of a branched pipeline system conveying high-pressurized unsteady internal flow. They found that the effect of the additional linear/nonlinear coupling terms in the present pipe dynamic theory could become significant at high fluid pressure and velocity. Biswas and Ahmed (2001) studied the optimum flow velocity in a pipeline for minimum flow-induced pipeline vibration. Koo and Park (1998) employed a periodic structural formulation to study the vibration reduction of a pipeline system using a periodic support.

In addition to the internal flow-induced vibration of onshore fluid carrying pipelines, subsea pipelines also subject to the excitation of external water flow. Vortex-induced vibration (VIV) is a particular concern of external flow-induced vibration which has attracted vast attentions from researchers and practitioners working in the field. A great number of publications have been produced in recent years. Sumer and Fredsøe (1995) provided a thorough review on the existing research of marine pipeline vibration and discussed the dominant factors concerning offshore pipeline vibration. Choi (2001) presented an analytical model to calculate the allowable free span length of a prestress pipeline to avoid the fatigue damage to the pipeline due to VIV by wave or current. The allowable free span length of offshore pipelines was also analyzed by Bakhtiary et al. (2007) taking into account the influence of the soil conditions of the supporting seabed. Xu et al. (1999) studied both free and forced in-line vibration of a multi-span offshore subsea pipeline due to vortex shedding. The coupling effect between pipeline vibration and seabed sand scour around the vibrating pipeline was studied by Gao et al. (2006). They showed that there are two phases in the process of sand scour around a vibrating pipeline. In the first phase, sand scour beneath pipeline without VIV. In the second phase, sand scour with VIV of the pipeline. They further illustrated that sand scour affects not only the vibration amplitude of the pipeline but also its vibration frequency. VIV and sand scour due to pipeline vibration were investigated experimentally by Yang et al. (2008). They observed that pipeline vibration and the development of local scour are two coupled physical processes. Zhao and Cheng (2010) presented a numerical study of local scour below a vibrating pipeline and found that pipeline vibration will increase the scour depth below the pipeline. Daei-Sorkhabi and Zehsaz (2009) studied the VIV of a thin symmetric airfoil by employing both numerical and experimental approaches. They showed that the response amplitude of VIV of the airfoil depends upon the initial flow condition. Van den Abeele et al. (2008) presented a numerical study of VIV of submarine pipeline and discussed the effectiveness of various vibration mitigation measures based on numerical simulation results.

Vibration will cause offshore pipelines to fail. However, there are numerous other factors affecting the health conditions of pipelines. Pipelines, like all other engineering assets, deteriorate and become more defective over time. To ensure their safety and reliability, we need effective maintenance strategy. Engineering asset maintenance can be defined as the combination of all technical and administrative actions used to maintain an engineering asset in, or restore it to its required functional state (British Standards Institution, 1984). Maintenance is commonly categorized into four policies: corrective, preventive, predictive, and proactive maintenances (PM) (Barnard, 2009; Moubray, 1997).

Corrective maintenance (CM) policy is the first generation of maintenance. It requires fixing an engineering asset only when it breaks. Therefore, it is often termed as breakdown maintenance. Preventive maintenance policy is the second generation of maintenance. It mainly consists of preventive overhauls of an engineering asset at scheduled and fixed intervals based on time or duty. Therefore, this policy is also called time/duty based maintenance policy. Predictive maintenance policy belongs to the third generation of maintenance. In this policy, the health condition of an engineering asset is monitored continuously. Once it reaches a predefined threshold, a preventive maintenance is carried out. As the maintenance decisions are made based on asset health condition, predictive maintenance is often termed as condition based maintenance (CBM). The fourth generation policy, PM policy, conventionally aims to avoid or to reduce the root causes of failures. Common causes of offshore pipeline failures include vibration, material defects, fatigue, weld crack, erosion, corrosion, improper installation and repairs, as well as third-party activities. As conventional PM policy requests the monitoring and correction of failure root causes, the policy can be termed as root cause based maintenance (RCBM) policy.

CM is generally not suitable for pipelines operating in harsh subsea environment even though it is still being used for low criticality pipelines. Pipelines would have various failures if they are not maintained properly. Oil leaking due to pipeline failures could have both severe financial and environmental consequences. An offshore pipeline failure could be an even worse scenario as repairing submarine equipment or facility could be a very costly and challenging task, and could take a long time to complete. For instance, fixing a leak of an offshore pipeline usually requires a large submarine vessel to the leakage location. The repair operation often takes a long period of time ranging from a few weeks to several months, and the technical challenge poised on the repair operation typically implies a very expensive repair cost tag.

Unlike CM policy, the time/duty based maintenance policy, the CBM policy, and the RCBM policy (i.e., conventional PM policy) all require to take maintenance actions prior to the occurrence of failures, either to avoid such an occurrence or to lower the occurrence probability and/or to reduce the failure consequences. In other words, time/duty based maintenance, CBM and RCBM, are all PM (Barnard, 2009), with different maintenance decision criteria.

Section 18.1 provides a background discussion of pipeline vibration and its causes, and highlights the need for proactive/preventive maintenance. The literature survey showed that most research works for pipeline vibration aimed to provide solutions to answer a particular phenomenon relating to pipeline vibration. For a better understanding of pipeline vibration problem and the important influence factors affecting the design of the allowable free span length of a pipeline, Section 18.2 aims to provide a systematic approach for pipeline vibration analysis and to summarize the most important factors affecting pipeline vibration. Some of the common practices to mitigate pipeline vibration are also discussed in this section. Section 18.3 will focus on the implementation and optimization of CBM for pipelines. The examples of disastrous incidents relating to oil and gas pipelines and offshore drilling rigs have highlighted the need of PM policies for maintaining the offshore pipelines and rigs. Among all PM policies, CBM is found to be the most suitable one for offshore pipeline maintenance, because the time/duty based maintenance is less reliable and often more expensive to implement, whereas the RCBM policy is usually costly and sometimes inapplicable (e.g., it is impossible to eliminate the effect of corrosive seawater). Both short-term maintenance strategy (i.e., deciding the next PM) and long-term maintenance strategy which involve determining multiple PM activities over a long time period are considered.

18.2 PIPELINE VIBRATION

To prevent structure fatigue caused by large vibration response of offshore pipeline, free span pipeline length has to be carefully chosen to avoid the coincidence between the natural frequencies of a pipeline and vortex shedding frequency of wave and current. Research shows that it is important to predetermine the free span length of a pipeline according to the flow velocity of the current.

18.2.1 Pipeline Vibration Analysis Using Beam Formulations

A simple and effective approach of analyzing pipeline vibration is to treat a pipeline as an elongated beam due to the typical large span length of pipeline segments between end supports. Examples of such approach can be found in the references (Biswas and Ahmed, 2001; Blevins, 1990; Choi, 2001; Koo and Park, 1998; Xu et al., 1999).

A pipeline segment supported by two end supports is shown in Figure 18.1. By considering the pipeline segment as an Euler–Bernoulli beam, the homogeneous governing equation of free flexural displacement of the pipeline (u) can be expressed as

$$EI \frac{\partial^4 u}{\partial x^4} + m_e \frac{\partial^2 u}{\partial t^2} = 0 \qquad (18.1)$$

where E and I are, respectively, the Young's modulus and moment of inertia of the pipe conduit, m_e is the effective mass per unit pipeline length. Equation (18.1) is derived by assuming that the beam is undergoing a small deflection.

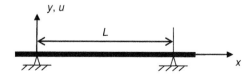

FIGURE 18.1   A typical pipeline segment and the associated coordinate system.

Equation (18.1) can be utilized to determine the natural frequencies of a pipeline, which is a primary concern of pipeline vibration. The fourth order differential equation given by Eq. (18.1) can be solved by the propagating wave approach as

$$u = A_1 e^{i(kx - \omega t)} + A_2 e^{-i(kx + \omega t)} + A_3 e^{(kx - i\omega t)} + A_4 e^{-(kx + i\omega t)} \tag{18.2}$$

where $k = \sqrt[4]{(m_e \omega^2 / EI)}$ is the flexural wave number of the pipe, ω is the angular frequency in radians per second, and i is the imaginary number. The first two terms in Eq. (18.2) are traveling wave components, and the last two terms are near field exponential decaying waves.

For a finite length pipeline segment, it is more convenient to express the solution with trigonometric functions. This is accomplished by invoking Euler's formula so that Eq. (18.2) becomes

$$u = [B_1 \sin(kx) + B_2 \sinh(kx) + B_3 \cos(kx) + B_4 \cosh(kx)]e^{-i\omega t} \tag{18.3}$$

Equation (18.3) can be used to determine the natural frequencies of a pipeline segment when the boundary conditions of the pipeline are specified.

18.2.2 Boundary Conditions and Natural Frequencies of a Pipeline Segment

In general, boundary conditions of a pipeline segment between two end supports can take the following forms:

a. Pined-Pined boundary conditions
b. Fixed-Fixed boundary conditions
c. Elastic supporting conditions

or in a mixture of the three boundary conditions depending upon the form of practical boundary supports. For example, if a pipe is supported by heavy concrete blocks at each end, the boundary conditions of the pipe can be considered as Fixed-Fixed conditions. If the pipe is supported by slender steel frames or rested on uneven seabed, the boundary conditions can be regarded as Pined-Pined conditions. If the effects of neighboring pipe segments are to be considered, the elastic supporting conditions could be applied.

18.2.2.1 Pined-Pined Boundary Conditions

For this boundary condition, both flexural displacement and bending moment at each end support of the pipe are zero, thus

$$u|_{x=0} = u|_{x=L} = EI \frac{\partial^2 u}{\partial x^2}\bigg|_{x=0} = EI \frac{\partial^2 u}{\partial x^2}\big|_{x=L} = 0 \tag{18.4}$$

Substituting Eq. (18.4) into Eq. (18.3), we have

$$B_2 = B_3 = B_4 = 0 \qquad (18.5a)$$

and the frequency equation,

$$\sin(kL) = 0 \qquad (18.5b)$$

From Eq. (18.5b), the flexural wave number of a Pined-Pined beam at resonance can be determined by

$$k_n = \frac{n\pi}{L}, \quad n = 1, 2, ..., \infty \qquad (18.6)$$

The natural frequencies of the pipe can now be obtained as

$$\omega_n = k_n^2 \sqrt{\frac{EI}{m_e}} \qquad (18.7)$$

and the mode shape function of the Pined-Pined pipe is given by

$$\phi_n(x) = \sin(k_n x) \qquad (18.8)$$

18.2.2.2 Fixed-Fixed Boundary Conditions

Both translational and angular displacements are zero for fixed boundary conditions, which are expressed by

$$u|_{x=0} = u|_{x=L} = \left.\frac{\partial u}{\partial x}\right|_{x=0} = \left.\frac{\partial u}{\partial x}\right|_{x=L} = 0 \qquad (18.9)$$

Substituting Eq. (18.9) into Eq. (18.3), we have

$$B_1 = B_3 = 0 \qquad (18.10a)$$

and the characteristic equation

$$\begin{bmatrix} \sinh(kL) - \sin(kL) & \cosh(kL) - \cos(kL) \\ \cosh(kL) - \cos(kL) & \sinh(kL) + \sin(kL) \end{bmatrix} \begin{Bmatrix} B_2 \\ B_4 \end{Bmatrix} = 0 \qquad (18.10b)$$

For a nontrivial solution, the determinant of Eq. (18.10b) has to be zero, which yields the following eigen-function for a Fixed-Fixed pipe:

$$\cosh(kL)\cos(kL) = 1 \qquad (18.11)$$

Equation (18.11) can be rewritten as

$$\tan\frac{kL}{2} = \pm \tanh\frac{kL}{2} \qquad (18.12)$$

Solution of Eq. (18.12) yields

$$k_n \cong \frac{\pi}{2L}[3.011, 5, 7, ..., (2n + 1)], \quad n = 1, ..., \infty \qquad (18.13)$$

The natural frequencies of the Fixed-Fixed pipe can once again be calculated by Eq. (18.7) using the modal wave number described by Eq. (18.13). The mode shape function of the Fixed-Fixed pipe is

$$\phi_n(y) = \sinh(k_n y) - \sin(k_n y) - \beta_n[\cosh(k_n y) - \cos(k_n y)] \tag{18.14}$$

where

$$\beta_n = \frac{\sinh(k_n L) - \sin(k_n L)}{\cosh(k_n L) - \cos(k_n L)} \cong \frac{\sinh(k_n L) - (-1)^n}{\cosh(k_n L)} \tag{18.15}$$

is the modal constant.

18.2.2.3 Elastic Supporting Conditions

If the effect and interaction of adjacent pipeline segments to a pipeline segment are to be taken into account, the boundary conditions of the pipeline segment can be expressed in terms of spring stiffness as

$$EI\frac{\partial^2 u}{\partial x^2}\bigg|_{x=0} = k_{rL}\frac{\partial u}{\partial x}\bigg|_{x=0}$$

$$EI\frac{\partial^2 u}{\partial x^2}\bigg|_{x=L} = k_{rR}\frac{\partial u}{\partial x}\bigg|_{x=L}$$

$$EI\frac{\partial^3 u}{\partial x^3}\bigg|_{x=0} = -k_{tL}u\big|_{x=0}$$

and

$$EI\frac{\partial^3 u}{\partial x^3}\bigg|_{x=L} = -k_{tR}u\big|_{x=L} \tag{18.16}$$

where k_{rL}, k_{rR}, k_{tL}, and k_{tR} are the equivalent rotational and translational spring stiffness at the left and the right ends of the pipe segment respectively. From Eqs (18.16) and (18.3), we have

$$\begin{bmatrix} kk_{rL} & kk_{rL} & EIk^2 & -EIk^2 \\ kk_{rR}\cos(kL)+EIk^2\sin(kL) & kk_{rR}\cosh(kL)-EIk^2\sinh(kL) & -kk_{rR}\sin(kL)+EIk^2\cos(kL) & kk_{rR}\sinh(kL)-EIk^2\cosh(kL) \\ -EIk^3 & EIk^3 & k_{tL} & k_{tL} \\ k_{tR}\sin(kL)-EIk^3\cos(kL) & k_{tR}\sinh(kL)+EIk^3\cosh(kL) & k_{tR}\cos(kL)+EIk^3\sin(kL) & k_{tR}\cosh(kL)+EIk^3\sinh(kL) \end{bmatrix}\begin{Bmatrix} B_1 \\ B_2 \\ B_3 \\ B_4 \end{Bmatrix} = 0 \tag{18.17}$$

By letting the determinant of the matrix in Eq. (18.17) equals zero for a nontrivial solution, the natural frequencies of the system can be determined.

For pipelines that undergo large flexural deflection, the shear deformation and rotational inertia effects of the pipe would need to be taken into account by using the so-called Timonshenko beam formulation. However, it is well known that Timonshenko beam formulation is most suitable to describe the flexural behavior of short beams or when a beam is subject to high-frequency excitation where the beam's bending wavelength is in the same magnitude scale of the beam thickness. For most pipelines, the pipeline span is usually substantially larger than the pipe's cross section. Furthermore, severe damage to pipelines is largely due to the low-frequency vibration. Therefore, pipeline vibration can be analyzed accurately and effectively by employing Euler–Bernoulli beam formulation. Timonshenko beam formulation for pipeline vibration is not in the interest of this text and will not be discussed further.

18.2.3 Free Span Length of a Pipeline

For safe operation of a pipeline, the free span length should be maintained within the allowable length to avoid the fatigue of the pipeline as a result of exceeding vibration amplitude causing by the resonance of the pipeline at different operating environments. This allowable pipeline span, according to DNV design guideline (Det Norske Veritas, 1998), is given by

$$L = \left(\frac{EI}{m_e}\right)^{1/4} \sqrt{\frac{CV_r D}{2\pi V_0}} \qquad (18.18)$$

where

$$V_r = \frac{2\pi V_0}{\omega_1 D} \qquad (18.19)$$

is the reduced velocity that is used to indicate the strength of pipeline interaction with the vortex shedding in its own wake. D is the external diameter of the pipeline, V_0 is the flow velocity of the external fluid normal to the pipeline axis, ω_1 is the fundamental natural frequency of the pipe segment, and C is the boundary coefficient to be determined from the end boundary conditions of the pipeline.

Equation (18.18) can also be obtained directly from the characteristic frequency equation of the pipeline (Eq. (18.7)) by letting

$$C = k_1^2 L^2 \qquad (18.20)$$

where k_1 is the modal wave number of the fundamental flexural vibration mode of the pipe, which is given for the three types of boundary conditions as:

1. For Pined-Pined boundary conditions

$$C = \pi^2 \qquad (18.21)$$

2. For Fixed-Fixed boundary conditions

$$C = (1.505\pi)^2 \qquad (18.22)$$

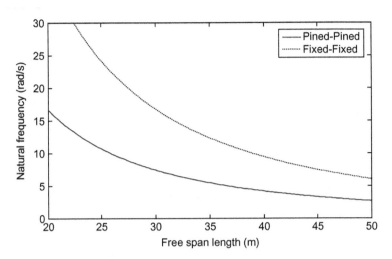

FIGURE 18.2　The fundamental natural frequency of a pipeline having Pined-Pined and Fixed-Fixed boundary conditions.

3. For elastic boundary conditions, C is obtained from Eq. (18.19) by first calculating k_1 from the determinant of the matrix in Eq. (18.17).

A major issue in pipeline design is thus concerning how to accurately modeling the end boundary conditions and determining the natural frequency of the pipe. Figure 18.2 shows the fundamental natural frequency as a function of the free span length of a pipeline for Pined-Pined and Fixed-Fixed boundary conditions. A steel pipe segment having Young's modulus $E = 2.1 \times 10^{11}$ Pa, density $\rho_s = 7800$ kg/m^3, and thickness of 0.02 m is used in this simulation. The external diameter of the pipe is assumed to be 0.5 m and the density of carrying fluid is $\rho_i = 1000$ kg/m^3.

18.2.4 Effects of Flowing Fluid on Pipeline Vibration

In the previous section, we discussed the free vibration of a pipeline by ignoring the effect of flowing fluid velocity on the pipeline vibration. The effect of flowing fluid on the vibration of a pipeline was discussed by several researchers (Blevins, 1990; Housner, 1952; Laura et al., 1987), which is described by the following equation:

$$\frac{\omega_n^F}{\omega_n} = \left(b - \sqrt{b^2 - 4\left[1 - \left(\frac{V_I}{V_c}\right)^2\right]\left[4 - \left(\frac{V_I}{V_c}\right)^2\right]} \right)^{1/2}, \quad n = 1, 2 \qquad (18.23)$$

where ω_n^F is the natural frequency of the fluid carrying pipe with internal fluid flow velocity V_I, ω_n is the corresponding natural frequency of the pipe

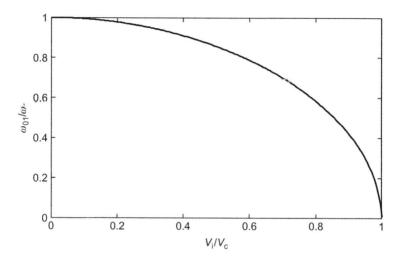

FIGURE 18.3 The effect of internal fluid flow on the fundamental natural frequency of a pipeline.

with stationary fluid (Eq. (18.7)). V_c is the critical flow velocity leading to pipeline buckling which is given by

$$V_c = \left(\frac{\pi}{L}\right)\sqrt{\frac{EI}{m_f}} \tag{18.24}$$

The parameter b is given by

$$b = 8.5 - \left(\frac{V_1}{V_c}\right)^2 \left(2.5 - \frac{128}{9\pi^2}\frac{m_f}{m_e}\right) \tag{18.25}$$

where m_f and m_e are, respectively, the fluid mass and effective mass per unit length of the pipe.

The effect of internal fluid flow velocity on the fundamental natural frequency of a pipeline is shown in Figure 18.3. In this figure, ω_{01} is the fundamental natural frequency of the pipeline carrying flowing fluid, ω_1 is the corresponding natural frequency of the pipeline with stationary fluid. It is illustrated that the velocity of the internal flowing fluid of a pipeline has substantial influence on the pipeline vibration. High fluid flow velocity can cause instability of a pipeline when the fluid flowing velocity approaches the critical velocity of the pipeline and therefore, it shall be avoided in the pipeline design.

18.2.5 Effects of Axial Force

Axial tension/compression force induced by residual tension during installation of a pipeline, internal pressure, temperature gradient, or pipe deflection could largely affect the flexural vibration of a pipeline. Therefore, the effect of axial force shall also be taken into account in calculating the free span and natural frequency of a pipeline.

By including the axial tension/compression in the formulation, Eq. (18.1) can be rewritten as (Xu et al., 1999)

$$EI\frac{\partial^4 u}{\partial x^4} + m_e\frac{\partial^2 u}{\partial t^2} - \Gamma\frac{\partial^2 u}{\partial x^2} = 0 \tag{18.26}$$

where Γ is the axial compression force. For axial tension force, the last term in Eq. (18.26) changes sign.

Equation (18.26) can be solved once again by the standing wave solution given by Eq. (18.3). Assuming Pined-Pined boundary conditions for the pipeline, the natural frequencies of the pipeline are obtained as

$$\omega_n^T = \omega_n\sqrt{\left(1 - \frac{\Gamma}{P_E}\right)} \tag{18.27}$$

where

$$P_E = \frac{n^2\pi^2 EI}{L_{eff}^2} \tag{18.28}$$

is Euler buckling load and ω_n is the angular natural frequency of the corresponding pipeline free of axial compression/tension force.

The effective length of a free span pipeline rested on uneven seabed, L_{eff}, is given by (Bakhtiary et al., 2007; Det Norske Veritas, 1998)

$$L_{eff} \cong \begin{cases} 1.12L & \frac{L}{D} \leq 40 \\ 1.12 - 0.001\left(\frac{L}{D} - 40\right) & 40 \leq \frac{L}{D} \leq 160 \\ 1.00 & \frac{L}{D} > 160 \end{cases} \tag{18.29}$$

Xu et al. (1999) also provided an analytical solution for the natural vibration of an elastically supported prestress pipeline based on a standing wave solution. However, results obtained from their solution were found to be rather different from the natural frequencies of offshore pipelines (Bakhtiary et al., 2007).

Equation (18.27) indicates that the natural frequency of a pipeline segment would decrease monotonously if it is under an axial compression force and would increase monotonously if it is subject to an axial tension force. DNV guideline for free spanning pipelines (Det Norske Veritas, 2008) also recommends a similar approximated formulation to calculate the effect

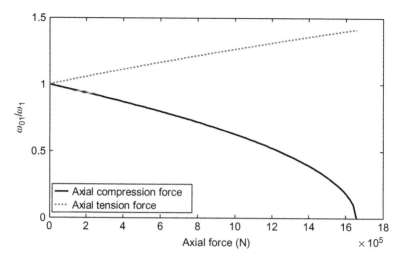

FIGURE 18.4 The effect of axial force on the fundamental natural frequency of a pipeline.

of axial force on the natural vibration of a pipeline. The effect of the pre-stress axial force on the first natural frequency of a pipeline is shown in Figure 18.4. It is illustrated that the system quickly becomes unstable when the axial compression force is close to the buckling load of the pipeline.

18.2.6 Free Vibration of Offshore Pipelines Considering the Added Mass Effect

The added mass effect of seawater on vibration of subsea pipelines should be included in the effective mass calculation of the pipeline free span length. The added mass, m_a, to an immersed pipeline is calculated by

$$m_a = \frac{\pi}{4} C_a \rho_0 D^2 \tag{18.30}$$

where C_a is the added mass coefficient and ρ_0 is the density of the seawater.

Combining Eq. (18.7) with Eqs (18.23), (18.27), and (18.30), the natural frequency of an offshore subsea pipeline can now be calculated as

$$\omega_n^0 = k_n^2 \sqrt{\frac{EI}{m_e + m_a} \left(b - \sqrt{b^2 - 4\left[1 - \left(\frac{V_1}{V_c}\right)^2\right]\left[4 - \left(\frac{V_1}{V_c}\right)^2\right]} \right)\left(1 - \frac{\Gamma}{P_E}\right)},$$

$$n = 1, 2, \ldots, \infty$$

$$\tag{18.31}$$

Similarly, the effective length of the pipeline (Eq. (18.18)) can be reexpressed as

$$L_{\mathrm{eff}} = \left(\frac{EI}{m_{\mathrm{e}} + m_{\mathrm{a}}}\right)^{1/4} \sqrt{\frac{C}{\omega_n^{\mathrm{o}}}} \qquad (18.32)$$

18.2.7 Flexural Vibration of Subsea Pipelines

A major concern of subsea pipeline vibration is the VIV. VIV has been widely recognized as one of the main causes to fatigue damage of pipelines (Blevins, 1990) and has been studied intensively by many researchers worldwide (Van den Abeele et al., 2008; Xu et al., 1999; Yang et al., 2008). An important parameter of VIV is vortex shedding frequency. Vortex shedding frequency relates to the dimensionless Strouhal number (S) by

$$f_{\mathrm{s}} = \frac{SV_0}{D} \qquad (18.33)$$

where V_0 is the velocity of external flow. Experiment shows that cross flow resonance of pipeline vibration occurs at vortex shedding frequency while in-line resonance occurs at twice the shedding frequency (Blevins, 1990).

Lock-in occurs when vortex shedding frequency is about half of the natural frequency of a pipeline. As the external flow velocity increases, cross flow vibration begins to occur and vortex shedding frequency locks into the natural frequency of the pipeline. Amplified vibration response due to resonance (a result of frequency matching between vortex shedding frequency and the natural frequency of a free span pipeline) can cause fatigue damage to the pipeline. Thus, it is important to determine the critical allowable pipeline span length under various flow conditions in the pipeline design.

Vortex shedding for a smooth circular pipeline in a subsonic flow is a function of Reynolds number, which is a ratio of the inertia force and the viscosity force of the flow. Vortex shedding can be grouped into four regions according to the Reynolds number:

1. Below subcritical region ($Re < 300$): In this region, vortices start to develop, increase linearly, and become turbulence when break away from the pipeline as Reynolds number increases.
2. Subcritical region ($300 < Re < 1.5 \times 10^5$): In this region, vortex shedding is strong and periodic.
3. Transition region ($1.5 \times 10^5 < Re < 3.5 \times 10^6$): In this region, the disruption of regular shedding process broadens the frequency band of the shedding frequency for smooth pipelines, and VIV is insignificant (Blevins, 1990).
4. Supercritical region ($Re > 3.5 \times 10^6$): In this region, regular shedding is reestablished with a turbulent cylinder boundary layer.

Another important parameter of offshore pipeline vibration is the stability parameter, which is defined by (Sumer and Fredsøe, 1995)

$$K_s = \frac{4\pi\zeta_s(m_e + m_a)}{\rho_0 D^2} \tag{18.34}$$

in which ζ_s is the structural damping coefficient. The stability parameter K_s is a measure determining uniquely the maximum amplitude of vortex-induced pipeline vibration.

Vortex shedding produces lift forces causing cross flow vibration. It also produces inertia and drag forces causing in-line vibration of the pipeline. Assuming that the lift force imposed on a pipeline per unit length by vortex shedding is harmonic in time, it can be written as

$$F_L = \frac{1}{2}C_L\rho_0 V_0^2 D \sin(\omega_s t) \tag{18.35}$$

where C_L is the lift coefficient and ω_s is the vortex shedding frequency in rad/s. Incorporating Eq. (18.35) into the right-hand side of Eq. (18.1), the formulation governing pipeline flexural vibration due to the lift force can be established. It can be subsequently solved by using propagating wave approach or standing wave solution once the boundary conditions of the pipeline are known.

Similarly, the inertia force and drag force in an oscillating flow are given by (Blevins, 1990)

$$F_I = \frac{\pi}{4}(1 + C_a)\rho_0 D^2 \dot{V}_0 \tag{18.36}$$

and

$$F_D = \frac{1}{2}C_D\rho_0 D\left|V_0 - \frac{\partial u}{\partial t}\right|\left(V_0 - \frac{\partial u}{\partial t}\right) \tag{18.37}$$

where C_D is the drag coefficient.

Incorporating Eqs (18.36) and (18.37) into the right-hand side of Eq. (18.1) and including the damping and axial prestress terms in the equation, Xu et al. (1999) presented an analytical model for the in-line vibration of a prestress pipeline as

$$EI\frac{\partial^4 u}{\partial x^4} + (m_e + m_a)\frac{\partial^2 u}{\partial t^2} + \zeta_s\frac{\partial u}{\partial t} - \Gamma\frac{\partial^2 u}{\partial x^2} = \frac{1}{2}\rho_0 DC_D\left|V_0 - \frac{\partial u}{\partial t}\right|\left(V_0 - \frac{\partial u}{\partial t}\right)$$
$$+ \frac{\pi}{4}C_M\rho_0 D^2\frac{\partial V_0}{\partial t} \tag{18.38}$$

where C_M is the inertia coefficient. In their solution, the coupled nonlinear drag force term on the right-hand side of Eq. (18.38) was first linearized to give a linear equation of motion, which was then solved in both time and frequency domains.

It is worth noting that formulations presented in this section are all based on one-dimensional (1D) beam models. To understand the full coupling effects between the disturbance of internal fluid flow and the structural and acoustic modes of a pipeline, a 3D cylindrical pipe model would be a more appropriate approach. This is a common practice in pipeline noise and vibration analysis when the L/D ratio of a pipe is relatively small (e.g., the noise and vibration analysis presented by Norton, 1989). Nevertheless, vibration of subsea pipelines can be analyzed efficiently by using 1D beam models with reasonable accuracy due to the elongated and slender nature of subsea pipelines.

18.2.8 Measures to Mitigate Pipeline Vibration

Measures should be taken to avoid or to mitigate the flow-induced vibration and VIV by modifying either the pipeline system or the flow condition. The simplest solution to mitigate pipeline vibration is to eliminate vibration at the source location. For example, vibration transmission to the pipeline due to imbalance of rotating machinery attached to the pipeline such as pump can be eliminated by rebalancing the centrifugal vane or shaft. Another example of eliminating vibration source from the pump is to ensure that the pump suction always provides an NPSH to avoid the development of cavitation when the fluid flows through the pump casing. However, eliminating vibration at the source location is quite often impractical or uneconomical. In such case, it would be a good practice to mitigate pipeline vibration along vibration propagating paths. For instance, the mechanical-induced pipeline vibration in the previous example can be reduced by installing a flexible coupling connecting the outlet of the pump and the pipe.

Pipeline vibration induced by internal fluid flow can be avoided or mitigated by careful arrangements of the pipeline system to optimize fluid flow velocity and to minimize flow disturbance due to change of flow direction (elbow, bend, Tee branch, etc.) or change in flow cross sections (valve, orifice, etc.). For example, an ideal inlet flow to a pump should be a continuous even flow through the pipe cross section. A 5D to 10D long straight pipe right before the fluid entering the pump inlet is recommended to reestablish an even flow. A smooth change of flow direction such as a large radius bend or elbow is also recommended to reduce the turbulence generated in the flow. The pressure pulsation generated in the flow through the pipeline due to reciprocating machineries in the system should be damped by using flow smoothing devices such as suction and discharge pulsation damper, accumulator, or surge drum. Vortex shedding due to branch outlet, valves, or attaching devices can also cause severe vibration in a pipeline when the frequency of pressure pulsation of the internal fluid flow matches the acoustic frequency of the pipeline. This type of vibration can be mitigated by eliminating the sharp branch entrance edge or valves that cause vortices in the pipeline.

Blevins (1990) has detailed a few measures to reduce vortex-induced pipeline vibration caused by external flow. Some of these measures are listed as follows: (1) Increasing structural damping of the pipeline by using materials of high damping coefficients; (2) keeping the reduced velocity below unity to avoid the resonant response of the pipeline; (3) a streamline cross-section design to minimize the vortex shedding and reducing the drag force; and (4) adding a vortex suppression device such as helical strake, perforated shroud, axial slats, streamlined fairing, splitter, ribbon, guiding vane, spoiler plates, and stepped cylinder. A detailed configuration and description of these devices can be found in Blevins's work (Blevins, 1990). The effectiveness of some of these devices in suppression of the vortex shedding was studied numerically by Van den Abeele et al. (2008).

18.3 CONDITION BASED MAINTENANCE

Figure 18.5 outlines the classification of engineering asset maintenance policies discussed in Section 18.1. It also highlights the focus of this section which is the implementation and optimization of CBM for pipeline structures.

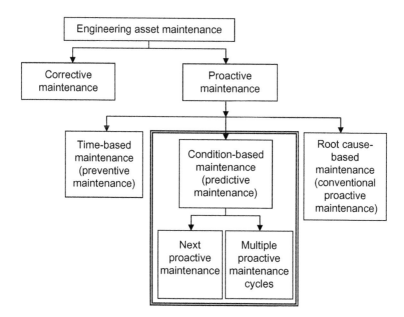

FIGURE 18.5   Engineering asset maintenance classification.

Engineering asset failures can be largely divided into two categories: (1) sudden failures and (2) gradual failures. An asset has a sudden failure indicates that it has a functional failure suddenly at a time point while it performs normally before this point; whereas a gradual failure is basically a degradation process, i.e., its performance degrades continuously over time until it cannot function satisfactorily. Offshore pipelines can suffer from both types of failures. Dents and costing crack caused by third-party activities are examples of sudden failures, whereas pipeline ruptures as a result of exceeding pipeline vibration and pipeline corrosion and erosion process are two examples of gradual failures. CBM is only suitable for continuous degradation process. For example, leaking due to internal erosion and corrosion is one of the common functional failures in offshore pipelines. To prevent the occurrence of a leakage, the inside of pipelines need be inspected regularly with certain condition monitoring tools. When a coating failure is detected, we can initiate a PM action to replace the deteriorated segment.

Implementing CBM involves the following procedure:

1. Determine asset health indicators or measurements such as efficiency, capacity (flow rate and pressure), corrosion rate, material loss, and hazard. Some of them can be measured directly while others need be calculated. Hazard is an example of the latter and will be further explained in Section 18.3.2.

2. Identify condition monitoring parameters that can reflect or indicate the changes of the selected health indicators or measurements. For instance, corrosion and erosion is one of the typical failure modes of steel offshore pipelines. Commonly used condition monitoring parameters include wall thickness, vibration, and acoustic emission. In practice, condition monitoring parameters can be the same as health indicators. Furthermore, several parameters can be used to monitor the same health indicator or measurement such as hazard rate.

    Note that condition monitoring data are often called covariates in reliability engineering and can be divided into two categories in terms of their relationship with asset health conditions. The first one is external covariates. The changes of these covariates will cause the characteristics of the health state of an engineering system which can be represented by the hazard of the system, to change. The changes of external covariates are normally independent of the changes of state of the system. Amount of erosive particles in pipelines is an example of an external covariate. The second category is internal covariates. These covariates are indicators of the health state of an engineering system. Their changes are dependent on the changes of the state of the system. Symptoms that reflect the deterioration of a system such as acoustic emission from deteriorated pipelines is one of internal covariates. Some condition monitoring data can have both roles.

3. Define PM thresholds. Threshold is a major decision variable in CBM decision making. Determination of optimal thresholds depends on

maintenance techniques, business objectives, cost and condition monitoring parameters. Different assets generally have different thresholds.

4. Identify the optimal inspection intervals which can be the same or different. Inspection interval is another major decision variable in CBM decision making. Various factors such as cost and the availability of condition monitoring devices can affect the decisions.

5. Monitor the changes of the health indicators. Once an indicator reaches its predefined threshold, a PM action is carried out (Figure 18.6).

18.3.1 Objectives of Optimizing CBM

Determining the optimal PM thresholds and inspection intervals is a major concern when applying CBM to offshore pipelines and therefore is the focus of the following sections. Although various factors can influence the selection of thresholds and inspection frequency, cost is often the most critical factor.

Cost associated with CBM of offshore pipelines can be classified into three categories: inspection cost, PM cost, and failure cost. They are all functions of threshold and inspection intervals, i.e.,

$$C(T, S) = C_I(T, S) + C_F(T, S) + C_P(T, S) \tag{18.39}$$

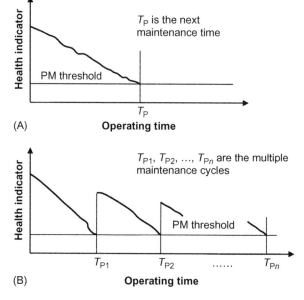

FIGURE 18.6 CBM decision making: (A) the next maintenance and (B) multiple maintenance cycles.

where $C(T,S)$ is the total CBM cost associated with a particular inspection interval T and threshold S. $C_I(T,S)$ is the cost for conducting inspections including inspectors' labor cost and fees for using inspections tools; $C_F(T,S)$ is the cost caused by a failure of a pipeline, which normally results in leakage of pollutant fluid to the sea, including cost to fix the failure, cost to clean up the pollution, production loss, penalty, compensation, and damage to the environment; $C_P(T,S)$ is the PM cost which includes the cost for the maintenance and the production loss due to the shutdown of operation for such a maintenance activity.

As pipelines are linear assets and often span long distances, cost analysis for a pipeline is generally conducted based on its segments instead of the entire pipeline, although proper segmentation of a pipeline can be also a challenge.

Thresholds and inspection intervals have different impacts on different types of costs. Thresholds set to an earlier stage at pipeline deterioration process can lower failure probability, correspondingly reducing expected failure cost, but will generally increase expected PM cost. While thresholds generally have little influence on inspection cost, inspection intervals affect all three types of costs. Increasing inspection frequency can increase the chance to detect potential failures or functional failures at an earlier time, consequently reducing the expected failure cost and increasing the expected PM cost. However, frequent inspections will increase the inspection cost (Figure 18.7). The objective of CBM decision making is therefore to decide the optimal threshold and inspection intervals to minimize the total cost. Various models have been developed to achieve this objective.

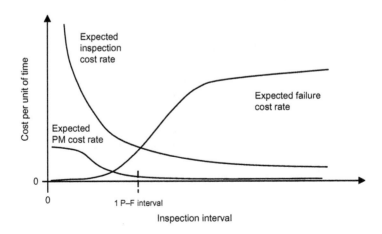

FIGURE 18.7   Costs versus inspection intervals.

18.3.2 Optimization of the Next Maintenance Time

Optimizing the next maintenance time of pipelines is often needed in pipeline management. Degradation curve or hazard functions of a pipeline can be utilized to decide when a preventive maintenance action of the pipeline is required.

18.3.2.1 P−F Curve Based Method

The concept of P−F curves was introduced by Moubray (1997) to describe the processes of some gradual failures. It implies that any of these failures is a degradation process consisting of two stages. In the first stage, the degradation develops slowly until a certain point where it becomes detectable. This point is defined as potential failure point "P." After this point, the degradation accelerates, finally leading to a functional failure at failure point "F" (Figure 18.8). Therefore, the failure process can be monitored and its P−F curves can be used for CBM decision making. The point "P" is the lead time for taking a maintenance action to prevent the occurrence of the failure, i.e., the health condition at the point "P" is a threshold for conducting PM. The time between points "P" and "F" is called P−F interval which is the maximum time allowing for such a PM.

   Some of offshore pipeline failures can also be described using P−F curves. One example is the pipeline failure caused by wave- and flow-induced vibration. Another example is the failure due to corrosion and erosion. The potential failure time T_P and functional failure time T_F shown in Figure 18.8 are deterministic. However, in reality, point "P" and point "F" both can vary over a large range. In this case, both T_P and T_F are random variables. For example, after the coating has been penetrated, different pipes may have different failure times which depend on the quality of materials, the thickness of wall, and the corrosiveness and erosiveness of the internal

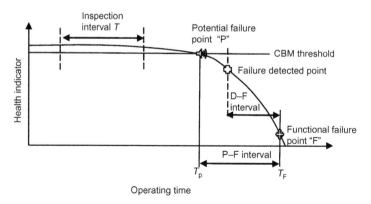

FIGURE 18.8   P−F curve and interval.

flowing fluid (Castanier and Rausand, 2006). Determining P—F curves needs a sound understanding of the failure process. The shapes of P—F curves depend on both failure modes and the adopted condition monitoring techniques. If condition indicators are different, P—F curves can be different.

The most important step in the application of the P—F based methods is to determine the optimal condition monitoring frequency since the maintenance thresholds are set to be the "P" points in such approaches. The objective of condition monitoring is thus to find this point in time so that a PM can be carried out to stop functional failures or to avoid the failure consequences. Identifying point "P" in time is crucial for an effective maintenance. However, cautions have to be exercised in determining the optimal inspection intervals since inspections of subsea pipelines are usually very costly. Too frequent inspections will result in a huge waste of money and other resources. On the other hand, insufficient inspection frequency with long intervals could miss the detection of the point "P," resulting in the occurrence of functional failures.

A simple rule recommended by Moubray (1997) states that a reasonable condition monitoring interval for a given P—F curve could be the half point of the P—F interval. To select an optimal interval, cost analysis is needed. An optimal condition monitoring interval should balance between condition monitoring cost and the benefits brought by avoiding the consequences of functional failures. Cost analysis is case dependent and the cost analysis model can be very complex in reality. The model developed by Castanier and Rausand (2006) is described in the following as an example.

Castanier and Rausand's model is used to determine the optimal inspection frequency of offshore steel pipelines subject to erosive—corrosive failures based on their P—F curves. The pipelines considered in their model were used to transport a mixture of oil, gas, and water. To protect the pipelines from internal corrosion, a coating or lining was adopted inside the pipelines. A major failure mode of the pipelines was leaking. The failure process could be described using a P—F curve as shown in Figure 18.9. At the beginning, the erosive material in the fluid gradually wears away the coating. Once the steel is exposed, it is subject to the combined effects of corrosion and erosion, and finally resulting in a hole in the pipe.

The following assumptions were made in their mathematical formulation:

1. Coating has a linear degradation rate and its hazard rate (the concept of hazard is explained in Section 18.3.2.2) is given by

$$h_c(t) = \lambda_c + \beta_c t \tag{18.40}$$

   where $h_c(t)$ is the hazard rate of coating at time t. λ_c and β_c are two constants.
2. The service life of the pipelines is infinite.
3. Corrosion rate is constant.

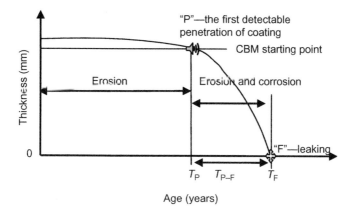

FIGURE 18.9   An example of P–F curves of offshore steel pipelines subject to erosive–corrosive failures.

4. Inspection intervals are the same.
5. The probability to detect the failure in the ith inspection is higher than in the $(i - 1)$th inspection ($i = 2, 3, 4...$).
6. When a coating failure is detected during an inspection, a new segment is ordered. As soon as the new segment is available, a preventive replacement is carried out.
7. A pipeline after a repair becomes as good as new.
8. After a maintenance action, no more than one segment will be damaged before the next maintenance.
9. The occurrence of a hole in the coating follows an exponential distribution.
10. The P–F interval is a random variable following a two-parameter Gumbel probability distribution.

Based on these assumptions, expected total cost per unit of time per segment of pipeline, $c(T)$, can be calculated using the following equation:

$$c(T) = \frac{E[C(T)]}{T_{br}(T)} \tag{18.41}$$

where $E[C(T)]$ is the expected total cost between two consecutive repairs for a segment, T is the inspection interval, and $T_{br}(T)$ is the mean time between repairs.

The expected total cost is given by

$$E[C(T)] = E[C_I(T)] + E[C_C(T)] + E[C_P(T)] \tag{18.42}$$

where $E[C_I(T)]$, $E[C_C(T)]$, and $E[C_P(T)]$ are the expected inspection cost, the expected corrective replacement cost, and the expected proactive replacement cost between two successive repairs for a segment, respectively.

Let $p_i(T)$ denote the probability that a coating failure is not detected during the ith inspection after the failure has become detectable.

Let T_V denote the time for transporting a maintenance vessel to the maintenance location.

Let $P_V(t)$ denote the probability distribution function of T_V.

Let T_D denote the time required for detecting a leakage.

Let $P_D(t)$ denote the probability that a functional failure is detected.

Let T_R denote the mean repair time.

Let T_{PF} denote the P–F interval.

Let $P_F(t)$ denote the probability that a functional failure occurs after the last inspection before point "P".

Let $P_{Fd}(t)$ denote the probability that a functional failure is detected after it has occurred for time t.

Let c_I denote unit inspection cost for a segment.

Let c_R denote unit repair cost for replacing a single segment.

Let c_L denote production loss rate during downtime.

Let c_V denote vessel cost rate.

The expected inspection cost, the expected corrective replacement cost, the expected proactive replacement cost, and the mean time between repairs can now be calculated respectively by

$$E[C_I(T)] = c_I \left\{ \frac{\exp(-\lambda_c T)}{1 - \exp(-\lambda_c T)} + \sum_{j=1}^{\infty} \left\{ \left[\prod_{i=0}^{j-1} p_i(T) \right] [1 - P_{Fd}(jT)] \right\} \right\}$$

$$(18.43)$$

$$E[C_P(T)] = C_{rt} P_p(T) \qquad (18.44)$$

$$E[C_C(T)] = C_{rt}[1 - P_P(T)] + \sum_{i=1}^{3} \{c_L[E_i(Y) + E_i(Z)]\} \qquad (18.45)$$

$$T_{br}(T) = \left\{ \frac{\exp(-\lambda_c T)}{1 - \exp(-\lambda_c T)} + E[N_{aP}(T)] \right\} \left\{ P_P(T) + \sum_{j=1}^{\infty} \{[1 - p_j(T)] \left[\prod_{i=0}^{j-1} p_i(T) \right] \right\}$$

$$\left\{ \int_0^{\infty} [P_F(jT + \tau) - P_{Fd}(jT + \tau)] dP_V(\tau) \} \right\} T + \left[\frac{1}{\lambda_c} + E(T_{PF}) + E(T_D) \right]$$

$$\left\{ \sum_{j=1}^{\infty} \left\{ \prod_{i=0}^{j-1} p_i(T)[P_{Fd}(jT) - P_{Fd}((j-1)T)] \right\} + \sum_{j=1}^{\infty} \{[1 - p_j(T)] \left[\prod_{i=0}^{j-1} p_i(T) \right] \right.$$

$$\left\{ \int_0^{\infty} P_{Fd}(jT + \tau) dP_V(\tau) - P_{Fd}(jT)] \} \right\} + \int_0^{\infty} [1 - P_V(\tau)] d\tau + T_R$$

$$(18.46)$$

where

$$C_{rt} = c_R + c_L T_R + c_V \left\{ \int_0^\infty [1 - P_V(\tau)]d\tau + T_R \right\}$$

$$P_P(T) = \sum_{j=1}^\infty \left\{ [1 - p_j(T)] \left| \prod_{i=0}^{j-1} p_i(T) \right| \int_0^\infty [1 - P_F(jT + \tau)]dF_V(\tau) \right\}$$

$$E_1(Y) = \sum_{j=1}^\infty \left\{ \prod_{i=0}^{j-1} p_i(T)[P_{Fd}(jT) - P_{Fd}((j-1)T)] \right\} \int_0^\infty [1 - P_V(\tau)]d\tau$$

$$E_2(Y) = \sum_{j=1}^\infty \left\{ [1 - p_j(T)] \left[\prod_{i=0}^{j-1} p_i(T) \right] \int_0^\infty \int_{jT}^{jT+\tau} (jT + \tau - t)dP_{Fd}(t)dP_V(\tau) \right\}$$

$$E_3(Y) = 0$$

$$E_1(Z) = \sum_{j=1}^\infty \left\{ \left[\prod_{i=0}^{j-1} p_i(\tau) \right] [P_{Fd}(jT) - P_{Fd}((j-1)T)] \right\} \int_0^\infty [1 - P_D(\tau)]d\tau$$

$$E_2(Z) = \sum_{j=1}^\infty \left\{ [1 - p_j(T)] \left[\prod_{i=0}^{j-1} p_i(T) \right] \left[\int_0^\infty P_{Fd}(jT + \tau)dP_V(\tau) - P_{Fd}(jT) \right] \right\}$$
$$\int_0^\infty [1 - P_D(\tau)]d\tau$$

$$E_3(Z) = \sum_{j=1}^\infty \left\{ [1 - p_j(T)] \left[\prod_{i=0}^{j-1} p_i(T) \right] \right.$$
$$\left. \left\{ \int_0^\infty \int_{jT}^{jT+\tau} (jT + \tau - t)[1 - P_D(jT + \tau - t)]dP_F(t)dP_V(\tau) \right\} \right\}$$

$$E[N_{aP}(T)] = \sum_{j=1}^\infty \prod_{i=0}^{j-1} p_i(\tau)$$

The original derivation of the above equations can be found in (Castanier and Rausand, 2006) for interested readers.

P–F curve based method requires a sound understanding of asset failure process, which can be difficult in reality. Moreover, the method is not suitable for failure processes that do not show clear P–F characteristics. Furthermore, point "P" may not necessarily be the optimal time for conducting a PM action.

18.3.2.2 Proportional Hazard Model Based Method

Proportional hazard model (PHM) based method enables PM decisions of an engineering asset to be made based on its hazard. This method can be used to deal with degradation processes which do not present P–F characteristics. Before moving to a detail description of the PHM method, we introduce the concept of hazard first.

Hazard is one of commonly used asset health measures in engineering asset management. Hazard function $h(t)$ is mathematically expressed as

$$h(t) = \lim_{\Delta t \to 0} \frac{F(t + \Delta t) - F(t)}{\Delta t \cdot R(t)} \qquad (18.47)$$

where $h(t)$ is the hazard function, $F(t)$ is the cumulative distribution function of failure, and $R(t)$ is the reliability function. $h(t)\Delta t$ approximately describes the conditional failure probability of an asset in the interval t to $t + \Delta t$ given it has survived until time t.

Reliability is another commonly used asset health measure. It is defined as the probability that an engineering system will function until time t:

$$R(t) = P(T_f > t) \qquad (18.48)$$

where T_f is the time to first failure.

The relationship among the reliability function, the cumulative distribution function, and hazard function is

$$R(t) = 1 - F(t) = \exp\left[-\int_0^t h(\tau)d\tau\right] \qquad (18.49)$$

Compared with reliability, hazard provides a clearer indication of the degradation process of engineering assets and is thus often used in CBM decision making.

Hazard functions are traditionally estimated using historical failure time data only. The traditional hazard model of an engineering asset cannot take into account its current conditions. PHM eliminates this limitation by including both the age of an asset and its condition in the calculation of the hazard of the asset. This model was initially developed by Cox and Oakes (1984) to assess the effects of external covariates on the hazard of a system but has also been used for hazard estimation using internal covariates (see, for example, Jardine et al., 1999).

In PHM, the hazard at time t of a system is modeled as a product of the baseline hazard function $h_0(t)$ and a positive function term $\Phi[\overrightarrow{Z}(t), \overleftarrow{\gamma}]$ as follows:

$$h(t) = h_0(t)\Phi[\overrightarrow{Z}(t), \overrightarrow{\gamma}] \qquad (18.50)$$

where $\overline{Z}(t)$ is a vector consisting of the covariates and $\overrightarrow{\gamma}$ is a vector of the weighting parameters used to combine the covariates together to form a composite indicator. Function $h_0(t)$ is the baseline hazard representing the hazard without influence of the covariates, i.e., the hazard when $\vec{Z}(t) = \vec{0}$. This implies

$$\Phi[\vec{0}, \vec{\gamma}] = 1 \qquad (18.51)$$

Equation (18.50) illustrates that the baseline hazard function is time-dependent only, the changes of which are not revealed by covariates in PHM. The equation also indicates that the hazard characteristics of a system change when its covariates change. In PHM, the covariates are explanatory variables and the hazard is a response variable. The values of covariates at time t can alter the hazard of the system at the same time point (Figure 18.10). PHM can be used to quantify the effects of external covariates to offshore pipelines. For example, the velocity of internal erosive fluid of a steel pipeline is a major factor influencing pipeline wall material loss. When the velocity of the internal fluid flow is faster, the pipeline will usually be eroded at a faster rate, leading to a higher hazard. In this case, the deterioration process is irreversible unless a maintenance action is carried out.

A commonly used form for $\Phi[\vec{Z}(t), \vec{\gamma}]$ is an exponential function:

$$\Phi[\vec{Z}(t), \vec{\gamma}] = \exp[\vec{\gamma} \cdot \vec{Z}(t)] = \exp\left[\sum_{i=1}^{n_c} \gamma_i Z_i(t)\right] \qquad (18.52)$$

where n_c is the number of covariate parameters.

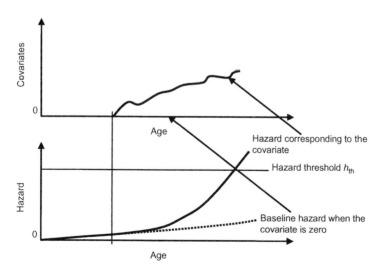

FIGURE 18.10   Concept of proportional hazard model.

Furthermore, the baseline hazard function is often assumed to have the form of two-parameter Weibull distribution such that

$$h(t) = \frac{\beta}{\alpha} \left(\frac{t}{\alpha}\right)^{\beta-1} \exp\left[\sum_{i=1}^{n_c} \gamma_i Z_i(t)\right] \tag{18.53}$$

where α and β are, respectively, the scale and shape parameters of Weibull distribution.

The scale and shape parameters as well as the weighting parameters can be estimated based on the following log-likelihood function (Liao et al., 2006):

$$L(\alpha, \beta, \vec{\gamma} | S_D) = \sum_{T_{Fk} \in S_F} \ln h(T_{Fk}) - \sum_{T_j \in \{S_F \cup S_s\}} \int_0^{T_j} h(t) dt \tag{18.54}$$

where S_D is the historical dataset including engineering asset ages, condition monitoring measurements, failure times, and surviving times if the data are censored. S_F is the dataset of failure times and S_S is the dataset of surviving times. T_{Fk} is the kth member in S_F. T_j is a member of S_F (failure time) or S_S (censored data). Estimations of the parameters α, β, and γ_i ($i = 1, 2,..., n_c$) can be made by maximizing the log-likelihood function (Eq. (18.54)).

When hazard based models are adopted, the focus would be to determine the optimal threshold even if optimizing inspection intervals is still a concern. Cost is the major factor to be considered in determining the optimal hazard threshold although various other factors also needed to be considered. Makis and Jardine (1992) provided the following formula for calculating an optimal proactive replacement threshold:

$$e_{op} = \frac{C_P + (C_F - C_P)P_F(e_{op})}{T_r(e_{op})} \tag{18.55}$$

and

$$h_{th} = \frac{e_{op}}{(C_F - C_P)} \tag{18.56}$$

where e_{op} is the optimal expected average cost and h_{th} is the optimal hazard threshold. C_p is the proactive replacement cost. C_F is the failure cost, $P_F(e_{op})$ is the probability that a failure will occur. $T_r(e_{op})$ is the expected time until a replacement which can be a proactive replacement or a corrective replacement.

The following assumptions were made in deriving Eq. (18.55):

1. Inspection is perfect, inspection time and cost are ignorable.
2. The asset degrades continuously but can have a failure at any time.
3. When the asset fails, a corrective replacement will be carried out.
4. Proactive replacement cost is lower than failure cost.
5. Decisions are only made at discrete inspection time points.

6. Covariates have nondecreasing processes and the asset hazard does not decrease when the values of covariates increase.

The optimal expected average cost, e_{op}, can be calculated using the following recursive equation:

$$e_n = \frac{C_P + (C_F - C_P)P_F(e_{n-1})}{T_r(e_{n-1})}, \quad (n = 2, 3, \ldots) \tag{18.57}$$

and

$$e_{op} = \lim_{n \to \infty} e_n \tag{18.58}$$

18.3.2.3 Proportional Covariate Model Based Method

In condition monitoring, we often observed that the internal covariates (i.e., symptoms) of a system change with the change of hazard of the system, which represents the alternations in the system health condition. A proportional covariate model (PCM) was proposed by Sun et al. (2006) to describe such observations (Figure 18.11). PCM assumes that the hazard of a system at time t, which represents the state of the system at time t, can be indicated by the condition indicators (internal covariates). Hence, the change of these

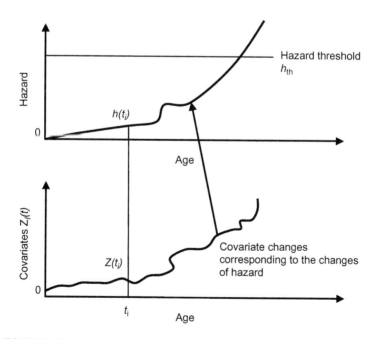

FIGURE 18.11 Concept of proportional covariate model.

indicators can be used to monitor the change of hazard. In PCM, the function of covariates $\Psi[Z_I(t)]$ is expressed as

$$\Psi[Z_1(t)] = C(t)h(t) \tag{18.59}$$

where $Z_1(t)$ is the covariate function describing the change of internal covariates, $C(t)$ is the baseline covariate function. The function $h(t)$ is the hazard function of the system. Equation (18.59) indicates that the covariates are response variables and the hazard is an explanatory variable.

Construction of a suitable function of covariates, $\Psi[Z_r(t)]$, plays an important role in improving the accuracy of hazard estimation, especially when multiple covariates are used in the estimation. In the following discussion, we only consider a simple scenario where a single covariate is used, and

$$\Psi[Z_1(t)] = Z_1(t) \tag{18.60}$$

Substituting Eq. (18.60) into Eq. (18.59) gives

$$Z_1(t) = C(t)h(t) \tag{18.61}$$

Estimating the baseline covariate function $C(t)$ is a critical step in PCM which can be conducted using the following two approaches:

18.3.2.4 Approach 1: Data Driven Approach

In this approach, the baseline covariate function is estimated using historical failure data and condition monitoring data.

Step 1. Estimate the initial hazard function of a system, $h_{in}(t)$, using historical failure times and surviving times if the data are censored.

Step 2. Generate a set of discrete baseline covariate function values using the following equation (Eq. (18.62)) providing that a set of historical covariate values $Z(t_i)$ $(i = 1,2,...,m)$ have been recorded at time $(t_1, t_2,...,t_m)$.

$$C_i = \frac{Z_1(t_i)}{h_{in}(t_i)} \quad (i = 1, 2, ..., m) \tag{18.62}$$

where m is the number of historical condition monitoring data and C_i is the baseline covariate value at time t_i. An average (or mean) covariate value should be used if several measurements are taken at the same time.

Step 3: Estimate the baseline covariate function $C(t)$ using the discrete dataset $\{C_i, t_i\}$ $(i = 1,2,...,m)$ and the regression analysis method.

18.3.2.5 Approach 2: Expert Knowledge Based Approach

When historical failure data are sparse or zero, the baseline covariate function $C(t)$ can be determined according to anecdotal experience of operators of plant and/or supplementary information such as data from accelerated life tests. Hence the hazard functions of engineering assets can still be estimated using PCM.

Once the baseline covariate function $C(t)$ is determined, the hazard of the engineering system at time t_n ($t_n = 1, 2,\ldots$ and $t_n > t_m$) could be estimated and updated using new condition monitoring data that act as covariates $Z_1(t_n)$ based on the following formula:

$$\tilde{h}(t_n) = \frac{Z_1(t_n)}{C(t_n)}, \quad (t_n = 1, 2,\ldots \text{ and } t_n > tm) \tag{18.63}$$

Hazard function $h(t)$ can be updated based on the dataset $\tilde{h}(t_n)$, generally using the regression technique and the maximum likelihood method, if its form has been assumed, e.g., Weibull distribution.

18.3.3 Optimization of Long-Term Maintenance Activities

In engineering asset management, we not only need to decide the next PM time of an asset but also need to decide multiple PM times over a long term (refer to Figure 18.6) which is defined as decision horizon T_h, to plan various resources such as finance, human resource, and logistics.

Pipelines are normally complex engineering asset systems. To improve the overall reliability of the entire pipeline system, CBM is needed to maintain the aged pipelines continuously to reduce unexpected failures and the associated resultant impacts. When a pipeline is preventively maintained, quite often only some of its segments rather than the whole pipeline are repaired or replaced. As a result, the health state of the pipeline after a PM action is better than old but not as good as new, i.e., the pipeline has an imperfect repair. The effects of imperfect repairs must be considered when optimizing a long-term CBM strategy which consists of multiple PM cycles.

Various methods/models have been developed to consider the influence of imperfect repairs on the reliability of an engineering system (Lin et al., 2000; Pham, 2003; Pham and Wang, 1996; Rausand and Hoyland, 2004). These models/methods can be largely classified into three categories: (1) state rules based models, (2) improvement factor methods, and (3) heuristic methods (Ebeling, 1997; Lewis, 1996; Sun et al., 2007a,b). Compared with the other two methods, the heuristic method can produce an intuitive and explicit prediction of reliability and is therefore suitable for offshore pipeline applications. Moreover, this type of methods directly links the hazard functions discussed previously to long-term PM decision making so that the decision can be updated using the latest condition monitoring information.

Among existing heuristic methods, split system approach (SSA) (Sun et al., 2007b) (also termed as system split model (SSM) (Sun et al., 2007a) can better deal with the reliability prediction of systems subject to CBM, in particular, reliability (or risk) based preventive maintenance (RBPM). SSA is different from the other methods in that it models the reliability of a system with multiple PM actions at the component level. This approach considers the effect of repaired components on the system reliability and allows the change of the reliability of the system after each PM action to be calculated rather than estimated by maintenance staff. SSA removes the assumptions on the probability of different states of a system after repairs. SSA is employed to calculate the reliability of the pipeline with multiple PM intervals in the following section. The outcomes can assist industrial personnel to determine the optimal PM strategy for pipelines.

18.3.3.1 Pipeline Reliability Prediction Based on SSA

When RBPM policy is applied, we preventively replace some pipes (or segments) in a pipeline whenever the reliability of the entire pipeline falls to a predefined reliability threshold R_{th}. Note that pipes scheduled for PM may or may not survive until their individual PM times. The survival probability of a system under the condition that its components have successfully been proactively maintained is defined as the conditional reliability of the system. The survival probability of the system that considers the survival probability of components scheduled for PM until their individual PM times is termed as overall reliability of the system. Both reliability concepts are useful in CBM decision making. The conditional reliability describes the reliability changes between two successive PM actions and is used to determine dynamic PM intervals; whereas the overall reliability illustrates the reliability changes of the system over the decision horizon which may cover a number of PM intervals and thus is used to evaluate PM strategies.

Let $R_s(t)_i$ and $R_{sc}(t)_i$ denote the overall reliability function and the conditional reliability function of a pipeline after the ith PM action, respectively. As a pipeline can usually be divided into a number of segments which are serially connected, according to Ebeling (1997) and Lewis (1996), the overall reliability and the conditional reliability of the pipeline has the following relationship:

$$R_s(t)_i = \prod_{k=0}^{i} R_c(T_{Pk} - T_{Pj})_j R_{sc}(t)_i, \quad (0 \le \tau < (T_{Pi+1} - T_{Pi}) \text{ for } i = 0, 1, 2, \ldots,$$

$$n - 1 \text{ and } 0 \le \tau < (T_h - T_{Pn}) \text{ when } i = n)$$

$$(18.64)$$

where n is the number of PM actions during a given decision horizon T_h, and T_{Pi} is the ith PM time ($i = 0, 1, 2, \ldots, n$). If we assume that repair times are ignorable, then T_{Pi} is also the start time for the pipeline to operate again after the ith PM action. The value $R_c(T_{Pi} - T_{Pj})_j$ is the reliability of the preventively repaired segment just before the ith PM action providing that it has been maintained at jth PM action ($0 \le j < i$). When $i = 0$, $\prod_{k=0}^{i} R_c(T_{Pk} - T_{Pj})_k$ is defined as one.

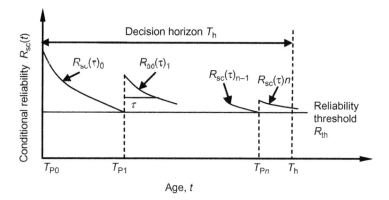

FIGURE 18.12   The conditional reliability of pipeline with multiple imperfect PM actions.

When only part of a pipeline is repaired or replaced in a PM action, the conditional reliability of the entire pipeline after this PM can be improved but lower than its original reliability, leading to an imperfect repair (Figure 18.12).

In Figure 18.12, R_{th} is the predefined reliability threshold for the entire pipeline. $R_{sc}(\tau)_i$ is the conditional reliability of the pipeline after the ith PM action in terms of the relative timescale τ which will be reset to zero after each PM action. Age t stands for absolute timescale, changing from zero to infinite. The relationship between the relative timescale and the absolute timescale is given by

$$\tau = t - T_{Pi} \quad (0 \leq \tau < (T_{Pi+1} - T_{Pi}) \quad \text{for } i = 0, 1, 2, \ldots, n-1 \text{ and } 0 \leq \tau$$
$$< (T_h - T_{Pn}) \text{ when } i = n)$$

$$(18.65)$$

Reliability Prediction When the Same Segments Repaired

To calculate the overall reliability of a pipeline according to Eq. (18.63), the conditional reliability has to be known, which is formulated in this section. For the model development, we first consider a simple case where the same segments in a single pipeline are proactively repaired (when we say repair, it also includes both repair and replacement) in all PM actions. In this case, the pipeline can be divided into two parts, one composes the repaired segments and the other consists of the remainder (Figure 18.13).

Following assumptions are made in formulating the reliability:

1. The failures of the repaired segments are independent of the unrepaired segments. This assumption means that when the pipeline is preventively maintained, only the reliability of the repaired segments changes. The

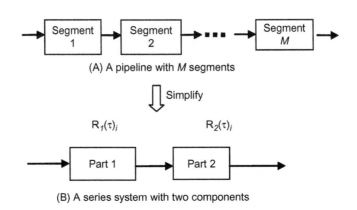

FIGURE 18.13   Simplification of pipelines with the same segments repaired.

reliability of the remainder of the pipeline before and after the PM action does not change.

2. The reliability function of the pipeline before any repair is known. The reliability functions of the repaired segments are also known.
3. PM time is a deterministic variable, which is calculated based on the pre-defined reliability threshold.

Let $R_{sc}(\tau)_i$, $R_1(\tau)_i$, and $R_2(\tau)_i$ denote the conditional reliability functions of the pipeline, the maintained Part 1 and the unmaintained Part 2 after the ith PM action, respectively. According to the reliability theory, we have the following relationship:

$$R_{sc}(\tau)_i = R_1(\tau)_i R_2(\tau)_i \quad (i = 0, 1, 2, \ldots, n) \tag{18.66}$$

The conditional reliability function of the pipeline before any repair equals to its original overall reliability function, i.e.,

$$R_{sc}(\tau)_0 = R_s(\tau)_0 \tag{18.67}$$

Substituting Eq. (18.67) into Eq. (18.66) gives the initial reliability function of Part 2:

$$R_2(\tau)_0 = \frac{R_s(\tau)_0}{R_1(\tau)_0} \tag{18.68}$$

where $R_1(\tau)_0 \neq 0$. The reliability functions for typical failure distributions such as exponential distribution and Weibull distribution all meet this requirement.

The first PM time is given by $R_{sc}(T_{P1})_0 = R_{th}$. It becomes obvious that a higher reliability threshold will result in a shorter T_{P1}. At time T_{P1}, the segments in Part 1 are proactively repaired in line with the CBM strategy. The reliability function of Part 1 after this PM action becomes $R_1(\tau)_1$, but the reliability function of Part 2 remains the same because it has not been repaired. However, the time for calculating the reliability of Part 2 changes to $\tau + T_{P1}$. Therefore, the conditional reliability of the pipeline after the first PM action becomes

$$R_{sc}(\tau)_1 = R_1(\tau)_1 R_2(\tau + T_{P1})_0 = \frac{R_1(\tau)_1 R_s(\tau + T_{P1})_0}{R_1(\tau + T_{P1})_0} \tag{18.69}$$

Equation (18.69) shows that if $R_1(\tau)_1 = R_1(\tau + T_{P1})_0$ (i.e., Part 1 after the PM is as bad as old), then $R_{sc}(\tau)_1 = R_s(\tau + T_{P1})_0$ (i.e., the state of the pipeline is also as bad as old). If $R_1(\tau)_1 = R_1(\tau)_0$ (i.e., Part 1 is proactively replaced with an identical new one), then $R_{sc}(\tau)_0 > R_{sc}(\tau)_1 > R_s(\tau + T_{P1})_0$ as $R_1(\tau)_0$ decreases monotonously. This result indicates that the reliability of the pipeline has been improved after the PM, but it does not become as good as new.

Substituting Eqs (18.65) and (18.69) into Eq. (18.64) and considering that $T_{P0} = 0$, the overall reliability function of the pipeline after the first PM action in terms of the absolute timescale is obtained as

$$R_s(t)_1 = \frac{R_1(T_{P1})_0 R_1(t - T_{P1})_1 R_s(t)_0}{R_1(t)_0} \tag{18.70}$$

If $R_1(t)_1 = R_1(t)_0 = \exp(-\lambda t)$, $R_s(t)_1 = R_s(t)_0$. This result indicates that if Part 1 is preventively replaced by an identical one with the same exponential failure distribution, the PM action is ineffective. After we have obtained the conditional reliability of the pipeline after the first PM action (Eq. (18.69)), the second PM time can be calculated by $R_{sc}(T_{P2})_2 = R_{th}$, and the conditional reliability of the pipeline after the second PM action can be obtained.

Continuing the above process, we can obtain the conditional reliability function of the pipeline after the ith PM action:

$$R_{sc}(\tau)_i = R_1(\tau)_i R_2(\tau + T_{Pi})_0 = \frac{R_1(\tau)_i R_s(\tau + T_{Pi})_0}{R_1(\tau + T_{Pi})_0}, \ (0 \le \tau < (T_{Pi+1} - T_{Pi}) \text{ for}$$

$$i = 0, 1, 2, \ldots, n - 1 \text{ and } 0 \le \tau < (T_h - T_{Pn}) \text{ when } i = n)$$

$$\tag{18.71}$$

The $(i + 1)$th PM time can be calculated by letting $R_{sc}(T_{Pi+1})_i = R_{th}$ $(i = 1, 2, \ldots, n - 1)$.

The overall reliability of the pipeline after the ith PM action is given by substituting Eqs (18.65) and (18.71) into Eq. (18.64):

$$R_s(t)_i = \frac{R_1(t - T_{Pi})_i R_s(t)_0}{R_1(t)_0} \prod_{k=0}^{i} R_1(T_{Pk} - T_{Pk-1})_k, (T_{Pi} \le t < T_{Pi+1} \tag{18.72}$$

for $i = 0, 1, 2, \ldots, n - 1$ and $T_{Pn} \le t < T_h$ when $i = n$)

where $\prod_{k=0}^{i} R_1(T_{Pk} - T_{Pk-1})_k = 1$, when $i = 0$.

Reliability Prediction When Various Segments Repaired

The model can be extended to the scenarios where different segments are repaired or replaced in multiple PM cycles (Sun et al., 2007b). In this case, a pipeline is assumed to be divided into M segments which are connected together serially (see the top diagram in Figure 18.13). All segments of the pipeline may or may not be the same. Among them, M_p ($M_p \leq M$) vulnerable segments are proactively repaired in n PM actions. In each PM action, only one segment whose hazard can be reduced most among all the segments by this action is repaired.

The reliability calculation of the pipeline after the first PM action is the same as in the previous text. However, it could be different after the second PM action because the proactively repaired segments may or may not be the same after the second PM action. For brevity, only the end equation is presented here without going into detailed derivation. Interested readers can refer to Sun et al. (2007b) for a detail deviation of the formulation.

The conditional reliability function of a system after the ith PM cycle is given by:

$$
R_{\text{sc}}(\tau)_i = \frac{R_{\text{s}}[\tau + \sum\limits_{k=1}^{i}(T_{\text{P}k} - T_{\text{P}k-1})]_0 \prod\limits_{u=1}^{M_p} R_u[\tau + \sum\limits_{k=l_u+1}^{i}(T_{\text{P}k} - T_{\text{P}k-1})]_{l_u}}{\prod\limits_{u=1}^{M_p} R_u[\tau + \sum\limits_{i=1}^{i}(T_{\text{P}k} - T_{\text{P}k-1})]_0},
$$

$$
0 \leq \tau < (T_{\text{P}i+1} - T_{\text{P}i})
$$

$$
\text{for } i = 0, 1, 2, \ldots, n-1 \text{ and } 0 \leq \tau < (T_{\text{h}} - T_{\text{P}n}) \text{ when } i = n
$$

$$
(18.73)
$$

where l_u indicates that the last PM of segment u ($u \leq M_p$) is carried out in the l_uth PM action ($l_u \leq i$ and $i = 1, 2, \ldots, n$), $\sum_{k=l_u+1}^{i}(T_{\text{P}k} - T_{\text{P}k-1}) = 0$ when $l_u + 1 > i$, and $\sum_{k=1}^{0}(T_{\text{P}k} - T_{\text{P}k-1}) = 0$.

Using SSA, the number of PM actions required to meet the predefined reliability demand and the corresponding PM times can be predicted. It is worth noting that in SSA, the PM times of a pipeline are predicted based on the reliability functions of the pipeline and its preventively repaired segments. These reliability functions can be updated using condition monitoring data as described in Sections 18.3.2.2 and 18.3.2.3. Thus, the predicted PM times can also be updated. This updating ability enables us to improve our predictions and PM decisions continuously.

18.3.3.2 Optimization of Reliability Threshold

Reliability threshold is the only decision variable in RBPM decision making. When determining an optimal threshold for a pipeline, we need to consider its failure risk and PM related cost. To meet a higher reliability requirement, more frequent PM actions are required. As a result, PM cost will increase. On the other hand, lowering the reliability threshold will result in the increase of failure probability and consequently increase the failure cost. Therefore, an optimal threshold exists.

Various asset maintenance cost models have been developed (see, for example, Blischke and Murthy, 2000; Kelly, 1984; Pham, 2003; Sun et al., 2008). In the following analysis, we assume that the failure related cost of a pipeline is proportional to its overall failure probability, and the PM related cost is proportional to the number of PM actions:

$$C_F(R_{th}) = \theta_F[1 - R_s(T_h, R_{th})] \tag{18.74}$$

and

$$C_P(R_{th}) = \theta_P N(T_h, R_{th}) \tag{18.75}$$

where T_h is decision horizon. $R(T_h, R_{th})$ is the overall reliability of the pipeline at time T_h. $N(T_h, R_{th})$ is the number of PM actions over the period of time T_h to maintain the conditional reliability of the pipeline above the threshold. Both R (T_h, R_{th}) and $N(T_h, R_{th})$ depend on the reliability threshold R_{th} and can be calculated using the SSA presented in the previous text. Parameter θ_F can be interpreted as the failure cost rate caused by each percentage of pipelines, whereas parameter θ_P can be interpreted as the cost rate for each PM action. Parameters θ_F and θ_P are assumed to be constant. This assumption implies that the reliability of the pipeline should not be lower than 0.9. In reality, most offshore pipelines can meet this requirement. $C_F(R_{th})$ and $C_P(R_{th})$ are respectively the expected failure cost and the expected PM cost of the pipeline. Then, the total expected cost of the pipeline is given by

$$C_T(R_{th}) = C_F(R_{th}) + C_P(R_{th}) \tag{18.76}$$

where $C_T(R_{th})$ is the total expected cost. The objective of optimizing the RBPM decision of pipelines is to find the optimal reliability threshold R_{th} which minimizes the total expected cost of the pipelines.

In practical situations, parameters θ_F and θ_P could be uncertain. Let Θ_F represent the cost of a PM action and Θ_P represent the cost due to the failures of a percentage of pipelines, then Θ_F and Θ_P are two random variables. The following assumptions are made in the subsequent formulation:

1. The changes of Θ_F and Θ_P are independent of the age of the pipeline and the number of PM actions.
2. Variable Θ_F takes values in $[0, \infty)$ with a distribution density function $p_F(\theta_F)$.
3. Variable Θ_P varies in $[0, \infty)$ with a distribution density function $p_P(\theta_P)$.

The failure cost conditional on $\Theta_F = \theta_F$ is given by

$$C_F[R(T_h, R_{th})|\Theta_F = \theta_F] = \theta_F[1 - R(T_h, R_{th})] \tag{18.77}$$

Removing the condition, we have

$$C_F(R_{th}) = \int_0^\infty \theta_F[1 - R(T_h, R_{th})]p_F(\theta_F)d\theta_F \qquad (18.78)$$

$$\text{Let } E[\Theta_F] = \int_0^\infty \theta_F p_F(\theta_F)d\theta_F \qquad (18.79)$$

Substituting Eq. (18.78) into Eq. (18.77), we obtain the expected failure cost:

$$C_F(R_{th}) = E[\Theta_F][1 - R(T_h, R_{th})] \qquad (18.80)$$

Similarly, we can calculate the expected PM cost:

$$C_P(R_{th}) = N(T_h, R_{th}) \int_0^\infty \theta_P p_P(\theta_P)d\theta_P = E[\Theta_P]N(T_h, R_{th}) \qquad (18.81)$$

REFERENCES

Bakhtiary, A.Y., Ghaheri, A., Valipour, R., 2007. Analysis of offshore pipeline allowable free span length. Int. J. Civil Eng. 5, 84–91.

Barnard, I., 2009. Engineering Asset Management: an insurance perspective, 1st ed. Reliabilityweb.com, Fort Myers, FL.

Biswas, S.K., Ahmed, N.U., 2001. Optimal control of flow-induced vibration of pipeline. Dyn. Control 11, 187–201.

Blevins, R.D., 1990. Flow-Induced Vibration. Van Nostrand Reinhold, New York, NY.

Blischke, W.R., Murthy, D.N.P., 2000. Reliability—Modelling, Prediction, and Optimization. John Wiley & Sons, New York, NY.

British Standards Institution, 1984. BS3811 Glossary of Maintenance Terms in Terotechnology. British Standards Institution, London.

Castanier, B., Rausand, M., 2006. Maintenance optimization for subsea oil pipelines. Int. J. Press. Vessels Piping 83, 236–243 (The 16th European Safety and Reliability Conference).

Choi, H.S., 2001. Free spanning analysis of offshore pipelines. Ocean Eng. 28, 1325–1338.

Cox, D.R., Oakes, D., 1984. Analysis of Survival Data. Chapman & Hall, London.

Daei-Sorkhabi, A.H., Zehsaz, M., 2009. Studies of vortex induced vibration on the NACA0015. Int. J. Mech. 393, 39–43.

Det Norske Veritas, 1998. Free spanning pipelines. Guidelines No. 14.

Det Norske Veritas, 2008. Free spanning pipelines. Recommended Practice, DNV-RP-F105.

Ebeling, C.E., 1997. An Introduction to Reliability and Maintainability Engineering. McGraw-Hill Company, Inc., New York, NY.

Gao, F.P., Yang, B., Wu, Y.X., Yan, S.M., 2006. Steady current induced seabed scour around a vibrating pipeline. Appl. Ocean Res. 28, 291–298.

Housner, C.W., 1952. Bending vibration of a pipe line containing flowing fluid. J. Appl. Mech. 19, 205–208.

Jardine, A.K.S., Joseph, T., Banjevic, D., 1999. Optimizing condition-based maintenance decisions for equipment subject to vibration monitoring. J. Qual. Maintenance Eng. 5, 192–202.

Kelly, A., 1984. Maintenance Planning and Control. Butterworth & Co Ltd, Cambridge.

Koo, G.H., Park, Y.S., 1998. Vibration reduction by using periodic supports in a piping system. J. Sound Vib. 210, 53–68.

Laura, P.A.A., Ficcadenti de Iglesias, G.M., Verniere de Irassar, L., 1987. A note on flexural vibrations of a pipeline containing flowing fluid. Appl. Acoust. 21, 191–198.

Lee, U., Kim, J., 1999. Dynamics of branched pipeline systems conveying internal unsteady flow. J. Vib. Acoust. 121, 114–122.

Lewis, A.L., Szenasi, F.R., Roll, D.R., 1998. Control-vale-induced vibration corrected by variable speed pump. Feature June, 27 31.

Lewis, E.E., 1996. Reliability Engineering. John Wiley & Sons, New York, NY.

Liao, H., Zhao, W., Guo, H., 2006. Prediction remaining useful life of an individual unit using proportional hazards model and logistic regression model. In: IEEE Annual Reliability and Maintainability Symposium, Newport Beach, CA.

Lin, D., Zhou, M.J., Yam, R.C.M., 2000. General sequential imperfect preventive maintenance models. Int. J. Reliab. Qual. Saf. Eng. 7, 253–266.

Makis, V., Jardine, A.K.S., 1992. Optimal replacement in the proportional hazards model. INFOR 30, 172–181.

Miller, H.L., 2001. Piping vibration involving control valves, Presented at International Joint General Conference and Exposition ASME International—Power Division, New Orleans, LA.

Moubray, J., 1997. Reliability Centred Maintenance. Industrial Press, New York, NY.

Norton, M.P., 1989. Fundamentals of Noise and Vibration Analysis for Engineers. Cambridge University Press, Cambridge (Chapter 7).

Pham, H. (Ed.), 2003. Handbook of Reliability Engineering. Springer, London.

Pham, H., Wang, H.Z., 1996. Imperfect maintenance. Euro. J. Operat. Res. 94, 425–438.

Rausand, M., Hoyland, A., 2004. System Reliability Theory: Models, Statistical Methods, and Applications. John Wiley & Sons, Hoboken, NJ.

Sumer, B.M., Fredsøe, J., 1995. A review of vibrations of marine pipelines. Int. J. Offshore Polar Eng. 5, 81–90.

Sun, Y., Ma, L., Mathew, J., Wang, W., Zhang, S., 2006. Mechanical systems hazard estimation using condition monitoring. Mech. Syst. Signal Proc. 20, 1189–1201.

Sun, Y., Ma, L., Mathew, J., 2007a. Reliability prediction of repairable systems for single component repair. J. Qual. Maintenance Eng. 13, 111–124.

Sun, Y., Ma, L., Mathew, J., 2007b. Prediction of system reliability for multiple component repairs. In: Helander, M., et al. (Eds.), Proceedings of the 2007 IEEE International Conference on Industrial Engineering and Engineering Management, Singapore.

Sun, Y., Ma, L., Mathew, J., 2008. Determination of optimal preventive maintenance strategy for serial production lines. Aust. J. Mech. Eng. 5, 97–103.

Van den Abeele, F., Vande Voorde, J. Goes, P., 2008. Numerical modelling of vortex induced vibrations in submarine pipelines. In: Proceedings of the COMSOL Conference, Hannover

Xu, T., Lauridsen, B., Bai, Y., 1999. Wave-induced fatigue of multi-span pipelines. Marine Struct. 12, 83–106.

Yang, B., Gao, F.P., Jeng, D.S., Wu, Y.X., 2008. Experimental study of vortex-induced vibrations of a pipeline near an erodible sandy seabed. Ocean Eng. 35, 301–309.

Zhao, M., Cheng, L., 2010. Numerical investigation of local scour below a vibrating pipeline under steady current. Coastal Eng. 57, 397–406.

Zhong, X., Wu, Y., Li, D., Li, Q., 2007. Effect of slug flow on pipe vibration in horizontal pipeline. In: Cai, X., et al. (Eds.), Proceedings of the Multiphase Flow: The Ultimate Measurement Challenge. America Institute of Physics, America Institute of Physics, Macau, China, 537–541.

Zou, G.P., Cheraghi, N., Taheri, F., 2005. Fluid-induced vibration of composite natural gas pipeline. Int. J. Solid Struct. 42, 1253–1268.

Gas–Liquid Multiphase Flow in Pipeline

A.1 INTRODUCTION

Offshore pipelines consist of export pipelines and infield pipelines. Export pipelines transport oil or gas from either platform or FPSO (floating production, storage, and offloading) to beach for further processing. The flow inside the export pipeline is usually gas–condensate flow or oil flow with a little amount of water. The infield pipelines transport wellstream either from the manifolds or from the wells to the platform or FPSO. The flow in the infield pipeline is usually gas–oil–water multiphase flow for oil fields or gas–condensate–water flow for gas fields. Thus, to properly design offshore pipelines, it is critical to understand the impacts of multiphase flow. All of the flow assurance issues associated with offshore pipeline operations are related to the multiphase flow inside the pipeline.

Pressure drop is one of the most critical parameters for pipeline sizing. For single-phase flow, pressure drop is mainly controlled by the Reynolds number that is a function of the fluid viscosity, fluid density, fluid velocity, and pipeline size. For gas–oil–water three-phase flow, the pressure drop inside the pipeline is governed by the flowing properties of all the fluids:

- density of oil, water, and gas;
- viscosity of oil, water, and gas;
- velocity of oil, water, and gas;
- volume fraction of oil, water, and gas;
- interfacial tension between fluids;
- system pressure and temperature.

When oil, water, and gas are flowing inside the pipeline simultaneously, the three phases can distribute in the pipeline in many configurations due to the density difference among the fluids. These phase configurations are called flow regimes or flow patterns, differing from each other in the spatial distribution of the interfaces of water–oil, water–gas, and oil–gas. Different fluid interfaces result in different hydrodynamics of the flow, as well as mechanisms of the momentum, heat, and mass transfer among the fluids. Since flow in different flow regimes may induce different pressure drops, for proper pipe line sizing, it is important to correctly predict the flow regime.

This appendix covers the fundamentals of multiphase flow. Common terminologies used in multiphase flow will be defined. Flow regimes for both horizontal and vertical pipelines will be classified. Flow regime transitions and flow modeling will also be discussed.

In the last decade or so, pipeline flow simulations using multiphase flow simulators have gained significant popularity. Both steady-state and transient simulators have been used for pipeline design and for pipeline operation simulations. This appendix addresses mathematical models used in pipeline simulations and discusses factors affecting the simulation accuracy.

A.2 MULTIPHASE FLOW CONCEPTS

This section introduces commonly used multiphase flow variables. Pressure loss and recovery concepts will also be discussed. Finally, water–oil emulsion viscosity will be briefly discussed.

A.2.1 Basic Flow Variables

A.2.1.1 Superficial Velocity
The superficial velocity of liquid or gas is defined as the ratio of the liquid or gas volumetric flowrate to the total pipeline cross-sectional area, i.e.,

$$U_{sl} = \frac{Q_l}{A_f} \tag{A.1}$$

$$U_{sg} = \frac{Q_g}{A_f} \tag{A.2}$$

where

U_{sl} = liquid superficial velocity,
U_{sg} = gas superficial velocity,
Q_l, Q_g = liquid and gas volumetric flowrate, respectively,
A_f = pipeline flow cross-sectional area.

A.2.1.2 Mixture Velocity
The fluid mixture velocity is defined as the sum of the superficial gas and liquid velocities.

$$U_m = U_{sl} + U_g = \frac{Q_l + Q_g}{A} \tag{A.3}$$

where

U_m = fluid mixture velocity.

A.2.1.3 Liquid Holdup

Liquid holdup is defined as the ratio of the liquid volume in a pipeline segment to the whole volume of the pipeline segment.

$$H_1 = \frac{V_1}{V} \tag{A.4}$$

where

H_1 = liquid holdup
V_1 = pipeline segment volume occupied by liquid,
V = whole pipeline segment volume.

Liquid holdup is a function of both space and time.

A.2.1.4 Gas Void Fraction

Gas void fraction is defined as the ratio of the gas volume in a pipeline segment to the whole volume of the pipeline segment:

$$\alpha_g = \frac{V_g}{V} \tag{A.5}$$

where

α_g = gas void fraction,
V_g = pipeline segment volume occupied by gas.

From the above two equations, it is obvious that the sum of the liquid holdup and gas void fraction equals one.

$$H_1 + \alpha_g = 1 \tag{A.6}$$

A.2.1.5 Average Gas and Liquid Velocities

If the superficial velocity and liquid holdup are known and the liquid holdup would not change longitudinally, the average gas and liquid velocities can be calculated as:

$$u_g = \frac{Q_g}{A_g} = \frac{Q_g}{A\alpha_g} = \frac{Q_g}{A(1 - H_1)} = \frac{U_{sg}}{1 - H_1} \tag{A.7}$$

$$u_1 = \frac{Q_1}{A_1} = \frac{Q_1}{AH_1} = \frac{U_{sl}}{H_1} = \frac{U_{sl}}{1 - \alpha_g} \tag{A.8}$$

where

u_1, u_g = average liquid and gas velocities, respectively,
A_1, A_g = pipeline cross-sectional area occupied by liquid and gas, respectively.

A.2.1.6 Slip Velocity

Due to the density difference, when gas and liquid flow simultaneously inside a pipeline, the gas phase tends to flow faster than the liquid phase. The gas is "slipping" away from the liquid. The slip velocity is defined as the difference of the average gas and liquid velocities.

$$u_s = u_g - u_l = \frac{U_{sg}}{1 - H_l} - \frac{U_{sl}}{H_l} \qquad (A.9)$$

In homogeneous gas and liquid two-phase flow, there is no slippage between gas and liquid, and the slip velocity equals zero. Then, the liquid holdup can be easily calculated as:

$$H_l = \frac{U_{sl}}{U_{sl} + U_{sg}} = \frac{Q_l}{Q_l + Q_g} \qquad (A.10)$$

A.2.1.7 Water Cut

In the oil industry, a parameter commonly used by the petroleum engineers is called water cut, which is defined as the ratio of the water volumetric flowrate to the total water and oil volumetric flowrates, i.e.,

$$f_w = \frac{Q_w}{Q_w + Q_o} = \frac{Q_w}{Q_l} \qquad (A.11)$$

where

f_w = water cut,
Q_o, Q_w = oil and water volumetric flowrate, respectively.

A.2.1.8 Mixture Density

The density of gas and liquid homogeneous mixture is expressed as:

$$\rho_m = \rho_l H_l + \rho_g (1 - H_l) \qquad (A.12)$$

where

ρ_m = gas–liquid mixture density,
ρ_l, ρ_g = liquid and gas density, respectively.

A.2.1.9 Mixture Viscosity

If the gas and liquid mixture is homogeneous, the viscosity of the mixture can be calculated by:

$$\mu_m = \mu_l H_l + \mu_g (1 - H_l) \qquad (A.13)$$

where

μ_m = gas–liquid mixture viscosity,
μ_l, μ_g = liquid and gas viscosity, respectively.

The liquid viscosity can be the viscosity of water, oil, or water−oil mixture. Normally, the water−oil mixture viscosity can be calculated based upon the water cut:

$$\mu_l = \mu_o(1 - f_w) + \mu_w f_w \tag{A.14}$$

A.2.2 Pressure Loss and Recovery

In single-phase flow, as the flow goes through upward and downward sections, pressure loss due to the elevation change is fully recovered when the flow goes through the downward section. As shown in Figure A.1A, when fluid flows from A to B, the pressure at B is lower than the pressure at A due to the elevation change (pressure loss). But as the flow continues from B to C, the pressure gained at C due to the elevation change is equal to the pressure loss from A to B. Therefore, with single-phase flow, the pressure lost in the upward flow can be fully recovered in downward flow.

But the same conclusion may not always hold true in gas−liquid two-phase flow. As shown in Figure A.1B, the flow regime in the upward flow section (from A to B) may not be the same as the flow regime in the downward flow section (from B to C). With different flow regimes, the liquid holdup in each section may not be the same. Thus, the pressure loss in the upward flow section may not be fully recovered in the downward flow section.

A.2.3 Water−Oil Emulsion Viscosity

When mixing inside the pipeline, water and oil, especially heavy oil, can form emulsion. Water−oil emulsion is a heterogeneous system which consists of either water droplets dispersed in a continuous oil phase (W/O) or oil droplets dispersed in a continuous water phase (O/W) (Becher, 2001).

For oil-in-water (O/W) emulsion, water is the continuous phase and the viscosity of O/W emulsion is dominated by water viscosity. For water-in-oil

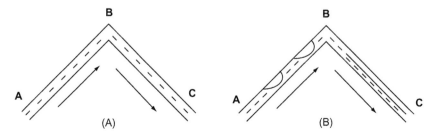

FIGURE A.1 Single-phase and multiphase flows through upward and downward pipe sections.

(W/O) emulsion, oil is the continuous phase and the viscosity of W/O emulsion is a strong function of the water cut and can be magnitudes higher than either the oil viscosity or water viscosity. There are quite a few parameters that can affect the emulsion viscosity. They are: oil and water viscosity, water cut, oil and water interfacial shear stress, water and oil physical properties, chemical surfactants, and any solid phase as wax or asphaltenes, and system temperature (Benayoune et al., 1998; Woelflin, 1942; Yan and Masliyah, 1993).

Over the years, quite extensive research has been conducted to try to develop simplified correlations for the water−oil emulsion viscosity (Becher, 2001). But since there are so many parameters that affect the emulsion viscosity, none of these correlations can be universally applied to engineering calculations. Instead, the best way to determine water−oil emulsion viscosity is to perform lab measurements of emulsions of different water cut at elevated pressure and temperature conditions. The oil sample should be from the live crude to be transported through the pipeline.

Only a couple of the correlations will be listed here for reference. For the diluted system with the concentration of the dispersed phase less than 10%, the classic Einstein equation can be used:

$$\mu_l = \mu_e(1 + 2.5 f_d) \tag{A.15}$$

where

μ_l = emulsion viscosity,
μ_e = viscosity of the continuous phase,
f_d = the volume fraction of the dispersed phase (less than 0.1).

For concentrated emulsions, Pal and Rhodes (1985) proposed the following correlation:

$$\mu_l = \mu_e \left[1 + \frac{f_d/f_c}{1.187 - f_d/f_c} \right]^{2.492} \tag{A.16}$$

where f_c = the concentration of the dispersed phase at which the emulsion viscosity μ_l is 100 times of the viscosity of the continuous phase μ_e.

A.3 FLOW REGIME CLASSIFICATIONS

Due to the physical property (mainly density) difference between liquid and gas, different flow patterns or flow regimes can occur when gas and liquid flow simultaneously inside the pipeline. The flow regimes differ from each other by having different gas−liquid interfaces. The mechanisms of mass, momentum, and energy transfer between phases are different in different flow regimes. Thus it is important to know the different flow regimes in both horizontal and vertical flows.

A.3.1 Horizontal Gas–Liquid Flow Regimes

The classification of flow regime is quite arbitrary, largely depending upon the individual's observations. For a given flow situation, different people may have different definitions of flow regime. Numerous flow regimes are defined in the literature. But for horizontal gas–liquid concurrent flow, the most widely accepted flow regimes (Bergles et al., 1981; Collier, 1972; Song, 1994) are shown in Figure A.2.

When gas and liquid are flowing concurrently inside a horizontal or near-horizontal pipeline, at low gas and liquid velocities, the gas and liquid will completely segregate from each other. The gas will flow on top of the liquid. The gas–liquid interface is smooth. This flow regime is called stratified smooth flow.

Starting from the stratified smooth flow, when gas flow and/or liquid flow increases, some waves will be generated at the gas–liquid interface. The gas–liquid interface becomes wavy. This flow is called stratified wavy flow.

If gas flow is further increased, the waves at the gas–liquid interface will grow. Some of the waves will be large enough to touch the upper inner pipe wall and block gas flow. Waves that are large enough to fill the pipe and block gas flow are called liquid slugs. This flow is defined as slug flow. In slug flow, the liquid inventory in the pipe is not uniformly distributed along

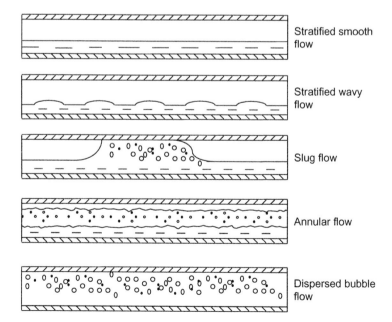

FIGURE A.2 Gas–liquid flow regimes in horizontal and near-horizontal pipeline.

the pipe axis, with slugs being separated by gas zones. The gas zones contain a stratified liquid layer flowing at the bottom of the pipe. The liquid slugs may be aerated by small gas bubbles.

If the gas flow is increased even further, the gas will flow as a core in the center of the pipe and the liquid will flow as a ring around the pipe wall. The liquid ring may not be uniform along the entire circumference but is thicker at the bottom of the pipe than at the top. Some small liquid droplets may be contained in the gas core. This flow is called annular flow.

With very low gas flow and high liquid flow, the gas will flow as discrete bubbles within a continuous liquid phase. The gas bubbles are usually not uniform in size and most of the bubbles flow at the upper portion of the pipe due to the buoyancy effects. This flow is called dispersed bubble flow.

A.3.2 Vertical Gas–Liquid Flow Regimes

The common flow regimes associated with upward vertical gas–liquid con-current flow are shown in Figure A.3. When gas flowrate is very low, the gas tends to flow as discrete bubbles in a continuous liquid phase. This flow is called bubble flow. The gas bubbles are not uniform in size and in shape and tend to flow in the center of the pipe with a zigzag pass. When the gas flowrate is increased, the gas bubble density becomes higher and some of the fast traveling larger bubbles catch up with the smaller ones and become even larger. At some point, large gas bubbles with a bullet front shape, called Taylor bubbles, appear as shown in Figure A.3. This flow is called slug

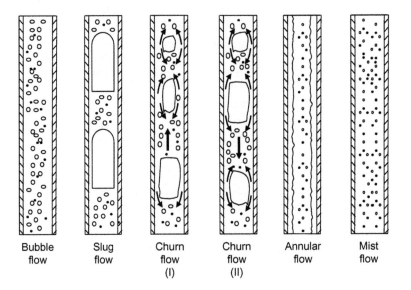

| Bubble flow | Slug flow | Churn flow (I) | Churn flow (II) | Annular flow | Mist flow |

FIGURE A.3   Gas–liquid flow regimes in upward vertical pipeline.

flow. Because the density of the gas is small, the pressure drop across the gas bubble is not enough to support the liquid film surrounding the gas bubble and the liquid film falls down. The falling liquid film is caught up by the liquid slugs which separate successive Taylor bubbles. In the liquid slugs, there may be small entrained gas bubbles. In slug flow, the liquid holdup along the pipe axis is not uniform but intermittent.

Another flow regime in vertical upward gas—liquid two-phase flow is called churn flow. Churn flow is formed by the breakdown of the Taylor bubbles in slug flow. As the Taylor bubbles flow upward, the bubbles become narrow and their shapes are distorted as shown in Figure A.3. At the same time, the liquid slugs between Taylor bubbles are penetrated by gas bubbles and start to fall downward. As the liquid falls down, the liquid forms a bridge at a lower position and is lifted again by the gas. This sequence repeats itself as fluids flow upward. Thus, in churn flow, the liquid slugs have oscillatory motions.

Even though churn flow is identified as one of the flow regimes in vertical two-phase flow, there is no existing model that can practically describe the chaotic physics. Thus, there is yet no practical usefulness in identifying the churn flow.

If the gas flowrate is very high, the gas will flow as a core in the center of the pipe and liquid will flow as a film along the pipe's inner wall. This flow is called annular flow. Inside the gas core, some liquid droplets are entrained. These entrained droplets enhance the interaction between gas and liquid. It is very important to predict the amount of liquid droplets that can be entrained in the gas core for thermal hydraulics analysis.

If the gas flowrate is even higher, the interfacial friction at the gas—liquid interface in the annular flow is so high that the liquid film will be destroyed by the gas. Thus, all the liquid flows as discrete droplets in the gas phase. This flow is called mist flow. Mist flow can be treated as homogeneous flow.

A.4 HORIZONTAL GAS–LIQUID FLOW REGIME MAPS

The preceding flow regime classifications are based upon visual observations of the phenomena occurring inside experimental pipelines. For engineering applications, visual observations may not always be available, and simple methods that can be used to predict flow regimes inside the pipeline for a given set of flow parameters are needed. Flow regime maps to define the various flow regime transitions were thus developed based upon either experimental data or mechanistic models.

For horizontal gas—liquid two-phase flow, Mandhane et al. (1974) developed a flow regime map using superficial gas and liquid velocities as coordinates. The map, as shown in Figure A.4, was based upon about 6000 experimental data points from pipelines of diameters between 1.27 and 16.51 cm.

With given flow conditions (pressure, temperature, volumetric flowrates, and pipeline sizes), the flow regime can be determined by using the map with superficial gas and liquid velocities. Again, the Mandhane map is good only for horizontal gas–liquid two-phase flow.

A much more widely used flow regime map for horizontal gas–liquid two-phase flow was developed by Taitel and Dukler (1976). The map was based upon mechanistic models, and the flow regime transitions are governed by different flow parameters in dimensionless form, as shown in Figure A.5.

In Taitel and Dukler's map, the flow regimes are defined by the transition curves. Curves A and B are defined by the coordinates F and X. Curve C is

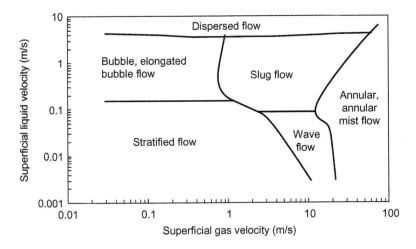

FIGURE A.4   Mandhane flow regime map for horizontal flow (Mandhane, 1974).

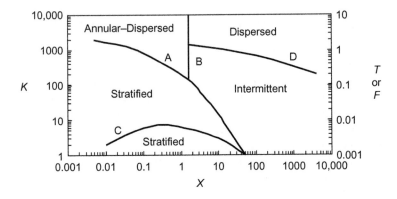

FIGURE A.5   Taitel–Dukler flow regime map for horizontal flow (Taitel and Dukler, 1976).

defined by the coordinates of K and X while Curve D by T and X. Parameters X, F, K, and T are defined as:

$$X = \left[\frac{(dP/dx)_l^s}{(dP/dx)_g^s} \right]^{1/2} \qquad F = \sqrt{\frac{\rho_g}{(\rho_l - \rho_g)}} \frac{U_{sg}}{\sqrt{D_g \cos \theta_p}}$$

$$T = \left[\frac{|(dP/dx)_l^g|}{(\rho_l - \rho_g)g \cos \theta_p} \right]^{1/2} \qquad K = \left[\frac{\rho_g U_{sg}^2 U_{sl}}{(\rho_l - \rho_g)g v_l \cos \theta_p} \right]^{1/2}$$

where

$(dP/dx)_g^s$ = pressure gradient for single-phase gas flow,
$(dP/dx)_l^s$ = pressure gradient for single-phase liquid flow,
θ_p = pipeline inclination angle,
v_l = liquid kinematic viscosity.

A.5 FLOW REGIME TRANSITIONS IN HORIZONTAL GAS–LIQUID FLOW

In offshore multiphase flow pipelines, flow regimes may vary spatially due to pipeline elevation changes. The flow regime may also change with time over the whole field life due to gas, oil, and water flowrate changes. It is possible that the flow changes from stable flow to unstable flow when flowrates decline over time. How to predict the flow regime transitions is one of the most important research topics in multiphase flow. Over the decades, numerous papers on flow regime transitions have been published (Andritsos and Hanratty, 1987; Andritsos et al., 1989; Bendiksen and Espedal, 1992; Bishop and Deshpande, 1986; Bontozoglou, 1991; Crowley et al., 1992; Fan et al., 1992; Johnston, 1984, 1985; Jones and Prosperetti, 1985; Kang and Kim, 1993; Kocamustafaogullari, 1985; Kordyban, 1961, 1977; Kordyban and Ranov, 1970; Kowalski, 1987; Lin and Hanratty, 1986, 1987; Mishima and Ishii, 1980; Ooms et al., 1985; Ruder et al., 1989; Taitel and Dukler, 1976; Wallis and Dobson, 1973).

A.5.1 Transition from Stratified Flow to Slug Flow

Taitel and Dukler (1976) presented a model to predict the transition from stratified flow to slug flow. In their model, the motion of the wave at the gas–liquid interface was neglected. Song (1994) stated, based upon experimental observations, that at the transition, the water depth inside the pipe can be higher than the pipe radius and the gas velocity is not very high. Thus, the velocity of the solitary wave at the interface can be substantial compared with the gas velocity at the transition. A transition model was then developed by modifying Taitel and Dukler's model without neglecting the motion of solitary waves at the interface.

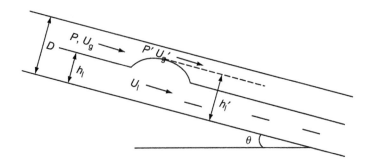

FIGURE A.6 Stratified wavy flow in slightly inclined pipe.

When a solitary wave is generated on the surface of a layer of motion-less water, as shown in Figure A.6, the wave velocity of this solitary wave can be expressed as (Friedichs and Hyers, 1954; Long, 1956; Stoker, 1957):

$$u_w = \sqrt{gh_1} \tag{A.17}$$

where

u_w = velocity of the solitary wave,
h_1 = equilibrium water depth.

As shown in Figure A.6, the pressure at the wave peak is lower than the pressure on the liquid surface where there is no wave due to the so-called Bernoulli effect. Thus, there is a pressure difference between the wave crest and flat liquid surface. This pressure difference tends to make the wave grow. On the other hand, the gravitational force of the wave tends to cause the wave to decay. Therefore, the condition for the wave to grow and to become a liquid slug is as follows:

$$p - p' \geq (h_1' - h_1)(\rho_1 - \rho_g)g \cos \theta_p \tag{A.18}$$

where

p = gas pressure on the flat liquid surface,
p' = gas pressure at the wave crest,
h_1' = liquid height at the wave crest,
θ_p = pipeline inclination angle.

Because of the wave motion, gas velocity relative to the wave will be reduced. By neglecting the gas gravitational force, the gas flow can be described as:

$$p - p' = \frac{1}{2}\rho_g \left[(u_g' - u_w)^2 - (u_g - u_w)^2 \right] \tag{A.19}$$

where

u'_g = gas velocity at the wave crest,
u_g = gas velocity on the flat liquid surface.

By assuming the gas density does not change significantly with time and with distance, the gas continuity equation can be expressed as:

$$u'_g - u_w = \frac{A_g}{A'_g}(u_g - u_w) \tag{A.20}$$

where

A_g = cross-sectional area occupied by gas at the equilibrium depth,
A'_g = cross-sectional area occupied by the gas at the wave crest.

From Eqs (A.19) and (A.20), one can get:

$$p - p' = \frac{1}{2}\rho_g\left(\left(\frac{A_g}{A'_g}\right)^2 - 1\right)(u_g - u_w)^2 \tag{A.21}$$

Eliminating the pressure from Eqs (A.18) and (A.21) yields,

$$(u_g - u_w)^2 \geq \frac{2(h'_1 - h_1)(\rho_1 - \rho_g)A'^2_g g \cos\theta_p}{(A^2_g - A'^2_g)\rho_g} \tag{A.22}$$

For small disturbances, neglecting the higher orders, one can get:

$$A'_g = A_g + \frac{dA_g}{dh_1}(h'_1 - h_1) \tag{A.23}$$

and

$$A'^2_g = A^2_g + 2A_g\frac{dA_g}{dh_1}(h'_1 - h_1) \tag{A.24}$$

Equation (A.24) can be expressed as:

$$A'^2_g - A^2_g = -2A_g\frac{dA_1}{dh_1}(h'_1 - h_1) \tag{A.25}$$

From Eqs (A.22) and (A.25), eliminating the liquid height at the wave crest gives:

$$(u_g - u_w)^2 \geq \frac{A'^2_g}{A^2_g}\left(\frac{(\rho_1 - \rho_g)A_g g \cos\theta_p}{\frac{dA_1}{dh_1}\rho_g}\right) \tag{A.26}$$

and the transition criterion can be expressed as:

$$u_g \geq \frac{A'_g}{A_g}\left(\frac{(\rho_1 - \rho_g)A_g \cos\theta_p}{(dA_1/dh_1)\rho_g}\right)^{1/2} + u_w \tag{A.27}$$

Equation (A.27) differs from the original Taitel–Dukler model by the solitary wave velocity.

Taitel and Dukler (1976) speculated the following relation:

$$\frac{A'_g}{A_g} = 1 - \frac{h_1}{D} \tag{A.28}$$

Thus, Eq. (A.27) becomes:

$$u_g \geq \left(1 - \frac{h_1}{D}\right)\left(\frac{(\rho_1 - \rho_g)A_g g \cos \theta_p}{(dA_1/dh_1)\rho_g}\right)^{1/2} + u_w \tag{A.29}$$

Equation (A.29) gives the criterion for the transition from stratified flow to slug flow.

Another popular model for the transition from stratified flow to slug flow is based on the classical Kelvin–Helmholtz instability (Song, 1994). When a gas flows parallel to a liquid surface and the interface becomes disturbed, the aerodynamic pressure will develop a component that is high at the troughs and low at the crests. When the gas velocity is high enough, the pressure difference will become sufficiently large enough to overcome the stabilizing effect of the gravity and the perturbation will grow and the flow will become unstable. This phenomenon is called the Kelvin–Helmholtz instability (Lamb, 1932; Milne-Thomson, 1968).

For long waves of small amplitude in horizontal flow, the criterion for the classical Kelvin–Helmholtz instability is given by:

$$u_g - u_1 \geq \left(\frac{(\rho_1 - \rho_g)h_g g}{\rho_g}\right)^{1/2} \tag{A.30}$$

where h_g is the height of the gas column.

Equation (A.30) is similar to Eq. (A.29).

A.5.2 Transition from Slug Flow to Annular Flow

Equation (A.29) shows the criteria for waves on the liquid surface to become unstable. If the equilibrium liquid level is high enough, there is sufficient liquid in the system for the slug to form. Otherwise, the waves will be swept up around the pipe wall to form annular flow (Taitel and Dukler, 1976). When the wave becomes unstable with increasing amplitude, more liquid is needed to sustain the wave. The liquid has to come from the equilibrium liquid film adjacent to the wave. For slug flow to form, the wave needs to touch the upper pipe wall to block the gas flow. Thus, the liquid level of the equilibrium liquid film has to be no less than the pipeline centerline. If the height of the liquid film is less than the centerline, the wave trough will reach the bottom before the wave peak can reach the top pipe wall and no slugs can form. Based upon the above reasoning, Taitel and Dukler (1976) proposed the following transition criteria between slug flow and annular flow:

$$h_1/D_i \geq 0.5 \tag{A.31}$$

where D_i is the pipeline's inner diameter.

Barnea et al. (1982) modified this criterion based upon the observations that liquid holdup inside the liquid slugs at the transition is less than 1.0 and proposed the new criterion as:

$$h_1/D_i \geq 0.35 \tag{A.32}$$

A.5.3 Transition from Stratified Smooth Flow to Stratified Wavy Flow

When the gas flowrate is high enough, waves will be generated on the liquid surface. How the waves are generated is very complicated and is not yet completely understood. Taitel and Dukler (1976) used the Jeffrey's theory (1925, 1926) for the transition between stratified smooth flow and stratified wavy flow to get:

$$(u_g - c)^2 c > \frac{4\nu_l g(\rho_l - \rho_g)}{s_c \rho_g} \tag{A.33}$$

where

c = wave propagation velocity,
ν_l = liquid kinematic viscosity,
s_c = a sheltering coefficient (0.01 is used by Taitel and Dukler).

The wave propagation velocity is much smaller than the gas velocity at the transition. The ratio of the wave velocity to the liquid velocity is a function of the Reynolds number of the liquid. At the transition, this ratio approaches 1.0–1.5. For simplicity, Taitel and Dukler used a ratio of 1.0 and assumed that the wave propagation velocity equals the liquid velocity. Thus, the transition between smooth flow and wave flow is given as:

$$u_g \geq \left[\frac{4\nu_l(\rho_l - \rho_g)g \cos \theta_p}{s \rho_g ul} \right] \tag{A.34}$$

A.5.4 Transition Between Slug Flow and Dispersed Bubble Flow

In slug flow, large gas bubbles exist between successive liquid slugs. If the liquid flow is high enough and the gas flow is low, the gas bubbles will shrink and the liquid level will approach the top pipe wall. When the liquid turbulence is large enough to overcome the buoyant force that tends to keep the gas at the top to form large bubbles, the transition to dispersed bubble flow will occur.

The buoyant force of the gas can be expressed as:

$$F_b = g \cos \theta(\rho_l - \rho_g)A_g \tag{A.35}$$

where F_b is the gas buoyant force.

Taitel and Dukler (1976) proposed the following equation for the estimation of the liquid turbulence force:

$$F_t = \frac{1}{2}\rho_l u_l^2 \left(\frac{f_l}{2}\right) s_i \tag{A.36}$$

where

F_t = liquid turbulence force,
f_l = liquid friction factor,
s_i = perimeter of gas−liquid interface.

At the transition, the turbulence force is much larger than the buoyant force. Thus, the criterion for the transition between slug flow and dispersed bubble flow can be expressed as:

$$u_l \geq \left[\frac{4A_g}{S_i} \frac{g \cos \theta_p}{f_l} \left(\frac{\rho_l - \rho_g}{\rho_l}\right)\right]^{1/2} \tag{A.37a}$$

A.6 MODELING OF MULTIPHASE FLOW IN HORIZONTAL PIPELINE

The major objectives of multiphase modeling are to calculate the pressure drop and liquid holdup inside the pipeline. After we define the common flow regimes and the flow regime transition criteria, we are ready to perform modeling calculations of multiphase flow in pipeline.

In all the modeling calculations, we assume the multiphase flow in the pipeline is in steady state and fully developed and all the flow parameters are independent of time.

A.6.1 Stratified Flow Model

In steady-state stratified flow, the equilibrium liquid level in the liquid film is a constant. The gas and liquid can be treated as two separate flows, and the so-called "two-fluid" model can be utilized.

Figure A.7 shows the horizontal gas−liquid stratified flow model. The gas is flowing on top of the liquid. From the control volume shown in

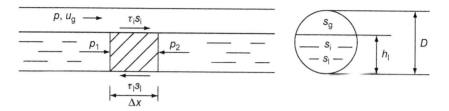

FIGURE A.7 Horizontal gas–liquid stratified flow.

Figure A.7, assuming the velocities are constant along the flowing direction, the momentum equation for the liquid can be expressed as (Song, 1994):

$$\left(p + \rho_{lg}\frac{h_1}{2}\right)A_1 - \left(p + \rho_{lg}\frac{h_1}{2}\frac{dp}{dx}\Delta x\right)A_1 + \tau_i s_i \Delta x - \tau_1 s_1 \Delta x = 0 \quad \text{(A.37b)}$$

where

p = gas pressure,
Δx = length of the control volume along the axis,
τ_i, τ_1 = shear stress at the interface and around the pipe wall occupied by the liquid, respectively,
s_i, s_1 = pipe perimeter at the interface and wetted periphery, respectively.

The above equation can be simplified as:

$$-A_1\frac{dp}{dx} + \tau_i s_i - \tau_1 s_1 = 0 \quad \text{(A.38)}$$

In the above equation, the shallow water approximation is assumed. Thus, the liquid pressure for the control volume is:

$$p_1 = p + \rho_{lg}\frac{h_1}{2} \quad \text{(A.39)}$$

Similarly, the momentum equation for the gas phase:

$$-A_g\frac{dp}{dx} - \tau_i s_i - \tau_g s_g = 0 \quad \text{(A.40)}$$

By eliminating the pressure from the two momentum equations, we have:

$$\tau_g\frac{s_g}{A_g} - \tau_1\frac{s_1}{A_1} + \tau_i s_i\left(\frac{1}{A_g} + \frac{1}{A_1}\right) = 0 \quad \text{(A.41)}$$

The shear stresses are evaluated using the following equations:

$$\tau_1 = f_1\frac{\rho_1 u_1^2}{2}, \qquad \tau_g = f_g\frac{\rho_g u_g^2}{2}$$

$$\tau_i = f_i\frac{\rho_g(u_g - u_1)^2}{2} \quad \text{(A.42)}$$

where f_l, f_g, f_i are the friction factors of liquid wall, gas wall, and at interface, respectively.

Based upon the suggestion by Taitel and Dukler (1976), the friction factors can be calculated as, for turbulent flow:

$$f_l = 0.046\left(\frac{D_l u_l}{\nu_l}\right)^{-0.2}, \quad f_g = 0.046\left(\frac{D_g u_g}{\nu_g}\right)^{-0.2} \tag{A.43}$$

$$f_i = f_g = 0.046\left(\frac{D_g u_g}{\nu_g}\right)^{-0.2}$$

where D_l and D_g are liquid and gas hydraulic diameters, respectively.

The hydraulic diameters are defined as:

$$D_l = \frac{4A_l}{s_l}, \quad D_g = \frac{4A_g}{s_g + s_i} \tag{A.44}$$

In the above analysis, it is assumed that the gas flows faster than the liquid. If the liquid flows faster than the gas, the term in Eq. (A.41), for the friction force at the interface will be negative. Thus, by taking into account both cases, we have:

$$\tau_g \frac{s_g}{A_g} - \tau_l \frac{s_l}{A_l} \pm \tau_i s_i \left(\frac{1}{A_g} + \frac{1}{A_l}\right) = 0 \tag{A.45}$$

where

$+$ = gas flows faster than liquid,
$-$ = liquid flows faster than gas.

Equation (A.45) can be solved for the equilibrium liquid depth for stratified flow. To simplify the calculation, the following dimensionless quantities are introduced:

$$h_{ld} = \frac{h_l}{D}, \quad s_{gd} = \frac{s_g}{D} = \cos^{-2}(2h_{ld} - 1)$$

$$s_{ld} = \frac{s_l}{D} = \pi - s_{gd} \quad s_{id} = \frac{s_i}{D} = [1 - (2h_{ld} - 1)^2]^{\frac{1}{2}}$$

$$A_{ld} = \frac{A_l}{D^2} = \frac{1}{4}\left[\pi - \cos^{-1}(2h_{ld} - 1) + (2h_{ld} - 1)[1 - (2h_{ld} - 1)^2]^{\frac{1}{2}}\right] \tag{A.46}$$

$$A_{gd} = \frac{A_g}{D^2} = \frac{\pi}{4} - A_{ld}$$

The shear stresses can also be expressed in these dimensionless parameters and the superficial velocities:

$$\tau_1 = 0.023 \left(\frac{\pi}{4} U_{sl} \right)^{1.8} \frac{\rho_1}{A_{ld}^2} \left[\frac{4D}{(\pi - s_{gd})\nu_1} \right]^{-1/5}$$

$$\tau_g = 0.023 \left(\frac{\pi}{4} U_{sg} \right)^{1.8} \frac{\rho_g}{A_{gd}^2} \left[\frac{4D}{(s_{id} - s_{gd})\nu_g} \right]^{-1/5} \quad\quad (A.47)$$

$$\tau_i = 0.023 \left(\frac{\pi}{4} \right)^{1.8} \left[\frac{4DU_{sg}}{(s_{id} + s_{gd})\nu_g} \right]^{-1/5} \rho \left[\frac{U_{sg}}{A_{gd}} - \frac{U_{sl}}{A_{ld}} \right]^2$$

Substituting Eqs (A.46) and (A.47) into Eq. (A.45), we have:

$$\rho_g U_{sg}^{1.8} [(s_{gd} + s_{id})\nu_g]^{\frac{1}{5}} \frac{s_{gd}}{A_{gd}^3} - \rho_1 U_{sl}^{1.8} \nu_1^{0.2} \frac{s_{ld}^{1.2}}{A_{ld}^3} \mp$$

$$\rho_g U_{sg}^{-0.2} [(s_{gd} + s_{id})\nu_g]^{\frac{1}{5}} s_{id} \left[\frac{U_{sg}}{A_{gd}} - \frac{U_{sl}}{A_{ld}} \right]^2 \left(\frac{1}{A_{gd}} + \frac{1}{A_{ld}} \right) = 0 \quad\quad (A.48)$$

This is the equation for horizontal, steady-state stratified flow with turbulent liquid flow. For given gas and liquid volumetric flowrates and pipeline diameter, the equilibrium liquid depth can be solved from the above equation. Once the liquid depth is known, the liquid holdup can be calculated. The pressure drop can be calculated using either Eq. (A.38) or (A.40).

If the liquid flow is laminar instead of turbulent, the friction factors can be evaluated as:

$$f_1 = 16 \frac{\nu_1}{D_1 u_1} \quad f_g = 16 \frac{\nu_{gl}}{D_g u_g} \quad f_i = f_g \quad\quad (A.49)$$

Equation (A.48) would change to:

$$p_g U_{sg} [(s_{gd} + s_{id})\nu_g] \frac{s_{gd}}{A_{gd}^3} - p_1 U_{sl} \nu_1 \frac{s_{ld}}{A_{ld}^3} \mp$$

$$p_g [(s_{gd} + s_{id})\nu_g] s_{id} \left[\frac{U_{sg}}{A_{gd}} - \frac{U_{sl}}{A_{ld}} \right] \left(\frac{1}{A_{gd}} + \frac{1}{A_{ld}} \right) = 0 \quad\quad (A.50)$$

Equation (A.50) is the stratified flow equation for laminar flow. When Eqs (A.48) and (A.50) are solved, multiple solutions for the liquid depth at low superficial velocities may exist. It is suggested that the minimum value should be used.

A.6.2 Slug Flow Model

Slug flow is one of the most complicated multiphase flows in pipeline. Extensive research has been conducted over the last few decades to try to develop mechanistic models for slug flow calculations. The first widely cited slug flow model was developed by Dukler and Hubbard (1975). Xiao et al. (1990) presented a comprehensive mechanistic model for slug flow. More recently, Zhang et al. (2003) presented a unified mechanistic model for gas–liquid slug flow in pipeline. The unified model is claimed to be applicable to all the pipeline inclining from −90° to +90° from horizontal. The approaches used by Zhang et al. (2003) will be presented here.

Figure A.8 shows a schematic of gas–liquid slug flow in inclined (near-horizontal) pipeline. There are a few assumptions associated with slug flow. They are as follows:

- The flow is steady state and fully developed.
- There is no gas bubble in the film region.
- Liquid droplets are entrained in the gas packet on top of the film region.
- There is no slippage between the gas and the liquid droplets in the gas packet.
- There is no slippage between the gas bubbles and the liquid in the slug body.

As described by Dukler and Hubbard (1975), when the liquid slug travels downstream, the slug will pick up fluids from the film region in front of it. When the slow-moving liquid film is overrun by the slug and is accelerated to the slug velocity, a mixing eddy forms in the slug front. Because of the mixing eddy, the fluid particle movement in the slug front is chaotic.

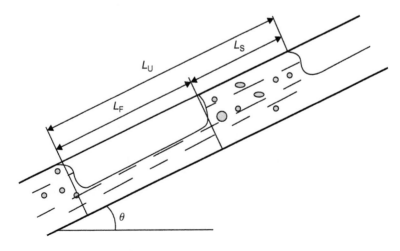

FIGURE A.8 Slug flow model.

Liquid also sheds away from the slug body at the back and forms the liquid film region behind the slug. The liquid in the film decelerates from the slug velocity and is picked up by the successive slug. With steady-state, fully established slug flow, fluids picked up from the front equal the fluids shedding away from the back. Because the slug picks up fluids from the front, the slug front velocity is higher than the average slug velocity. The slug front velocity is also called translational velocity, u_T.

Gas is trapped inside the slug due to the mixing process at the slug front. The higher the gas velocity, the more the gas is entrained. Thus, the liquid holdup inside the slug body is usually less than one.

Slug flow can be divided into two regions, as shown in Figure A.8. One region is called the film region and consists of liquid film at the bottom and gas packet at the top. The other region is called the slug region and consists of the slug body and gas bubbles in the body. The sum of the film region and the slug region is called the slug unit. If it is treated as a series of slug units flowing together, slug flow is a normal "continuous" flow (continuous from unit to unit), just like any other flow patterns. But at any point, slug flow, in nature, is intermittent with gas packets and liquid slugs alternating.

Because the slug front travels at the translational velocity, at one stationary point, one would observe only this velocity, not the fluid particle velocity. Thus, with a coordinate system moving at the translational velocity, continuity equations can be applied. Choosing a control volume consists of the film region, the gas and liquid continuity equations can be expressed as:

$$(1 - H_{IS})(u_T - u_S) = (1 - H_{IF} - H_{IC})(u_T - u_C) \quad (A.51)$$

where

H_{IS} = liquid holdup inside the slug body,
H_{IF} = liquid holdup inside the liquid film,
H_{IC} = liquid holdup inside the gas core or gas packet,
u_T = slug translational velocity,
u_S = slug velocity which equals to the mixture velocity,
u_C = gas core velocity.

$$H_{IS}(u_T - u_S) = H_{IC}(u_T - u_C) + H_{IF}(u_T - u_F) \quad (A.52)$$

where u_F is the liquid film velocity.

By combining Eqs (A.51) and (A.52), one gets

$$u_S = H_{IF}u_F + (1 - H_{IF})u_C \quad (A.53)$$

By definition, the slug unit length equals to the sum of film region length and slug length.

$$l_U = l_F + l_S \tag{A.54}$$

and

$$l_U = u_T t_U \tag{A.55}$$

where t_U is the time required for the slug to travel a distance of slug unit.

$$l_S = u_T t_S \tag{A.56}$$

where t_S is the time required for slug to travel a distance of the slug region.

$$l_F = u_T t_F \tag{A.57}$$

where t_F is the time required for the slug to travel a distance of the film region.

Again, using a coordinate that moves at the translational velocity, the liquid mass balance can also be expressed as:

$$Q_l t_U = A H_{lS} u_S t_S + A(H_{lF} u_F + H_{lC} u_C) t_F \tag{A.58}$$

Based upon Eqs (A.54)–(A.58), one can get:

$$l_U U_{sl} = l_S H_{lS} u_S + l_F(H_{lF} u_F + H_{lC} u_C) \tag{A.59}$$

Similarly, for the gas phase:

$$l_U U_{sg} = l_S(1 - H_{lS}) u_S + l_F(1 - H_{lF} - H_{lC}) u_C \tag{A.60}$$

The liquid holdup in the gas core is related to the liquid entrainment fraction which is defined as:

$$f_E = \frac{H_{lC} u_C}{H_{lF} u_F + H_{lC} u_C} \tag{A.61}$$

Again using the whole film region (both gas packet and liquid film) as the control volume, one can write the momentum equation for the liquid based upon the momentum into the control volume and momentum out of the control volume. The momentum equation for the liquid film is expressed as:

$$\frac{(p_2 - p_1)}{l_F} = \frac{\rho_l(u_T - u_F)(u_s - u_F)}{l_F} + \frac{\tau_i s_i - \tau_F s_F}{H_{lF} A} - \rho_l g \sin \theta_p \tag{A.62}$$

where p_1 and p_2 are the pressures at the right and left boundaries of the film region, respectively.

Similarly, the momentum equation for the gas packet is:

$$\frac{(p_2 - p_1)}{l_F} = \frac{\rho_C(u_T - u_C)(u_s - u_C)}{l_F} - \frac{\tau_i s_i + \tau_C s_C}{(1 - H_{lF}) A} - \rho_C g \sin \theta_p \tag{A.63}$$

where ρ_C is the gas core density.

The gas core density is related to the gas and liquid densities through,

$$\rho_C = \frac{\rho_g(1 - H_{IF} - H_{IC}) + \rho_l H_{IC}}{1 - H_{IF}} \tag{A.64}$$

By eliminating the pressures from Eqs (A.62) and (A.63), one can get the combined momentum equation:

$$\frac{\rho_l(u_T - u_F)(u_S - u_F) - \rho_C(u_T - u_C)(u_S - u_C)}{l_F} - \frac{\tau_F s_F}{H_{IF}A} +$$

$$\frac{\tau_C s_C}{(1 - H_{IF})A} + \tau_i s_i \left(\frac{1}{H_{IF}A} + \frac{1}{(1 - H_{IF})A} \right) - (\rho_l - \rho_C)g \sin \theta = 0 \tag{A.65}$$

The above equations are the governing equations for slug flow. Before the equations can be solved to get the liquid holdup, pressure drop, and other slug flowing parameters, a few closure equations are still needed.

The shear stresses can be evaluated as:

$$\tau_F = f_F \frac{\rho_l u_F^2}{2}, \tau_C = f_C \frac{\rho_g u_C^2}{2}, \tau_i = f_i \frac{\rho_C(u_C - u_F)|u_C - u_F|}{2} \tag{A.66}$$

Zhang et al. (2003) presented an equation to link the shear stress at the pipe wall with the shear stress at the gas−liquid interface:

$$\tau_F = \frac{3\mu_l u_F}{h_F} - \frac{\tau_i}{2} \tag{A.67}$$

where h_F is the average liquid film height in the film region, and is defined as

$$h_F = \frac{2AH_{IF}}{s_F + s_i} \tag{A.68}$$

The friction factors at the pipe wall can be estimated as:

$$f = m\, Re^{-n} \tag{A.69}$$

For laminar flow (Reynolds number less than 2000), $m = 16$ and $n = 1$. For turbulent flow (Reynolds number larger than 3000) and smooth pipe wall, $m = 0.046$ and $n = 0.2$.

The Reynolds numbers for the film and gas core are defined as:

$$Re_F = \frac{4H_{IF}Au_F\rho_l}{s_F\mu_l} \tag{A.70}$$

$$Re_C = \frac{4(1 - H_{IF})Au_C\rho_g}{(s_C + s_i)\mu_g} \tag{A.71}$$

The liquid holdup in the gas core is neglected in Eq. (A.71).

For the pipe perimeter calculations, Zhang et al. (2003) proposed the following geometry relations.

$$\Theta_1 = \Theta_{10}\left(\frac{\sigma_{\text{water}}}{\sigma}\right)^{0.15} + \frac{\rho_g}{\rho_1 - \rho_g}\frac{1}{\cos\theta}\left(\frac{\rho_1 U_{sl}^2 D}{\sigma}\right)^{0.25}\left(\frac{U_{sg}^2}{(1-H_{1F})^2 gD}\right)^{0.8} \quad \text{(A.72)}$$

where

Θ_1, Θ_{10} = pipe wall fraction wetted by liquid with curved and flat gas−liquid interfaces, respectively,

σ_{water}, σ = water and liquid surface tension, respectively.

and

$$s_F = \pi D\Theta_1 \quad \text{(A.73)}$$

$$s_i = \frac{s_F\left(\frac{D^2}{4}\left(\pi\Theta_1 - \frac{\sin(2\pi\Theta_1)}{2}\right) - H_{1F}A\right)H_{1F}AD\sin(\pi\Theta_1)}{\frac{D^2}{4}\left(\pi\Theta_1 - \frac{\sin(2\pi\Theta_1)}{2}\right)} \quad \text{(A.74)}$$

Another closure equation on the liquid entrainment in the gas core is needed. The correlation by Oliemans et al. (1986) for vertical pipe is presented here:

$$\frac{f_E}{1-f_E} = 10^{\beta_0}\rho_1^{\beta_1}\rho_g^{\beta_2}\mu_1^{\beta_3}\mu_g^{\beta_4}\sigma^{\beta_5}D^{\beta_6}U_{sl}^{\beta_7}U_{sg}^{\beta_8}g^{\beta_9} \quad \text{(A.75)}$$

Equation (A.75) is based upon data regression and all the coefficients are constants and are given below:

$\beta_0 = -2.52,$ $\beta_1 = 1.08,$ $\beta_2 = 0.18,$ $\beta_3 = 0.27,$ $\beta_4 = 0.28$
$\beta_5 = -1.80,$ $\beta_6 = 1.72,$ $\beta_7 = 0.70,$ $\beta_8 = 1.44,$ $\beta_9 = 0.46$

Zhang et al. (2003) proposed a model to calculate the holdup inside the slug body. The simpler correlation by Gregory et al. (1978) is presented here.

$$H_{1S} = \frac{1}{1 + \left(\frac{u_S}{8.66}\right)^{1.39}} \quad \text{(A.76)}$$

The slug velocity is the sum of the superficial gas and liquid velocities.

Another parameter that must be calculated independently is the slug translational velocity. Zhang et al. (2003) proposed the following equation based upon Bendiksen's work in 1984.

$$u_T = Cu_S + 0.54\sqrt{gD}\cos\theta_p + 0.35\sqrt{gD}\sin\theta_p \quad \text{(A.77)}$$

Zhang et al. (2003) also proposed the following equation for the slug length:

$$l_S = (32\cos^2\theta_p + 16\sin^2\theta_p)D \quad \text{(A.78)}$$

From the above equation, for horizontal flow, the slug length equals 32 times the pipe diameter, and for vertical flow, the slug length is 16 times the pipe diameter. Slug flow parameters can now be solved from all the above equations. Trial and error must be used. It is suggested to start with guessing a value for the film length.

A.6.3 Annular Flow Model

In steady-state annular flow, the majority of the gas and liquid are segregated with gas flows as a core in the center of the pipe and liquid flows as a film around the pipe wall, as shown in Figure A.9. Some liquid droplets are entrained in the gas core. As for stratified flow, annular flow can be calculated by using the two-fluid model. It is also assumed that there is no gas bubble in the film region.

Similar to Eqs (A.38) and (A.40), the momentum equation for the film and for the gas core can be written as:

$$-A_F \frac{dp}{dx} + \tau_i s_i - \tau_F s_F - A_F \rho_l g \sin \theta_p = 0 \tag{A.79}$$

and

$$-A_C \frac{dp}{dx} - \tau_i s_i - A_C \rho_C g \sin \theta_p = 0 \tag{A.80}$$

Since the gas core is surrounded by the liquid film, there is no gas wall shear stress term in the momentum equation for the gas core.

Equations (A.79) and (A.80) can be combined by eliminating the pressure:

$$\tau_F \frac{s_F}{A_F} - \tau_i s_i \left(\frac{1}{A_F} + \frac{1}{A_C}\right) + (\rho_l - \rho_C)g \sin \theta_p = 0 \tag{A.81}$$

The shear stresses are defined in Eqs (A.66), (A.69)–(A.71). For annular flow:

$$s_F = \pi D, \quad s_i = \pi(D - 2\delta_l) \tag{A.82}$$

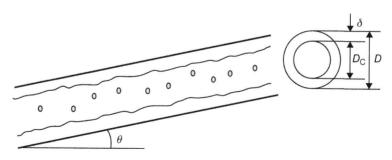

FIGURE A.9 Schematic for annular flow.

where δ_1 is the average film thickness.

Since it is assumed that there is no gas bubble in the liquid film, the film liquid holdup can be calculated as:

$$H_{\mathrm{lF}} = \frac{4\delta_1(D - \delta_1)}{\pi D^2} \tag{A.83}$$

And the liquid holdup in the gas core can be calculated as:

$$H_{\mathrm{lc}} = \frac{f_{\mathrm{E}} U_{\mathrm{sl}}(1 - H_{\mathrm{lF}})}{U_{\mathrm{sg}} + f_{\mathrm{E}} U_{\mathrm{sl}}} \tag{A.84}$$

Equation (A.84) is based upon the definition of the liquid droplet entrainment fraction. Thus, the gas core velocity can be calculated as:

$$u_{\mathrm{C}} = \frac{U_{\mathrm{sg}}}{1 - H_{\mathrm{lF}} - H_{\mathrm{lC}}} \tag{A.85}$$

The liquid film velocity can be calculated as:

$$u_{\mathrm{F}} = \frac{U_{\mathrm{sl}} - H_{\mathrm{lC}} u_{\mathrm{C}}}{H_{\mathrm{lF}}} \tag{A.86}$$

The last correlation needed to solve for annular flow is the friction factor at the interface. The correlation suggested by Xiao et al. (1990) is presented here:

$$f_{\mathrm{i}} = f_{\mathrm{c}}\left[1 + 2250 \frac{(\delta_1/D)}{((\rho_{\mathrm{C}}(u_{\mathrm{C}} - u_{\mathrm{F}})^2 \delta_1)/\sigma)}\right] \tag{A.87}$$

The core friction factor can be calculated using Eq. (A.69). Thus, the description of annular flow is completed. To solve for the flow parameters, it is suggested to start by guessing an average film thickness, δ_1. With the guessed average film thickness, if Eq. (A.81) is satisfied, the guessed film thickness is the right one and the corresponding flow velocities, liquid holdups, and pressure drops can be calculated using the proper equations.

A.6.4 Dispersed Bubble Flow Model

In dispersed bubble flow, the gas flows as discrete bubbles in a continuous liquid phase. If it is assumed that there is no slippage between the gas bubbles and the liquid, dispersed bubble flow can be treated as a pseudo single-phase flow. The parameters, such as density and viscosity, can be evaluated using the nonslip liquid holdup as discussed in Section A.2. Once the parameters are evaluated, single-phase flow equations can be used for pressure drop calculations.

REFERENCES

Andritsos, N., Hanratty, T.J., 1987. Interfacial instabilities for horizontal gas—liquid flows in pipelines. Int. J. Multiphase Flow 13 (5), 583—603.

Andritsos, N., Williams, L., Hanratty, T.J., 1989. Effect of liquid viscosity on the stratified—slug transition in horizontal pipe flow. Int. J. Multiphase Flow 15 (6), 877—892.

Barnea, D., Shoham, O., Taitel, Y., 1982. Flow pattern transition for vertical downward inclined two-phase flow: horizontal to vertical. Chem. Eng. Sci. 37 (5), 735—740.

Becher, P., 2001. Emulsions—Theory and Practice. Oxford University Press, New York.

Benayoune, M., Khezzar, L., Al-Rumhy, M., 1998. Viscosity of water in oil emulsion. Petrol. Sci. Technol. 16 (7—8), 767—784.

Bendiksen, K.H., 1984. An experimental investigation of the motion of long bubbles in inclined tubes. Int. J. Multiphase Flow 10 (4), 467—483.

Bendiksen, K.H., Espedal, M., 1992. Onset of slugging in horizontal gas—liquid pipe flow. Int. J. Multiphase Flow 18 (2), 237—247.

Bergles, A.E., Collier, J.G., Delhaye, J.M., Hewitt, G.F., Mayinger, F., 1981. Two-Phase Flow and Heat Transfer in the Power and Process Industries. McGraw-Hill, Washington.

Bishop, A.A., Deshpande, S.D., 1986. Interfacial level gradient effects in horizontal Newtonian liquid—gas stratified flow—I. Int. J. Multiphase Flow 12 (6), 957—975.

Bontozoglou, V., 1991. Weakly nonlinear Kelvin—Helmholtz waves between fluids of finite depth. Int. J. Multiphase Flow 17 (4), 509—518.

Collier, J.G., 1972. Convective Boiling and Condensation, second ed. McGraw-Hill.

Crowley, C.J., Wallis, G.B., Barry, J.J., 1992. Validation of a one-dimensional wave model for stratified-to-slug flow regime transition, with consequences for waves growth and slug frequency. Int. J. Multiphase Flow 18 (2), 249—271.

Dukler, A.E., Hubbard, M.G., 1975. A model for gas—liquid slug flow in horizontal and near horizontal tubes. Ind. Eng. Chem. Fundam. 14 (4), 337—347.

Fan, Z., Jepson, W.P., Hanratty, T.J., 1992. A model for stationary slugs. Int. J. Multiphase Flow 18 (4), 477—494.

Friederichs, K.O., Hyers, D.H., 1954. The existence of solitary waves. Commun Pure Appl Math 7, 517—550.

Gregory, G.A., Nicolson, M.K., Aziz, K., 1978. Correlation of the liquid volume fraction in the slug for horizontal gas—liquid slug flow. Int. J. Multiphase Flow 4 (1), 33—39.

Jeffreys, H., 1925. On the formation of water waves by wind. Proc. R. Soc. A 107, 189—205.

Jeffreys, H., 1926. On the formation of waves by wind. Proc. R. Soc. A 110, 341—347.

Johnston, A.J., 1984. An investigation into the interfacial shear stress contribution in two-phase stratified flow. Int. J. Multiphase Flow 10 (3), 371—383.

Johnston, A.J., 1985. Transition from stratified to slug flow regime in countercurrent flow. Int. J. Multiphase Flow 11 (1), 31—41.

Jones, A.V., Prosperetti, A., 1985. On the suitability of first-order differential models for two-phase flow prediction. Int. J. Multiphase Flow 11 (2), 133—148.

Kang, H.C., Kim, M.H., 1993. The relation between the interfacial shear stress and the wave motion in a stratified flow. Int. J. Multiphase Flow 19 (1), 35—49.

Kocamustafaogullari, G., 1985. Two-fluid modeling in analyzing the interfacial stability of liquid film flows. Int. J. Multiphase Flow 11 (1), 63—89.

Kordyban, E.S., 1961. A flow model for two-phase slug flow in horizontal tubes. J. Basic Eng.

Kordyban, E.S., 1977. The transition to slug flow in the presence of large waves. Int. J. Multiphase Flow 3 (6), 603−607.

Kordyban, E.S., Ranvo, T., 1970. Mechanism of slug formation in horizontal two-phase flow. J. Fluids Eng 92 (4), 857−864.

Kowalski, J.E., 1987. Wall and interfacial shear stress in stratified flow in a horizontal pipe. AIChE J. 33 (2), 274−281.

LAMB, H., 1932. Hydrodynamics. Cambridge University Press.

Lin, P.Y., Hanratty, T.J., 1986. Prediction of the initiation of slugs with linear stability theory. Int. J. Multiphase Flow 12 (1), 79−98.

Lin, P.Y., Hanratty, T.J., 1987. Detection of slug flow from pressure measurements. Int. J. Multiphase Flow 13 (4), 549−563.

Long, R.R., 1956. Solitary Waves in the One- and Two-Fluid Systems. Tellus 8 (4), 460−471.

Mandhane, J.M., Gregory, G.A., Aziz, K., 1974. A flow pattern map for gas−liquid flow in horizontal pipes. Int. J. Multiphase Flow 1 (4), 537−553.

Milne-Thomson, L. M. 1968. Theoretical Hydrodynamics. Macmillan.

Mishima, K., Ishii, M., 1980. Theoretical prediction of onset of horizontal slug flow. J. Fluids Eng. 102, 441−445.

Oliemans, R.V., Pots, B.F.M., Trompe, N., 1986. Modeling of annular dispersed two-phase flow in vertical pipes. Int. J. Multiphase Flow 12 (5), 711−732.

Ooms, G., Segal, A., Cheung, S.Y., Oliemans, R.V.A., 1985. Propagation of long waves of finite amplitude at the interface of two viscous fluids. Int. J. Multiphase Flow 11 (4), 481−502.

Pal, R., Rhodes, E., 1985. A novel viscosity correlation for non-Newtonian concentrated emulsion. J. Colloid Interface Sci. 107 (2), 301−307.

Ruder, Z., Hanratty, P.J., Hanratty, T.J., 1989. Necessary conditions for the existence of stable slugs. Int. J. Multiphase Flow 15 (2), 209−226.

Song, S.H., 1994. Characterization and Metering of Multiphase Mixtures from Deep Subsea Wells, Ph.D. dissertation. The University of Texas at Austin, TX.

Stoker, J.J., 957. Water Waves. Interscience.

Taitel, Y., Dukler, A.E., 1976. A model for predicting flow regime transitions in horizontal and near horizontal gas−liquid flow. AIChE J. 22 (1), 47−55.

Wallis, G.B., Dobson, J., 1973. The onset of slugging in horizontal stratified air−water flow. Int. J. Multiphase Flow 1 (1), 173−193.

Woelflin, W., 1942. The viscosity of crude-oil emulsions. Drilling and Production Practice. API Annual.

Xiao, J.J., Shoham, O., Brill, J.P., 1990. A comprehensive mechanistic model for two-phase flow in pipelines. Presented at the Sixty-Fifth Annual Technical Conference and Exhibition of the Society of Petroleum Engineers, 23−26 September 1990, New Orleans, LA. SPE Paper 20631.

Yan, Y., Masliyah, J.H., 1993. Effect of oil viscosity on the rheology of oil-in-water emulsions with added solids. Can. J. Chem. Eng. 71 (6), 852−858.

Zhang, H.Q., Wang, Q., Sarica, C., Brill, J.P., 2003. Unified model for gas−liquid pipe flow via slug dynamics—part 1: model development. J. Energy Res. Technol. 125 (4), 266−273.

Steady and Transient Solutions for Pipeline Temperature

B.1 ASSUMPTIONS

The following assumptions are made in model formulation:

1. Friction-induced heat is negligible.
2. Heat transfer in the radial direction is fully controlled by the insulation fluid.
3. Specific heat of fluid is constant.

B.2 GOVERNING EQUATION

Figure B.1 depicts a small element of pipe with an insulation layer. Such a pipe can be a well tubing or pipeline.

Consider the heat flow during a time period of Δt_f. Heat balance is given by:

$$q_{in} - q_{out} - q_R = q_{acc} \tag{B.1}$$

where

q_{in} = heat energy brought into the pipe element by fluid due to convection, J,

q_{out} = heat energy carried away from the pipe element by fluid due to convection, J,

q_R = heat energy transferred through the insulation layer due to conduction, J,

q_{acc} = heat energy accumulation in the pipe element, J.

These terms can be further formulated as:

$$q_{in} = \rho_f C_p v A_f T_{f,L} \Delta t_f \tag{B.2}$$

$$q_{out} = \rho_f C_p v A_f T_{f,L+\Delta L} \Delta t_f \tag{B.3}$$

367

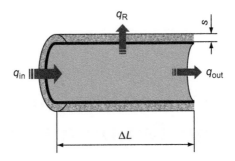

FIGURE B.1 Sketch illustrating convection and conduction heat transfer in a pipe.

$$q_R = 2\pi R_n k_n \Delta L \frac{\partial T_f}{\partial r} \Delta t_f \qquad \text{(B.4)}$$

$$q_{acc} = \rho_f C_p A_f \Delta L \Delta \overline{T}_f \qquad \text{(B.5)}$$

where

ρ_f = fluid density, kg/m³,
C_p = specific heat at constant pressure, J/kg °C,
v = the average flow velocity of fluid in the pipe, m/s,
A_f = cross-sectional area of pipe open for fluid flow, m²,
$T_{f,L}$ = temperature of the flowing-in fluid, °C,
Δt_f = flow time, s,
$T_{f,L+\Delta L}$ = temperature of the flowing-out fluid, °C,
R_n = inner radius of insulation layer, m,
k_n = thermal conductivity of the insulation layer, W/m °C,
ΔL = length of the pipe segment, m,
$\dfrac{\partial T_f}{\partial r}$ = radial-temperature gradient in the insulation layer, °C/m,
$\Delta \overline{T}_f$ = the average temperature increase of fluid in the pipe segment, °C.

Substituting Eqs (B.2) through (B.5) into Eq. (B.1) gives:

$$\rho_f C_p v A_f \Delta t (T_{f,L} - T_{f,L+\Delta L}) - 2\pi R_n k_n \Delta L \frac{\partial T_f}{\partial r} \Delta t_f = \rho_f C_p A_f \Delta L \Delta \overline{T}_f \qquad \text{(B.6)}$$

Dividing all the terms of this equation by $\Delta L \Delta t_f$ yields:

$$\rho_f C_p v A_f \frac{(T_{f,L} - T_{f,L+\Delta L})}{\Delta L} - 2\pi R_n k_n \frac{\partial T_f}{\partial r} = \rho_f C_p A_f \frac{\Delta \overline{T}_f}{\Delta t_f} \qquad \text{(B.7)}$$

For infinitesimal of ΔL and Δt_f, this equation becomes:

$$v\frac{\partial T_f}{\partial L} + \frac{\partial T_f}{\partial t_f} = -\frac{2\pi R_n k_n}{\rho_f C_p A_f}\frac{\partial T_f}{\partial r} \tag{B.8}$$

The radial-temperature gradient in the insulation layer can be formulated as:

$$\frac{\partial T_f}{\partial r} = \frac{T_f - (T_{f,0} - G\cos(\theta)L)}{s} \tag{B.9}$$

where

$T_{f,0}$ = temperature of the medium outside the insulation layer at $L = 0$, °C,
G = geothermal gradient, °C/m,
θ = inclination time, degree,
s = thickness of the insulation layer, m,

Substituting Eq. (B.9) into Eq. (B.8) yields:

$$v\frac{\partial T_f}{\partial L} + \frac{\partial T_f}{\partial t_f} = aT_f + bL + c \tag{B.10}$$

where

$$a = -\frac{2\pi R_n k_n}{\rho_f C_p s A_f} \tag{B.11}$$

$$b = aG\cos(\theta) \tag{B.12}$$

and

$$c = -aT_{f,0} \tag{B.13}$$

B.3 SOLUTIONS

Three solutions are sought in this study:

Solution A: Steady flow;
Solution B: Transient flow with static fluid as the initial condition;
Solution C: Transient flow with steady flow as the initial condition.

Solution A gives temperature profile during normal operation conditions; Solution B simulates temperature change during a start-up process; and Solution C yields temperature trend during a shutdown process.

B.3.1 Steady Heat Transfer

If the mass flowrate is maintained for a significantly long time, a steady heat transfer condition between the system and its surroundings is expected. Under steady flow conditions, the temperature at any point in the system is time independent. Therefore, Eq. (B.10) becomes:

$$v\frac{dT_f}{dL} = aT_f + bL + c \qquad (B.14)$$

This equation can be solved with boundary condition:

$$T = T_{f,s} \quad \text{at } L = 0 \qquad (B.15)$$

To simplify the solution, Eq. (B.14) is rearranged as:

$$\frac{dT_f}{dL} + \alpha T_f + \beta L + \gamma = 0 \qquad (B.16)$$

where

$$\alpha = -\frac{a}{v} \qquad (B.17)$$

$$\beta = -\frac{b}{v} \qquad (B.18)$$

and

$$\gamma = -\frac{c}{v} \qquad (B.19)$$

Let

$$u = \alpha T_f + \beta L + \gamma \qquad (B.20)$$

then

$$T_f = \frac{u - \beta L - \gamma}{a} \qquad (B.21)$$

and

$$\frac{dT_f}{dL} = \frac{1}{\alpha}\frac{du}{dL} - \frac{\beta}{\alpha} \qquad (B.22)$$

Substituting Eqs (B.21) and (B.22) into Eq. (B.16) gives:

$$\frac{1}{\alpha}\frac{du}{dL} - \frac{\beta}{\alpha} + u = 0 \qquad (B.23)$$

Integration of this equation with the method of separation of variables yields:

$$-\frac{1}{\alpha}\ln(\beta - \alpha u) = L + C \tag{B.24}$$

where C is a constant of integration. Substituting Eq. (B.20) into Eq. (B.24) and rearranging the latter result in:

$$T_f = \frac{1}{\alpha^2}\left[\beta - \alpha\beta L - \alpha\gamma - e^{-\alpha(L+C)}\right] \tag{B.25}$$

Applying boundary condition (B.15) to Eq. (B.25) gives the expression for the integration constant:

$$C = -\frac{1}{\alpha}\ln(\beta - \alpha^2 T_{f,s} - \alpha\gamma) \tag{B.26}$$

B.3.2 Transient Heat Transfer During Starting-Up

The temperature profile along the pipe during the starting-up process can be obtained by solving Eq. (B.10) with the method of characteristics, subject to the initial condition:

$$T_f = T_{f,0} - G\cos(\theta)L \quad \text{at } t = 0 \tag{B.27}$$

Consider a family of curves defined by the equation:

$$dt_f = \frac{dL}{v} = \frac{dT_f}{aT + bL + c} \tag{B.28}$$

The characteristics are:

$$L = vt_f + K \tag{B.29}$$

We also have from Eq. (B.28):

$$\frac{dT_f}{dL} = \frac{aT_f + bL + c}{v} \tag{B.30}$$

Using notations (B.17)–(B.19), Eq. (B.30) becomes:

$$\frac{dT_f}{dL} + \alpha T_f + \beta L + \gamma = 0 \tag{B.31}$$

which is exactly Eq. (B.16). Its solution is the same as Eq. (B.25), i.e.,

$$T_f = \frac{1}{\alpha^2}\left[\beta - \alpha\beta L - \alpha\gamma - e^{-\alpha(L+A)}\right] \tag{B.32}$$

where A is an arbitrary constant of integration. This constant is different on each characteristic curve. Further, each characteristic curve has a different value of K. Hence, as K varies, A varies, and we may write $A = f(K)$, where f is an arbitrary function to be determined. Writing $A = f(K)$ in Eq. (B.32) yields:

$$T_f = \frac{1}{\alpha^2}\left[\beta - \alpha\beta L - \alpha\gamma - e^{-\alpha(L+f(k))}\right] \tag{B.33}$$

Eliminating K using Eq. (B.29) gives:

$$T_f = \frac{1}{\alpha^2}\left[\beta - \alpha\beta L - \alpha\gamma - e^{-\alpha[L+f(L-vt_f)]}\right] \tag{B.34}$$

Now applying the initial condition (B.27) gives:

$$T_{f,0} - G\cos(\theta)L = \frac{1}{\alpha^2}\left[\beta - \alpha\beta L - \alpha\gamma - e^{-\alpha[L+f(L)]}\right] \tag{B.35}$$

which gives:

$$f(L) = -L - \frac{1}{\alpha}\ln\left[\beta - \alpha\beta L - \alpha\gamma - \alpha^2(T_{f,0} - G\cos(\theta)L)\right] \tag{B.36}$$

Therefore,

$$f(L - vt_f) = -(L - vt_f) - \frac{1}{\alpha}\ln[\beta - \alpha\beta(L - vt_f)$$

$$- \alpha\gamma - \alpha^2[T_0 - G\cos(\theta)(L - vt_f)]] \tag{B.37}$$

Substituting Eq. (B.37) into Eq. (B.34) results in the solution to Eq. (B.10) subject to the initial condition (B.27). This solution is valid for $L - vt_f > 0$. For points at which $L - vt_f < 0$, $L - vt_f = 0$ should be used.

B.3.3 Transient Heat Transfer During a Flowrate Change

The temperature trend along the pipe during a flowrate change (shutting down is a special case) process can be obtained by solving Eq. (B.10) with a new velocity v' corresponding to a new flowrate. The general solution is still given by Eq. (B.34) with new parameters corresponding to the low velocity, i.e.,

$$T_f = \frac{1}{\alpha'^2}\left[\beta' - \alpha'\beta'L - \alpha'\gamma' - e^{-\alpha'[L+f(L-v't_f)]}\right] \tag{B.38}$$

where

$$\alpha' = -\frac{a}{v'} \tag{B.39}$$

$$\beta' = -\frac{b}{v'} \tag{B.40}$$

and

$$\gamma' = -\frac{c}{v'} \tag{B.41}$$

The initial condition is defined by Eq. (B.25), i.e.,

$$T_{\mathrm{f}} = \frac{1}{\alpha^2}\left[\beta - \alpha\beta L - \alpha\gamma - \mathrm{e}^{-\alpha(L+C)}\right] \quad \text{at } t_{\mathrm{f}} = 0 \tag{B.42}$$

where the constant C is given by Eq. (B.26).

Now applying the initial condition (B.42) to Eq. (B.38) gives:

$$\frac{1}{\alpha^2}\left[\beta - \alpha\beta L - \alpha\gamma - \mathrm{e}^{-\alpha(L+C)}\right] = \frac{1}{\alpha'^2}\left[\beta' - \alpha'\beta'L - \alpha'\gamma' - \mathrm{e}^{-\alpha'[L+f(L)]}\right] \tag{B.43}$$

which yields:

$$f(L) = -L - \frac{1}{\alpha'}$$

$$\times \ln\left\{\beta' - \alpha'\beta'L - \alpha'\gamma' - \left(\frac{\alpha'}{\alpha}\right)^2\left[\beta - \alpha\beta L - \alpha\gamma - \mathrm{e}^{-\alpha(L+C)}\right]\right\} \tag{B.44}$$

Therefore,

$$f(L - vt_{\mathrm{f}}) = -(L - vt_{\mathrm{f}}) - \frac{1}{\alpha'}\ln\left\{\beta' - \alpha'\beta'(L - vt_{\mathrm{f}}) - \alpha'\gamma' - \left(\frac{\alpha'}{\alpha}\right)^2\right.$$

$$\left.[\beta - \alpha\beta(L - vt_{\mathrm{f}}) - \alpha\gamma - \mathrm{e}^{-\alpha[(L-vt_{\mathrm{f}})+C]}]\right\} \tag{B.45}$$

Substituting Eq. (B.45) into Eq. (B.38) results in the solution to Eq. (B.10) subject to the initial condition (B.42).

Strength De-Rating of Old Pipelines

C.1 INTRODUCTION

The strength of old pipelines declines because of a number of reasons, with corrosion being the major one. This is especially true when the pipeline is not well corrosion protected. Corrosion mechanisms include electrochemical corrosion, chemical corrosion, and stress-promoted corrosion. Pipeline pitting (cavity) due to failure of corrosion protection is most common. Figure C.1 shows a typical cavity due to corrosion. Stress concentration around the pitted area results in degradation of pipeline strength (pressure rating). This document presents expressions of stress concentration factors (SCFs) around spherical pits of various geometries based on Sun's work (2003). The de-rated strength of pipeline is equal to the strength of new pipeline divided by the SCF.

C.2 CLASSIFICATION OF CAVITIES

To simplify the stress concentration analysis of corrosion cavity on pipeline walls, spherical surface cavities are classified into three categories: shallow, medium, and deep cavities. A simplified diagram for a spherical surface cavity is shown in Figure C.2 where t is the wall thickness of the pipeline; d and h are the diameter and depth of the corrosion cavity, respectively. When h is equal to $d/2$, the cavity is exactly hemispherical and is classified as medium cavity. Shallow cavity is defined as a cavity with h value being less than $d/2$, while deep cavity is defined as the one with h value being greater than $d/2$.

C.3 ANALYTICAL SCF MODEL FOR MEDIUM CAVITY

Using the same approach adopted by Wang (2001) for SCF analysis, Sun (2003) derived the following expression for pipe with a hemispherical cavity:

$$\text{SCF} = \frac{(27 - 15v)/(14 - 10v)}{1 - k_1 \times \dfrac{a^3}{t^3} - k_2 \times \dfrac{a^5}{t^5}} \tag{C.1}$$

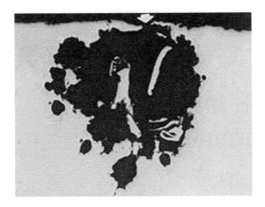

FIGURE C.1 A typical spherical cavity on a pipeline wall.

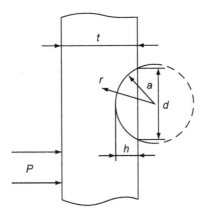

FIGURE C.2 Simplified diagram for a spherical surface corrosion cavity.

where the constants k_1 and k_2 are given by:

$$k_1 = -\frac{27 - 15v}{14 - 10v} \cdot \frac{5 - 4v^2}{(6 - 4v)(1 + v)} + 2.5 \qquad \text{(C.2)}$$

and

$$k_2 = \frac{27 - 15v}{14 - 10v} \cdot \frac{5 - 4v^2}{(6 - 4v)(1 + v)} - 1.5 \qquad \text{(C.3)}$$

C.4 ANALYTICAL SCF MODEL FOR SHALLOW CAVITY

Sun (2003) derived the following expression for pipe with a shallow spherical cavity:

$$\text{SCF} = \frac{b^2 \cdot \gamma - (a-h) \cdot \sqrt{b^2 - (a-h)^2}}{b^2\gamma - a^2\beta - (a-h)^2(tg\gamma - tg\beta) + (a^3(4m - 5mv + 3a^2n)/(7-5v))} \cdot \frac{27 - 15v}{14 - 10v} \tag{C.4}$$

where the constants are given by:

$$a = \frac{d^2 + 4h^2}{8h} \tag{C.5}$$

$$b = a - h + t \tag{C.6}$$

$$m = \frac{\sin\gamma - \sin\beta}{a - h} + \frac{\beta}{a} - \frac{\gamma}{b} \tag{C.7}$$

$$n = \frac{3\sin\gamma - 3\sin\beta - \sin^3\gamma + \sin^3\beta}{3(a-h)^3} + \frac{\beta}{a^3} - \frac{\gamma}{b^3} \tag{C.8}$$

$$\beta = \arccos\left(\frac{a-h}{a}\right) \tag{C.9}$$

$$\gamma = \arccos\left(\frac{a-h}{b}\right) \tag{C.10}$$

C.5 ANALYTICAL SCF MODEL FOR DEEP CAVITY

Sun (2003) derived the following expression for pipe with a deep spherical cavity:

$$\text{SCF} = \frac{\left(\pi \cdot b^2 - b^2 \cdot \psi + (h-a) \cdot \sqrt{b^2 - (h-a)^2}\right) \cdot [(27 - 15v)/(14 - 10v)]}{(b^2 - a^2) \cdot (\pi - \psi) - a^2(\psi - \omega) + (h-a)^2(tg\psi - tg\omega) + (a^2(4w - 5vw + 3a^3u)/(7-5v))} \tag{C.11}$$

where

$$a = \frac{d^2 + 4h^2}{8h} \tag{C.12}$$

$$b = a - h + t \tag{C.13}$$

$$u = \frac{3\sin\omega - 3\sin\psi + \sin^3\psi - \sin^3\omega}{3(h-a)^3} + \frac{\pi - \psi}{a^3} - \frac{\pi - \psi}{b^3} \tag{C.14}$$

$$w = \pi - \omega - \frac{a(\sin\psi - \sin\omega)}{h - a} - \frac{a(\pi - \psi)}{b} + \frac{3(\psi - \omega)}{4 - 5v} \tag{C.15}$$

$$\psi = \arccos\left(\frac{h-a}{b}\right) \tag{C.16}$$

$$\omega = \arccos\left(\frac{h-a}{a}\right) \tag{C.17}$$

C.6 ILLUSTRATIVE EXAMPLE

Consider an X-60 steel pipeline with an outside diameter of 30 in. and wall thickness of 1 in. The pipeline is classified as a thin-wall pipe because the diameter/thickness ratio ($D/t = 30$) is greater than 20. According to Eq. (6.1), the yield strength of 60 ksi corresponds to a net internal pressure of:

$$P = \frac{2tS_y}{D}$$

$$= \frac{(2)(1.0)(60,000)}{(30)}$$

$$= 4000 \text{ psi}$$

Suppose a spherical cavity were present on the surface of the pipeline due to corrosion pitting. Assume the cavity had an open diameter of $d = 0.1$ in. and depth of $h = 0.75$ in. Since $h > d/2$, the cavity is classified as a deep cavity. The following calculations are made to determine SCF.

$$a = \frac{d^2 + 4h^2}{8h}$$

$$= \frac{(0.1)^2 + 4(0.075)^2}{8(0.075)}$$

$$= 0.054 \text{ in.}$$

$$b = a - h + t$$
$$= 0.054 - 0.075 + 0.5$$
$$= 0.48 \text{ in.}$$

$$\psi = \arccos\left(\frac{h-a}{b}\right)$$

$$= \arccos\left(\frac{0.075 - 0.054}{0.48}\right)$$

$$= 1.53 \text{ rad}$$

$$\omega = \arccos\left(\frac{h-a}{a}\right)$$

$$= \arccos\left(\frac{0.075 - 0.054}{0.054}\right)$$

$$= 1.18 \text{ rad}$$

$$u = \frac{3\sin\psi - 3\sin\omega + \sin^3\psi - \sin^3\omega}{3(h-a)^3} + \frac{\pi-\psi}{a^3} - \frac{\pi-\psi}{b^3}$$

$$= \frac{3\sin(1.53) - 3\sin(1.18) + \sin^3(1.53) - \sin^3(1.18)}{3(0.075-0.054)^3} + \frac{3.14-1.53}{(0.054)^3} - \frac{3.14-1.53}{(0.48)^3}$$

$$= 9505 \text{ in.}^{-3}$$

$$w = \pi - \omega - \frac{a(\sin\psi - \sin\omega)}{h-a} - \frac{a(\pi-\psi)}{b} + \frac{3(\psi-\omega)}{4-5v}$$

$$= 3.14 - 1.18 - \frac{\sin(1.53) - \sin(1.18)}{0.075 - 0.054} - \frac{0.054(3.14-1.53)}{0.48} - \frac{3(3.14-1.18)}{4-5(0.3)}$$

$$= 2.007$$

$$(27 - 15v)/(14 - 10v) = (27 - 15(0.3))/(14 - 10(0.3))$$
$$= 2.045$$

$$\frac{a^2(4w - 5wv + 3a^3u)}{(7 - 5v)} = \frac{(0.054)^2(4(2.007) - 5(2.007)(0.3) + 3(0.054)^3(9505))}{(7 - 5(0.3))}$$

$$= 0.005$$

$$\text{SCF} = \frac{\left(\pi \cdot b^2 - b^2 \cdot \psi + (h - a) \cdot \sqrt{b^2 - (h-a)^2}\right) \cdot [(27 - 15v)/(14 - 10v)]}{(b^2 - a^2) \cdot (\pi - \psi) - a^2(\psi - \omega) + (h-a)^2(tg\psi - tg\omega) + (a^2(4w - 5wv + 3a^3u)/(7 - 5v))}$$

$$= \frac{(3.14(0.48)^2 - (0.48)^2(1.53) + (0.075 - 0.054)\sqrt{(0.48)^2 - (0.075 - 0.054)^2}[2.045]}{((0.48)^2 - (0.054)^2)(3.14 - 1.53) - (0.054)^2(1.53 - 1.18) - (0.075 - 0.054)^2(tg(1.53) - tg(1.18)) + 0.005}$$

$$= 2.055$$

Therefore, the pipe will start to yield near the corrosion pit at a net internal pressure of:

$$P_{\text{yield}} = \frac{P}{\text{SCF}}$$

$$= \frac{4000}{2.055}$$

$$= 1946 \text{ psi}$$

REFERENCES

Sun, K., 2003. Casing strength degradation due to corrosion—applications to casing pressure assessment, MS Thesis. University of Louisiana at Lafayette, Lafayette, LA.

Wang, Q.Z., 2001. Simple formulae for the stress concentration factor for two- and three-dimensional holes in finite domains. J. Strain Anal. 37 (3), 259–264.

Index

Note: Page numbers followed by "*f*" and "*t*" refer to figures and tables, respectively.

Printed and bound by CPI Group (UK) Ltd, Croydon, CR0 4YY

03/10/2024

01040422-0012